L. Poppe, L. Novák

Selective Biocatalysis

VCH

©VCH Verlagsgesellschaft mbH. D-6940 Weinheim (Federal Republic of Germany), 1992

Distribution:
VCH, P.O. Box 10 11 61, D-6940 Weinheim (Federal Republic of Germany)
Switzerland: VCH, P.O. Box, CH-4020 Basel (Switzerland)
United Kingdom and Ireland: VCH (UK) Ltd., 8 Wellington Court, Wellington Street, Cambridge CB1 1HZ (England)
USA and Canada: VCH, Suite 909, 220 East 23rd Street, New York, NY 10010-4606 (USA)

ISBN 3-527-28372-2 (VCH, Weinheim)   ISBN 1-56081-188-9 (VCH, New York)

L. Poppe, L. Novák

Selective Biocatalysis

A Synthetic Approach

Weinheim
New York
Cambridge
Basel

CHEMISTRY

Dr. László Poppe
Central Research Institute for Chemistry,
Hungarian Academy of Sciences
H-1025 Budapest
Pusztaszeri út 59-67
Hungary

Prof. Lajos Novák
Institute of Organic Chemistry
Technical University Budapest
H-1521 Budapest
XI., Gellért tér 4
Hungary

This book was carefully produced. Nevertheless, authors, editors and publisher do not warrant the information contained therein to be free of errors. Readers are advised to keep in mind that statements, data, illustrations, procedural details or other items may inadvertently be inaccurate.

Published jointly by
VCH Verlagsgesellschaft mbH, Weinheim (Federal Republic of Germany)
VCH Publishers, Inc., New York, NY (USA)

Editorial Director: Dr. Christina Dyllick-Brenzinger
Production Manager: Dipl.-Ing. (FH) Hans Jörg Maier

Library of Congress Card No.: 92-9284

British Library Cataloguing-in-Publication Data: A catalogue record for this book is available from the British Library.

Deutsche Bibliothek Cataloguing-in-Publication Data:
Poppe, László:
Selective biocatalysis: a synthetic approach / L. Poppe; L. Novák. - Weinheim; New York; Cambridge; Basel: VCH, 1992
ISBN 3-527-28372-2
NE: Novák, Lajos:

Composition: Ulrich Hellinger, D-6901 Heiligkreuzsteinach. Printing: betz-druck gmbh, D-6100 Darmstadt 12. Bookbinding: J. Schäffer GmbH & Co.KG, D-6718 Grünstadt.
Printed in the Federal Republic of Germany.

Preface

Increasingly today, one of the main challenges facing organic chemistry is the rational synthesis of an ever growing number of complex, optically active natural products and their analogues. There is an increasing demand not only for individual stereoisomers but also for higher and higher standards of purity of the product. It is thus understandable that the stereoselective elaboration of chiral centers has become a central issue, most importantly in pharmaceutical research and drug manufacturing.

The response to this challenge has been an uprecedented development in the methodology of synthetic organic chemistry. The use of metallo-organic compounds, metal catalyzed homogeneous and heterogeneous catalysis, phase transfer catalysis, electrochemical and photochemical techniques, reagents on solid supports and other up-to-date methods have all contributed to the armory of stereoselective synthesis. At the same time biocatalysis i.e. the use of enzymes and microorganisms to facilitate the reaction process has achieved prominent significance, most notably in the preparation of key chiral intermediates. For synthetic organic chemistry, both on the laboratory and the industrial scale, enzyme catalyzed transformations offer manifold possibilities. For almost every type of synthetic reaction there exists an enzymic counterpart. Indeed, there are a number of important points worth considering in favour of enzyme catalyzed reactions, such as their efficiency, their selectivity, and topically, their avoidance of toxic by-products which means they have a low level environmental impact.

Many of the enzymes currently used in organic chemistry are commercially available, and in fact some of them can be obtained in an immobilized form too. Several easily applicable microorganisms are also available in large quantities, and well established methods are available to facilitate their immobilization. An almost inexhaustible source of rarely used microorganisms, however, is typed culture collections which provide a rich supply of novel enzymes. This apart, the fields of biotechnology and genetic engineering are constantly tuning up promising new ways to obtain modified or even completely new enzymes. Thus, in the planning of synthesis strategies one can rely upon enzymes as one might on any other novel catalyst or reagent. If a biocatalytic process is available as an alternative to a chemical method, the decision in favour of one or the other should be based on a cost-benefit analysis. Nowadays the most up-to-date synthetic pathways are regarded as being those that can most efficiently blend whatever modern synthetic techniques are available, and biocatalysis is most firmly included amongst these.

As a result of intensive research over recent decades, a large body of information about biocatalytic processes applicable to organic synthesis is available to us. In amongst all this information are a number of topics which are quite important in their own right, areas such as the manufacturing and transformation of natural products by fermentation, and the production of amino acids, peptides, nucleotides and nucleic acids or isotopically labelled compounds by biotechnological methods. Unfortunately, these cannot be discussed in detail in this work. In addition, enzyme catalyzed reactions which have as yet found no, or only very restricted synthetic application, or those which involve sophisticated fermentational and microbiological methods have only been presented briefly.

Our main intention in this book was to present through the eyes of an organic chemist, detailed and practically useful information on those enzymes and cell systems which have established themselves recently as "reagents". In doing this, we wish to contribute to their further acceptance by the chemical community. The most important biocatalysts with a "reagent status" are hydrolases, oxido-reductases, several lyases, and among intact microorganisms, baker's yeast.

It is a pleasure to acknowledge the contribution to this book by our colleagues at the Technical University, Budapest.

We are very grateful to Professor Mihály Nógrádi who both provided the excellent translation of the text and contributed many valuable suggestions.

We also wish to thank Katalin Recseg and Andrea Farkas for their careful assistance in the editing and correction of the manuscript.

Finally, our special thanks go to our families for their helpful patience and understanding.

Budapest, October 1991   L. Poppe
                                                                                    L. Novák

Contents

List of Abbrevations*

AA	Acylase I from *Aspergillus* sp.
AAT	Amino acid transferase
AC	Acetyl
AChE	Acetylcholine esterase
AcL	*Achromobacter* sp. lipase
ADH	Alcohol dehydrogenase
AldDH	Aldehyde dehydrogenase
AnL	*Aspergillus niger* lipase
ATP	Adenosine triphosphate
Bn	Benzyl
BsP	Protease, from *Bacillus stearotermophilicus*
Bu	Butyl
Bz	Benzoyl
CcL	*Candida cylindracea* lipase
CfADH	Culvularia falcata alcohol dehydrogenase
ChE	Cholesterol esterase
CoA	Coenzyme A
CTR	α-Chymotrypsin
CvL	*Chromobacterium viscosum* lipase
DH	Dehydrogenase
DMF	Dimethyl formamide
DMSO	Dimethyl sulfoxide
EH	Epoxy hydrolase
Et	Ethyl
FAD	Flavin adenine dinucleotide
FDH	Formate dehydrogenase
FDPA	Fructose-1,6-diphosphate aldolase, from rabbit muscle
FMN	Flavin mononucleotide
GK	Glycerokinase
HKA	Acylase I from hog kidney
HLADH	Horse liver alcohol dehydrogenase
HLE	Horse liver esterase
HrL	*Humicola ramingera* lipase

---

* Abbreviations used more than once in this monograph have been collated here. Please refer to Tables 3.1 and 3.2 for abbreviations of hydrolases not included here.

HSDH	Hydroxysteroid dehydrogenase
LDH	Lactate dehydrogenase
Me	Methyl
MEH	Epoxy hydrolase, mitochondrial
MjADH	*Mucor javonicus* alcohol dehydrogenase
MmL	*Mucor meihei* lipase
NAD	Nicotinamide adenine dinucleotide
NADH	Nicotinamide adenine dinucleotide, reduced
NADP	Nicotinamide adenine dinucleotide phosphate
NADPH	Nicotinamide adenine dinucleotide phosphate, reduced
PA	Penicillin acylase
PAF	Platelet activating factor
PaL	*Pseudomonas aeruginosa* lipase
PfL	*Pseudomonas fluorescens* lipase
Ph	Phenyl
PLADH	Pig liver alcohol dehydrogenase
PLE	Pig liver esterase
PP	Papain
PPL	Porcine pancreatic lipase
Pr	Propyl
PsL	*Pseudomonas* sp. lipase
RaL	*Rhisopus arrhisus* lipase
RAMA	Fructose-1,6-diphosphate aldolase, from rabbit muscle
RjL	*Rhisopus japonicus* lipase
SgP	Protease, from *Streptomyces griseus*
TbADH	*Thermoanaerobium brocki* alcohol dehydrogenase
THF	Tetrahydrofuran
TR	Trypsin
TvL	*Trichoderma viridae* lipase
STC	*Subtilisin Carlsberg*
YADH	Yeast alcohol dehydrogenase

1 Biotransformation and Biocatalysis

The chemical potential of biochemical systems[1] was recognized by mankind very early on, the best known examples of practical exploitation of this being alcohol fermentation and vinegar production. Many biologically active natural products have been and are being produced or modified by fermentational or microbiological methods [1,2,3]. Scientists working on the borderline of biology and chemistry have often used biochemical transformations in their research, studying enzyme catalyzed reactions or applying microorganisms and enzymes for the solution of stereochemical or mechanistic problems.

Within recent decades reactions carried out with the aid of enzymes or intact microorganisms have been used with ever increasing frequency and ever increasing success to catalyze synthetic chemical reactions [4-16]. The range of such operations extends from analytical procedures in the lab [17] to industrial processes carried out on a multiton scale such as the production of amino acids [18, 19], of high fructose corn syrup [20-22], or the transformation of penicillin G to 6-aminopenicillanic acid and other penicillin derivatives with penicillin acylase enzyme [23-25] etc. Gradually, a technique called biotransformation or biocatalysis has taken shape which can be clearly distinguished from traditional fermentation. In the course of a biotransformation the selective, enzymatic transformation of a well-defined substrate is taking place, whilst in fermentation, the life processes of a living microorganism are exploited in a complex, multistep metabolic process, coupled with the utilization of inexpensive carbon and nitrogen sources. The product of a fermentation is always a natural product. Biotransformation, in turn, is often but a single step in a synthetic sequence composed mainly of purely chemical reactions. The first practical example of the use of such a biotransformation was the preparation of D-ephedrine [26]. Here the key step was the fermentation of benzaldehyde with baker's yeast in the presence of carbohydrates which provided (*R*)-1-phenyl-1-hydroxy-2-propanone. From this product D-ephedrine could then be obtained by more traditional chemical means.

Nowadays, a large proportion of the newly discovered catalysts able to be used in organic syntheses are biocatalysts [27]. Table 1.1 lists the most important systems which have been used as biocatalysts with their scope of application.

---

[1]   In this book acronyms will be used for some notations and for the names of the more common enzymes and microorganisms (see List of Abbreviations).

Table 1.1. Types of biocatalysts.

Whole Cells	Cell Free Preparations
Types	*Types*
a) growing cultures	a) cell free extracts
b) resting cells	b) purified enzymes
c) lyophilized cells	c) treated or modified enzymes
d) treated or modified cells	d) multienzyme systems
Forms	*Forms*
a) free cells	a) free form
b) microcapsules, microemulsions	b) microcapsules, microemulsions
c) immobilized cells	c) immobilized form
Environment	*Environment*
a) aqueous solution	a) aqueous solution
b) aqueous solution containing organic cosolvent	b) aqueous solution containing organic cosolvent
c) water-organic solvent biphasic system	c) water-organic solvent biphasic system
d) cell-containing preparations in organic solvent	d) water-restricted organic solvent

Other Promising Systems
a) semisynthetic enzymes
b) enzymes by genetic engineering
c) catalytic antibodies
d) synthetic enzymes (non-proteins)

In the biocatalytic process, both intact cells and cell-free enzyme preparations can be used in practically any combination of types, modes of application and conditions. Immobilized resting cells, for example, can be used either in aqueous suspension containing a cosolvent [28] or by applying the purified free enzyme in an organic solvent [29]. Clear distinction between individual types, however, is sometimes difficult. Thus lipase activity is exhibited by a resting or freeze-dried cell mass, a crude, cell-free preparation or by a purified enzyme, but can be equally exploited for practical purposes from either. Similarly, the boundaries between different modes of application are also ill-defined: a suspension of a free enzyme in an organic solvent may behave similarly to an immobilized enzyme, for example it can often be removed by filtration and recycled from either preparation. Research scientists in diverse fields are constantly on the lookout for new ways to exploit enzymatic properties, and some novel biocatalysts showing promise are enzymes that have been chemically or genetically modified [30, 31] and catalytic antibodies produced by immune systems [32, 190]. Fully synthetic enzyme models can also be regarded as a new class of catalysts [33-36].

Some characteristics of these individual systems and certain features of their application will be discussed in the following chapters.

1.1 Characteristics of Enzyme Catalyzed Processes

The principal component of all known enzymes is protein[2]. Proteins are linear polymers of naturally occurring amino acids which are, except for glycine, all chiral. The sequence of the amino acids (the primary structure), by virtue of the interaction of the side chains striving to achieve an energy minimum, determines a relatively well-defined three-dimensional structure (the secondary and tertiary structure). Often, two or more polypeptide chains may agglomerate by means of physical and chemical forces (quaternary structure) behaving as sub-units within the framework of a larger structural array. Due to a variety of functional groups attached to the side chains (amino, carboxy, hydroxy, mercapto, etc.) proteins are potentially acidic, basic or nucleophilic catalysts. Substrates are bound by chemical or hydrogen bonds, electrostatic or other forces to specific (and not necessarily adjacent) sites on the polypeptide chain collectively called the active site, and as such substrates are necessarily complementary to the three-dimensional topology of the active site. Subsequent transformation of substrate bound in this way to the enzyme is carried out by functional groups in the side chains situated in a favorable steric disposition at the surface of the active site.

In nature enzymes can rarely be found outside cells, most of them exerting their catalytic activity inside the cell integrated into various cell components. Some enzymes completely lose their activity when separated from their environment and therefore cannot be used in a cell-free medium, an example being enzymes embedded into membranes which are usually inactive in the absence of membrane components. Thus, enzyme functions operative in the intact biological system cannot always be utilized in cell free systems.

Enzymes, as any other catalyst, accelerate reactions by lowering the activation energy of the reaction. In enzyme catalyzed reactions, the most important source of activation energy is the conformational change of enzyme protein. This is in contrast to most of the non-enzyme catalyzed reactions in which activation energy is supplied by the thermal energy of the environment. Enzyme catalyzed reactions can not only be a real alternative to well established chemical methods but also may enable the realization of reactions which are difficult or impossible to carry out by purely chemical means. Before considering the synthetic application of enzymes [37-45] some of their special features should be reviewed.

Amongst the many advantages of enzymes the following, from our point of view, are the most important:

---

[2] The recently discovered catalytical activity of some polynucleotides will not be discussed here.

i)   Of nearly 3000 enzymes that have been described [46-48], many are available commercially, several also in an immobilized form [49] (Table 1.2).

ii)   Enzymes are capable of very efficiently catalyzing a wide selection of reactions. Enzyme catalyzed reactions may be as much as 10-12 orders of magnitude faster than the same reaction under identical conditions but without enzyme catalysis.

iii) Enzymes work under relatively mild conditions, i.e. at room temperature and nearly neutral pH. This is of prime importance with products which tend to isomerize or are otherwise unstable.

iv)   Enzyme catalyzed reactions are generally highly selective both regarding the type of reaction catalyzed and the substrate required, also taking into account the nature of its stereostructure. Enzymes, as chiral macromolecules are capable of catalyzing asymmetric transformations - that is, they may be both enantiomer and enantiotope selective.

v)   Some enzymes have the capacity to introduce functional groups at positions otherwise unreactive towards common organic reagents.

vi)   The functioning, activity and selectivity of enzymes may be influenced by several factors: concentration of both substrate and product, pH, temperature, and also by other molecules present in the system. These effects can be exploited for directing enzyme catalyzed processes.

vii) With enzyme catalyzed reactions associated environmental hazards are generally small.

The utilization of enzyme catalyzed reactions for synthetic purposes involves also a number of disadvantages, the most important ones being as follows:

i)   Owing to their natural origin, enzymes usually operate in aqueous media. This may cause difficulties when the substrate or the product is either insoluble in or sensitive to water. Some techniques gaining ground recently e.g. the use of organic solvents [29, 50-53], or gases in a supercritical state [59], reverse micelles [54-57] or emulsions made with liquid crystals [58] may offer solutions to these problems.

ii) Pure enzymes are relatively expensive (Table 1.2). This drawback can often be avoided by using relatively crude, less costly enzyme preparations or even intact cells, provided that contaminants do not much interfere with the main process. It must be kept in mind, however, that pure, protease free enzyme preparations usually have a longer active lifetime. In the preparation of optically active products, contaminating enzymes may catalyze the same transformation, but with different kinetics and (sometimes opposite) stereoselectivity [61]. Thus it is important to determine that particular degree of purity which guarantees the required product purity in an economically acceptable way.
Immobilization [62-77] may help to reduce the operational costs although an additional possibility for obtaining cheaper enzymes is their production using biotechnology processes and genetic engineering.

iii) Some enzymes are rather unstable, but their lifetime may be prolonged in several ways, for example, by the elimination of proteolytic impurities, immobilization, addition of stabilizers or by chemical or genetic modification. Research on the isolation of highly stable enzymes from heat and salt resistant microorganisms is being actively pursued.

iv) Some enzymes are capable of transforming only a very narrow selection of substrates. The ideal candidates for synthetic applications are those enzymes which are able to transform a relatively broad range of unnatural substrates whilst at the same time maintaining the high degree of stereoselectivity shown towards the natural substrates. Although these requirements seem to be contradictory, they are surprisingly well fulfilled by a number of mainly mammalian enzymes. The substrate tolerance of microbial enzymes however is generally narrower, but this is in some ways compensated for by their much greater variety.

1.1.1   Classification and Nomenclature of Enzymes

Since enzymes are highly complex macromolecules the exact chemical structure of only a fraction of them is known and attempting to describe even those by systematic chemical names would be totally impractical. Enzymes are usually discovered by recognizing the transformation they catalyze and this is the reason why their classification and nomenclature is based on functional, rather than on structural features [46,78]. Accordingly, the recommendations of the Nomenclature Committee of the International Union of Biochemistry [46] decreed that the name of an enzyme should not define chemical structure(s) but rather a definite catalytic process. Thus, it may arise that the same name is used for enzymes from different sources and having different structures because they catalyze the same reaction. In turn, enzymes catalyzing more than one transformation may be recorded under more than one name.

Depending on the type of reaction they catalyze, enzymes are classified by the International Enzyme Committee into six main groups (EC 1.-EC 6.) [46]:

1. *Oxidoreductases*

These catalyze oxidations and reductions, such as the transformations C-H $\Leftrightarrow$ CH-OH, CH-OH $\Leftrightarrow$ C=O, and CH-CH $\Leftrightarrow$ C=C.

2. *Transferases*

These catalyze the transfer of functional groups, e.g. of acyl, glycosyl and phosphate groups from one substrate to another.

3. *Hydrolases*

These catalyze the hydrolysis of esters, glycosides, amides, peptides, acid anhydrides, epoxides and other compounds.

4. *Lyases*

These catalyze the addition of most often H-X type compounds onto C=C, C=O, and C=N double bonds, as well as the reverse process.

5. *Isomerases*

These are enzymes that catalyze various isomerizations, e.g. double bond migrations, Z-E and *cis-trans* isomerizations, racemizations, and epimerizations.

6. *Ligases*

Which catalyze at the expense of ATP energy the formation of C-O, C-S, C-N, C-C, and other bonds.

These six main groups were further divided into subgroups and sub-subgroups, which is how the code composed of the prefix EC and four numbers separated by periods came into being. The first number refers to one of the six main groups, the second and third ones to sub- and sub-subgroups, whilst the fourth one is a serial number within the subsubgroup. For example, with hydrolases (EC 3) the second number informs about the type of bond affected (esterases - EC 3.1, glycosidases - EC 3.2, etc.), and the third number refers to the type of substrate (e.g. carboxylic ester hydrolases - EC 3.1.1, thiolester hydrolases - EC 3.1.2).

At present in synthetic organic chemistry hydrolases and oxidoreductases are the most frequently used, although there are many examples also for the application of lyases, transferases and isomerases.

1.1.2   Coenzymes

Although, with the exception of some nucleic acids, all known enzymes are proteins, some of them will only function in association with non-proteinic cofactors called coenzymes. In such cases, the catalytically active species is a complex of protein and coenzyme called holoenzyme and cofactors bound to the protein by stable covalent bonds are often called prosthetic groups. Coenzymes may also exist as solutes in the cytoplasm and become bound to the protein only in the course of the catalytic process. From this point of view then, enzymes can be classified into two main groups.

The first group covers enzymes which work without cofactors (Fig. 1.1.a) and within this group are the hydrolases, and most of the lyases and isomerases. Synthetic utilization of such enzymes is relatively simple. Also, the handling of enzymes in which the cofactor is tightly bound to the protein is similar, e.g. as in certain metalloproteins and pyridoxal dependent enzymes, since with such species the regeneration of cofactors can be achieved in an autocatalytic way by the enzyme itself (Fig. 1.1.b).

The second large group comprises enzymes which require added cofactors such as NAD(P)H, ATP, and CoA-SH, for their activation (Fig. 1.1.c). Since cofactors may be quite expensive, their use in stoichiometric quantities is not economically viable for synthetic applications. Thus, when the application of such an enzyme is appropriate, the problem of *in situ* regeneration of the cofactor must be solved [79-83]. Cofactor dependent enzymic transformations usually require careful purification of the protein component in order to avoid degradation of the cofactor by contaminants.

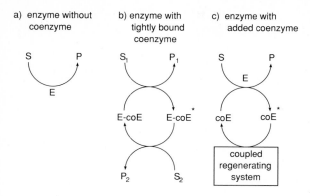

a) enzyme without coenzyme

b) enzyme with tightly bound coenzyme

c) enzyme with added coenzyme

coupled regenerating system

Fig. 1.1.   Types of enzyme-coenzyme relations

1.1.3 Enzyme Kinetics

A thorough understanding of the mechanism [85,191] and kinetics of enzymic reactions [78,86-91] as well as of factors influencing their action [92] is essential for their successful application to synthetic processes, primarily when the intention is for the optimization of an already working biocatalytic process or the assessment of its economic viability. This said, however, these details are not indispensable in a preliminary study of a new application.

The majority of enzyme catalyzed reactions follow s.c. Michaelis-Menten kinetics (Fig. 1.2) [93], a model which has remained to the present day the starting point for almost all studies on enzyme catalyzed processes. According to this model the enzymic process involves two (or more) elementary steps. Initially, in a fast equilibrium process, enzyme (E) and substrate (S) associate to form an enzyme-substrate complex (ES). Transformation to the product (P) takes place within this complex, and finally in a slow process the product is liberated whilst the enzyme (E) is recovered.

$$E + S \quad \underset{k_2}{\overset{k_1}{\rightleftharpoons}} \quad ES \quad \overset{k_3}{\longrightarrow} \quad E + P \tag{1}$$

With the progress of the transformation, both substrate and product concentration ([S] and [P]) change continuously and the enzyme-substrate complex is in a steady-state equilibrium with the other components of the system (provided that [S]>>[E]), thus:

$$k_1([E_0] - [ES])\,[S] = k_2[ES] + k_3[ES] \tag{2}$$

a)

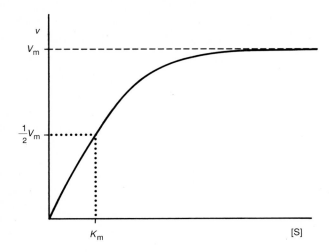

b)

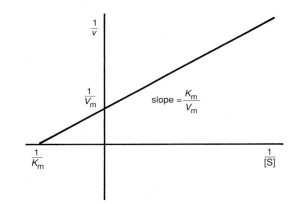

Fig. 1.2. Representations of Michaelis-Menten kinetics: a) dependence of the reaction rate (v) on the substrate concentration ([S]) for a reaction following Michaelis-Menten kinetics [93]; b) double reciprocal plot for linear representation (Lineweaver-Burk plot [94])

which after rearrangement gives

$$\frac{([E_0] - [ES])[S]}{[ES]} = \frac{k_2 + k_3}{k_1} = K_m \tag{3}$$

where K_m is the Michaelis-Menten constant. This constant provides information about the stability of the enzyme-substrate complex (ES): the smaller its value the more stable the complex. Characteristics of the enzyme system can be established by varying the concentration of the substrate and measuring the steady state rate (v) of the reaction at a fixed enzyme concentration ($[E_0]$) (Fig. 1.2.a) such that:

$$v = \frac{d[P]}{dt} = k_3[ES] = k_3\frac{[E_0]\,[S]}{K_m + [S]} \tag{4}$$

At low substrate concentrations increasing [S] results in a steep and initially linear rate increase followed by an attenuated rate increase which finally approximates to a maximum (V_m) at high substrate concentrations. At this point the enzyme becomes saturated, i.e. completely transformed to the complex (ES). Consequently, V_m is dependent on the quantity of the enzyme present:

$$V_m = k_3[ES_{max}] = k_3[E_0] \tag{5}$$

Combination of Eqs. 4 and 5 provides the Michaelis-Menten equation as expressed by K_m and V_m, thus:

$$v = V_m\frac{[S]}{K_m + [S]} \tag{6}$$

For given conditions, the rate and profile of an enzyme reaction is characterized by both K_m and V_m. (A change of conditions results, of course, in different constants.) This was extended by Lineweaver and Burk [94], who said that the reciprocal form of the Michaelis-Menten equation lends itself to the practical determination of the two constants, thus:

$$\frac{1}{v} = \frac{K_m + [S]}{V_m[S]} = \frac{K_m}{V_m}\frac{1}{[S]} + \frac{1}{V_m} \tag{7}$$

Accordingly a plot of $1/v$ vs. $1/[S]$ yields a straight line provided the process follows Michaelis-Menten kinetics (Fig. 1.2.b). Thus, simply by taking a few experimental points the values of K_m and V_m can be determined. This, and similar linear plots may also, of course, help to identify enzymic processes deviating from Michaelis-Menten kinetics [78,86-91].

The latter in fact, are not rare and are often met in multiphase systems in which the reaction kinetics are complicated by transport processes across a phase boundary. A prime example of this are lipase catalyzed hydrolyzes which proceed at the boundary of aqueous and organic phases. Important kinetic limitations may be set by various forms of inhibition [78,86,88,90], perhaps the most inconvenient of which is product inhibition, especially when it is non-competitive (Fig. 1.3). Special strategies may help to avoid it [95] but there is no general solution to the problem.

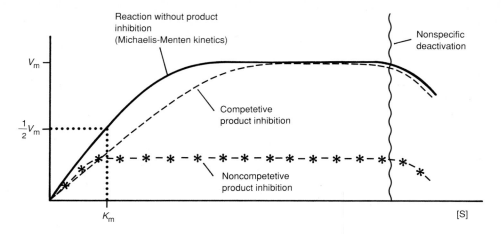

Fig. 1.3.   Types of product inhibition

1.1.4   Enzyme Immobilization

The synthetic application of enzymes carries a number of inherent disadvantages. Enzymes used in synthesis are water soluble and relatively unstable proteins. This can lead to difficulties in the isolation of the product and in addition, an enzyme once dissolved usually cannot be recovered intact. Since pure enzymes are still relatively expensive (see Table 1.2) applications on a large scale can become uneconomical. Because of this, soluble enzymes are only used in batch processes, which means that in the isolation step following the transformation a dissolved enzyme most often becomes denatured.

In the past 25 years efficient techniques for enzyme immobilization have been developed (Table 1.3) [62-77] in order to solve the problems mentioned above.

Immobilized enzymes have several advantages, the most important of which are listed below:

The biocatalyst can be easily removed from the reaction medium and recycled.

Continuous processes become possible, whereby operational costs decrease and a higher degree of control is facilitated.

Immobilization may enhance stability, activity and selectivity of enzymes.

Storage and handling of the immobilized protein is often easier.

The practical value and characteristics of immobilized enzymes are much influenced by the method of fixation and the inherent properties of the carrier. On the one hand, a properly immobilized enzyme may even be more active than its soluble counterpart, whilst on the other hand activity can also be seriously impaired or even completely destroyed if, for example, a covalent bond is established between the carrier and a vital group in the active center.

Table 1.2. Prices of commercial enzymes[a].

Enzyme	E.C. Number	Specific Activity[b] [U/mg protein]	Price [US $/1000 U]
HLADH	1.1.1.1	2	150
YADH	1.1.1.1	400	0.4
TBADH	1.1.1.2	90	260
Glycerol DH	1.1.1.6	60	70
L-Lactate DH	1.1.1.27	1200	0.7
D-Lactate DH	1.1.1.28	2000	4
Glucose DH	1.1.1.47	200	360
3α-HSDH	1.1.1.50	30	1600
20β-HSDH	1.1.1.53	10	2300
7α-HSDH	1.1.1.159	20	1200
Glucose oxidase	1.1.3.4	200	0.3
Formate DH	1.2.1.2	3	770
Aldehyde DH	1.2.1.5	25	240
Catalase	1.11.1.6	65000	0.001
Peroxidase	1.11.1.7	1200	1
Lipoxidase	1.13.11.12	5	4
Hexokinase	2.7.1.1	300	7
Glycerokinase	2.7.1.30	70	25
Pyruvate kinase	2.7.1.40	500	2
Acetate kinase	2.7.2.1	600	300
PLE	3.1.1.1	200	4
PPL	3.1.1.3	50	0.0001
Lipases	3.1.1.3	50-400000	0.001-80
Phospholipase A	3.1.1.4	600	5
Cholinesterase, acetyl	3.1.1.7	300	13
Cholesterol esterase	3.1.1.13	20	40
Lipoprotein lipase (CvL)	3.1.1.34	1300	0.5
Lipoprotein lipase (PsL)	3.1.1.34	2100	0.5
Phosphatase, alkaline	3.1.3.1	1500	15
Phosphatase, acidic	3.1.3.2	1	40
Phosphodiesterase	3.1.4.17	25	120
α-Amylase	3.2.1.1	1000	0.015
β-Glucosidase	3.2.1.21	6	9
β-Galactosidase	3.2.1.23	900	8
α-Chymotrypsin	3.4.21.1	50	0.2
Trypsin	3.4.21.4	10000	0.001
Protease (Subtilisin)	3.4.21.14	15	6
Papain	3.4.22.2	3	0.09
Thermolysin	3.4.24.4	100	1.4
Acylase I (Aspergillus)	3.5.1.14	2.3	0.5
Penicillinase	3.5.2.6	200	5
Hydroxymandelonitrile lyase	4.1.2.11	100	20
Aldolase (rabbit muscle)	4.1.2.13	15	10
N-Acetylneuraminic acid aldolase	4.1.3.3	30	12000
Aspartase	4.3.1.1	5	1800
Fumarase	4.2.1.2	350	10
Glucose isomerase	5.3.1.5	30	0.005
Pyruvate carboxylase	6.4.1.1	10	1100

[a]Price quotations for 1990 [49]; [b]One International Unit (IU) is the amount of catalytic activity in 1 mg of enzyme which converts 1 μmol of standard substrate to product in 1 minute.

Table 1.3. Techniques for enzyme immobilization.

I) *Cross-linking*
    1) without spacer
    2) with an inert spacer

II) *Binding to solid support*
    1) by physical adsorption
    2) by ionic interactions
    3) by metal-ligand attachment
    4) by covalent bonds

III) *Entrapping*
    1) gel entrapping
        a) by polymerization of monomers
        b) by cross-linking of linear prepolymers
        c) by gelation of long-chain polymers
    2) fiber entrapping

IV) *Retention by semipermeable membranes*
    1) microencapsulation
        a) within semipermeable polymeric membrane systems
        b) within liquid surfactant-membrane systems
    2) macroscopic membrane systems
        a) ultrafiltration membrane systems
        b) hollow fiber devices

For further details about immobilization techniques and the properties of fixed enzymes you are referred to the relevant literature given above.

1.1.5 Effect of Operational Conditions on Enzyme Activity

The catalytic activity of enzymes depends, in addition to substrate concentration, on several other factors. The basic kinetic constants (e.g. K_m and V_m) are modified by external effects such as temperature, pH, ion milieu, ionic strength, cosolvents, solvents and others. A thorough knowledge of these effects enables one to control, modify and optimize a given biotransformation.

Temperature
When plotted against temperature the rate of enzyme catalyzed processes passes through a maximum (Fig. 1.4.a). This maximum has been called, mainly in the older literature, the temperature optimum of enzyme activity. It must be borne in mind,

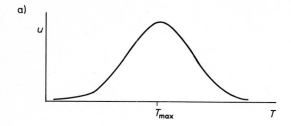

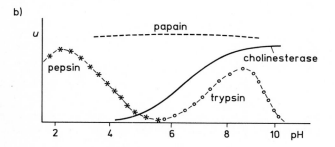

Fig. 1.4.   Dependence of the apparent enzyme activity (U): a) on temperature (T); b) on pH

however, that this temperature dependence arises as the resultant of several factors: the rate constant of the catalytic step increases with increasing temperature (the ascending leg of the curve), but simultaneously the steric structure of the enzyme also undergoes change. With rising temperature, the ordering forces maintaining this steric structure gradually lose their grip and as a consequence initially reversible, and later irreversible denaturation takes place (the descending leg of the curve). Denaturation diminishes enzyme activity to the point of total inactivation. Therefore when activity is considered over an extended period of time the term "optimum temperature" becomes meaningless.

The enzymes of some thermophilic organisms adapted to high temperature show resistance to thermal denaturation up to 60-80 °C. As a result of this high thermal stability, much interest has been shown in trying to isolate these species and their proteins and utilization of such enzymes and organisms as biocatalysts holds much promise for the future [96-98]. Enzymes may be active at low temperatures, e.g., proteases efficiently catalyze peptide synthesis even in frozen aqueous solution [192].

pH

The pH dependence of the rate of enzyme catalyzed processes usually follows a bell-shaped curve too (Fig. 1.4.b), and it is thus possible to define a pH optimum for a given process. pH dependence reflects the interplay of several factors, namely:
– changes in the dissociation of ionizable groups in the active center;

–   changes in the dissociation of other ionizable groups of the enzyme which induces a
    modification of the overall steric structure of the enzyme which in turn also affects
    the active center;
–   changes in the state of the eventual coenzyme.

Since it is controlled by many factors, the pH sensitivity of individual enzymes is
very different. Some remain active over a wide range of pHs, while others require more
careful pH control (Fig. 1.4.b).

In addition, *ion milieu, ionic strength* [99-101] or *pressure* [102] can also effect enzyme
catalysis.

Organic solvents and cosolvents

The use of organic solvents either as cosolvents, in a two-phase water-solvent system or
even alone is gaining ever wider acceptance as a component in the synthetic application of
biocatalytic processes [29,50-53,103-105]. Solvents promote the solubilization of water
insoluble substrates and as such can influence both selectivity and the state of equilibria.
Some transformations which are impossible to implement in aqueous media may be
possible in organic solvents [29,106] (Table 1.4). Examples of such applications will be
quoted later.

Table 1.4. Potential advantages of organic media over aqueous media in biotransformations.

I) *From the enzyme side:*
  1) Enzymes are insoluble in organic solvents, therefore
     a) the enzyme can be separated by filtration,
     b) immobilization can be carried out by simple adsorption.
  2) Enzyme stability increases, which means
     a) increased thermal stability,
     b) decreased proteolysis,
     c) danger of microbial infection eliminated.

II) *From the side of the enzyme catalyzed process:*
  · 1) Transformations requiring water are suppressed, that is:
     a) equilibria involving addition - elimination of water
        are shifted to the side of elimination,
     b) side reactions involving water addition are suppressed.
  2) Non-polar, hydrophobic substrates (i.e. most of the
     non-natural ones) are soluble in organic solvents.
  3) Selectivity often increases in organic solvents as
     compared to aqueous media.

III) *From an operational point of view:*
  1) Work-up and product isolation becomes simpler.
  2) Concurrent realization of enzyme catalyzed and chemical
     transformations becomes possible.

1.2 Practical Aspects of Transformations Carried Out with Microorganisms

Microorganisms are extensively used for the manufacture of a variety of useful products, such as foodstuffs, solvents, biochemicals, simple organic intermediates or even complex natural products [107-111].

Microorganisms can be applied in two different, relatively distinct ways to generate these products, namely by fermentation [3, 107, 108] or by microbial transformations [109,112-116].

In a *fermentation process* enzymes already present or inducible in a microorganism convert simple compounds in a cascade of well organized metabolic steps to the end product, which is a natural compound. Fermentations usually operate with cell populations in their growth phase. The process is relatively long, and a frequently low final concentration of the target metabolite means that the isolation of the product is often tedious.

In *microbial transformations*, in contrast, a well defined reaction is performed by one or a few enzymes of the microorganism on an added substrate (which may be a natural product or not); the product does not undergo further change, accumulates in the system and can ultimately be isolated. Microbial transformations can be carried out both with growing and resting cultures, as well as with immobilized ones. They generally take less time than fermentations, yielding a product of higher purity and higher concentration which in turn facilitates product isolation.

Scientists have been aware of microbial transformations for many years [1,2,117], some early examples being the microbial oxidation of ethanol to acetic acid by Pasteur [118] or the transformation of glucose to gluconic acid, first reported in 1880 [119]. It was, however, not until about fifty years ago when the potential of this sort of process to transform steroids was recognized that the method became really important and found large scale application. Ever since it has been regularly employed for the preparation of steroids [120,121], alkaloids [121,122], amino acids [19,62], nucleic acids and their analogues [123,124], carbohydrates [125] and relatively recently it has found its way to being used for the preparation of general organic intermediates and fine chemicals [10,14,107,116].

When considering the synthetic application of microorganisms, it is useful to keep in mind the following points:

i)   Availability. Several microorganisms can be purchased cheaply and in large quantities as resting cell mass; firstly baker's yeast (*Saccharomyces cerevisiae*) but also various other strains of beer and vine yeasts and other species used in the food industry are readily available. A large number of strains are accessible as samples of pure cultures from national typed culture collections. In this case, the necessary quantity of cell mass must be prepared by propagation of the inocculum, and this may require proficiency in special microbiological techniques [126,127]. Constancy and ease of handling of the culture are also important.

ii)   Each microorganism produces a multitude of enzymes, some of them are present permanently, whilst others are inducible, thus more than one type of reaction can be realized with a single species of microorganism (e.g. redox reactions, hydrolyzes, addition or condensation reactions). This then highlights the problem of selectivity for a given microbial transformation. Was the substrate transformed only to the desired product, or did it also undergo an undesirable reaction? Further secondary transformation of the product must also be considered and it is additionally possible that the same transformation (e.g. carbonyl reduction) can be catalyzed by different enzymes with opposite stereoselectivity [128-130].

iii)   Since microbial transformations are all enzyme catalyzed processes, most of the features already discussed for enzymic reactions (selectivity, chiral catalysis, mild conditions, kinetics, the role of external parameters) also apply, with certain limitations to microbial transformations.

iv)   With transformations requiring cofactors which are troublesome to regenerate or which require enzymes that are unstable when isolated, the use of intact microorganisms must be preferred, since in this case enzymes work under *in vivo* conditions and regeneration of cofactors is taken care of by the cells' own natural enzyme system. Similarly, with multistep transformations needing several enzymes, the use of intact living cells is recommended since here, not only are all the necessary enzymes present but they are present under more or less optimum conditions, whereas in an *in vitro* multienzymic procedure [72] optimization may require some degree of effort, even in the case of two cooperating enzymes [131].

v)   Ease of handling and stability of the microorganism are important. It is convenient if no special conditions (anaerobic or sterile operation, expensive additives, etc.) need be provided and the stability of the active cell mass is preserved for a sufficiently long time. It is also an asset if the active cell mass can be stored in some form, e.g freeze dried, for a long period.

vi)   Most preferable are microorganisms which can be conveniently immobilized [67,70,71,73,77,132-138]. Such systems can offer simple product recovery, recycling, and increased stability of the biocatalyst, as well as the potential of continuous operation.

vii)   Unfortunately there are no golden rules available for selecting the microorganism best suited for a given transformation. A common practice is to screen pure cultures selected by way of analogy and a number of rough guidelines exist to aid the preselection of species to be tested [113]. Yeasts are most effective for hydrolysis and the reduction of carbonyl compounds, fungi are suitable for hydroxylations, while bacteria exhibit the highest oxidative potential (they may oxidize organic molecules to carbon dioxide and water). Whilst hydrolyses are generally fast (a few hours), bacterial oxidations and hydratations may be equally rapid, reductions carried out with yeasts are more lengthy. Transformations by fungi require the longest time of all. The judicious use of known analogies [109,112,115,121] may help in cutting down on the screening period.

If readily available microorganisms fail to produce results, or it is the isolation of new strains which provides the main purpose of the study, the candidates must be collected from natural sources (from soil, or from places exposed to extreme conditions such as high temperature or salt concentration).

viii) The efficiency of microbial transformations can be characterized by a s.c. productivity number (PN) [80] defined by:

$$PN = n_{prod}/m_{dry} \cdot t$$

where n_{prod} is the mass of product (mmol), m_{dry} the mass of the dry cell mass (kg) and t the time of transformation (h). This number resembles specific activity as defined for pure enzymes, but includes in its definition in addition to the specific activity of the enzyme, several other important factors (the effects of inhibitions, transport processes, concentrations) too. The higher the PN value for a given transformation, the higher the output per unit volume and time. A high PN value is important for the economic isolation of the target product since it means less cell mass and volume per unit product.

1.2.1   How to Carry Out Microbial Transformations

Microbial transformations can be realized with cells at different stages of their life cycle, i.e. with growing cultures, resting cell mass, or spores. In all of these cases immobilization can also be effected. In addition, the combined use of more than one microorganism for multistep processes is possible [14,109].

With biotransformations carried out with *growing cultures* the substrate is added at a specific stage of microbial growth. Thus, optimal timing and mode of addition are important and it is helpful if the substrate is an inducer of the very enzyme responsible for the desired transformation. Conversely, substrate inhibition of cell growth may be a nuisance. In growing cultures, transformation of the substrate and cell growth run hand in hand, and as a result such cultures are often preferred because the whole operation takes less time. Similarly, addition of the substrate directly to growing cultures is also the method of choice in screening experiments.

When working with *resting cultures* preparation of the cell mass and biotransformation take place in separate operations. Initially, cells are grown on a favorable medium and subsequently, after separation by centrifuging or filtration, the cells are suspended in a much simpler s.c. transformation medium to which finally the substrate is added and biotransformation accomplished. In this way it becomes possible to optimize both steps independently and thus the best substrate - cell mass ratio can be easily achieved. In addition, isolation of the product may be easier from a less complex mixture and since cell nutrients are scarce or absent in the transformation medium, so predisposition to infection by alien organisms is much reduced, often to an extent where non-sterile operation is feasible.

Spores of fungi have often been employed in biotransformations [139,140]. For the preparation of spores fungi are cultivated on special media which promote sporulation. The same transformations can be carried out with spores as with miceliums, but the use of spores may offer some advantages. Firstly, the productivity number of spores often surpasses that of miceliums, however the main asset of spores is stability, since they can be stored in large quantities for long periods of time.

Biotransformations consisting of two or more steps and requiring more than one microorganism can be carried out in a single operation working with different microorganisms simultaneously, or with a combination of microorganisms and enzymes. The advantage of such complex systems [141] is that there is no need for the isolation of intermediates. Two examples of such combined microbial transformations are the preparation of L-lysin from α-amino-ε-caprolactam by enantiomer selective hydrolysis and racemization of the unchanged lactame [124] or the preparation of L-alanine from fumarate using microorganisms exhibiting both aspartase and aspartate decarboxylase activity [143].

Reports of detailed studies on transformations with microorganisms in different stages of their life cycle and their optimization are available in the relevant literature [109,126,127,144,145].

1.2.2   Immobilized Cells

Systems working with intact cells, both living and dead, may use them in an immobilized form [70,71,73,77,132-135,137,138] (Table 1.5).

Most of the methods currently employed for immobilizing cells are merely modified and extended versions of procedures originally developed for fixing enzymes. There are but a few of them that can only be used for the immobilization of cells and that are unsuited to the fixation of the much smaller enzyme molecules. Among the methods listed in Table 1.4, inclusion into polymer matrices (e.g. synthetic prepolymers [146], natural polysaccharide aggregates [147]) are the most wide-spread although in large scale operations membrane reactors [74] are also often used.

Immobilization of cells offers many advantages. For example, using these techniques the tedious and expensive procedures of enzyme isolation and purification can be avoided. Within cells, enzymes are in a natural environment, therefore their stability, temperature and pH tolerance, as well as resistance toward toxic effects is higher than in an isolated form. Advantages are even more pronounced with multienzyme transformations or when cofactors are needed (*in vivo* cofactor regeneration). Immobilized cells can also be used in continuous operation, whereby higher cell density and flow rates can be realized, product isolation becomes much simpler and repeated recycling of the biocatalyst is also possible.

Table 1.5. Techniques for cell immobilization.

I) *Carrierless immobilization*
  1) natural flocculation
  2) flocculation by pH-shift, heat shock, detergents, metal
     ion treatment or other effects

II) *Cross-linking with bifunctional reagent*
  1) without spacer
  2) with an inert spacer (e.g. gelatin, albumin)

III) *Immobilization on solid support*
  1) adsorption onto solid support
    a) by electrostatic interactions
    b) by ionic interactions
    c) by biospecific adsorption to an immobilized ligand
  2) covalent attachment to solid supports

IV) *Entrapment in gels/polymers*
  1) entrapment in synthetic polymers
    a) polymerization of monomers
    b) cross-linking of preformed linear polymers
  2) entrapment in natural polymers
    a) thermal gelation (e.g. agar, κ-carrageenan, gelatin)
    b) ionotropic gelation (e.g. alginate, chitosan)

V) *Entrapment in membrane systems*
  1) cells retained by semipermeable membrane
  2) cells on and/or within the matrix of membrane
  3) liquid membrane systems

VI) *Combinations of the above methods*

The less advantageous aspects of cell immobilization must also be included in this discussion. Problems may arise with keeping immobilized cells alive and supplying them with nutrients and energy sources. One of the main causes of activity decline in immobilized cell systems is cell death. Similarly, in systems with cells enclosed into polymeric matrices or semipermeable membranes, factors affecting material transport, permeability and diffusion may also impose limitations by hampering the take up of both nutrients and substrate, as well as the release of product. For this reason, the catalytic capacity of immobilized cells can rarely be fully exploited and optimum conditions can only be secured in a thin surface layer. In the case of aerobic cells diffusion control of oxygen uptake may be a severe limiting factor and finally, one cannot exclude from immobilized cells the problems caused by side reactions catalyzed by alien enzymes.

For successful optimization, a number of special features associated with immobilization should be considered [148].

1.3 Novel Types of Biocatalysts

The rapidly advancing methodologies of biotechnology have enabled the introduction of several novel and promising types of biocatalysts. By means of chemical, biochemical or genetic approaches it has become possible to modify enzymes or even cells. Thus, knowing the mechanism of action of an enzyme, semisynthetic and fully synthetic "enzymes" can be prepared which mimic the natural enzyme and even more exciting, with the aid of immunology and gene technology biopolymers other than enzymes exhibiting catalytic activity can be obtained.

1.3.1 Biotransformations with Cells of Higher Order Animals and Plants

Biotransformations can also be carried out with animal and plant cell cultures [149]. Since animal cells are more sensitive and are more difficult to culture, initially only cultures of plant cells were used for syntetic purposes [150, 151].

Plant cell cultures are mainly used to prepare secondary plant metabolites such as alkaloids, flavonoids, terpenes, polyphenols, etc. [152] although other kinds of biotransformations are often performed with such systems too. For example, methyldigitoxin can be converted to methyldigoxin by cell cultures of *Digitalis lanata* [153] and similarly codeinon to codein by cell cultures of *Papaver somniferum* [154].

Plant cells are also amenable to stereoselective transformations. Using immobilized cells of *Nicotiana tabacum*, ethyl acetoacetate can be reduced to ethyl *(S)*-3-hydroxybutanoate [155].

If forced to compare, the main advantage of plant cell cultures over microorganisms may be their longer lifetime, whilst as disadvantages slow growth and lower productivity may be cited.

1.3.2 Artificial Metabolism, Artificial Cells

A number of novel types of biotransformation are made possible by the use of multienzymic methods, often called artificial cells or artificial metabolism [72]. Multienzymic systems can be useful for the recovery of cofactors [14,81].

Complex biotransformations can be realized with these systems encapsulated into a semipermeable membrane, the so called artificial cell. It is such a system for example, that facilitates the preparation of L-glutamic acid from α-ketoglutaric acid (1.1) [156] (Fig. 1.5).

It is not unfeasible with multienzyme systems to realize, *in vitro*, the metabolic processes of living organisms or analogues transformations. For example, consider the preparation of *N*-acetyllactosamine [157] which requires the cooperation of six distinct enzymes involved in the biosynthesis of oligosaccharides (Leloir pathway).

Similarly, the manufacture of ethanol from glucose by means of a complex enzyme system comprising not fewer than 12 enzymes [81,158] is, of course not a new concept but clearly exemplifies that it is possible to operate such a highly complex system under *in vitro* conditions. Furthermore, ribulose-2,5-diphosphate is prepared on a several hundred grams scale again in a multienzyme system containing not less than six different enzymes [159].

There is no reason why multienzyme systems cannot, of course, be composed of enzymes which do not coexist in nature and this approach may gain increasing importance in the future.

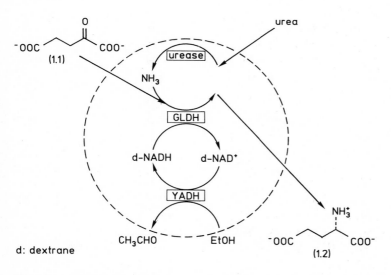

Fig. 1.5. Artificial cells: multienzyme transformation of α-keto glutarate to L-glutamic acid within microcapsules [d-NADH: NADH bound to dextran]

1.3.3   Modified Enzymes and Cells

The catalytic properties of enzymes and microbial cells can be significantly modified by chemical and genetic methods in order to create novel biocatalysts.

I) Modified enzymes

Modification of enzymes can be achieved both through chemical means and by genetic engineering. Modification may affect solely the active center or alternatively it could alter the complete structure of the enzyme as a whole.

i) An example of such *modification of overall structure* can be seen with the attachment of polyethyleneglycol chains to enzyme, whereby the enzyme becomes soluble in organic solvents [29,56].

  Gene manipulations can be used to enhance enzyme stability, for example by introducing additional disulfide bridges through directed mutagenesis[3] [160].

  This was neatly demonstrated in an impressive application of site-directed mutagenesis involving six parallel mutations, whereby it was possible to improve the stability of the enzyme subtilisin by a factor of 50-100 while its specificity remained unchanged. Computer aided molecular modeling was used to select the sites of modifications [161].

ii) Catalytic properties, such as pH profile and substrate selectivity can be modified by *changing the active center.*

  By using *the chemical modification of the active center* in pig pancreatic lipase (PPL), for example, the activity of the enzyme in hydrolyzing triglycerides was enhanced by reductively alkylating the lysine and histidine side chains in the enzyme [164]. Furthermore, subtilisin was modified by converting serine units to cysteines. The resulting product, called thiol-subtilisin, has a very weak protease activity and is thus highly amenable to the enzyme coupling of peptide units [165]. Similarly, protease activity in α-chymotripsin can be suppressed by methylation of the nitrogen in histidine-57 [166,193].

  Using *genetic modification of the active center* it has been possible to prepare several variants of subtilisin with different catalytic activities [167], another example of this being the alteration of the substrate specificity of L-lactate dehydrogenase from *Bacillus stearothermophilus* by site-directed mutagenesis [168].

II) Modified cells

The modification of cells can be effected by physical, chemical, and genetic methods.

i) One way of modifying cells by chemical and physical means is immobilization. Through physical treatment the permeability of cell walls and/or the cell aggregation properties can be changed.

ii) Modification of the genetic material of microorganisms (mutations) may increase the efficiency and/or selectivity of the required biocatalytic process or may even permit novel biotransformations. Thus a mutant of *Pseudomonas putida* in which the enzyme

---

[3] Directed mutagenesis of enzymes [147,30] is the replacement of a single amino acid in the protein chain by appropriate modification of the genetic code. Comparison of the mutant enzyme with the original one provides information about the function of the amino acid in question in the enzyme as a whole.

system for further transformation of *cis*-1,2-diols (1.4) is missing, can be used for the conversion of benzene and substituted benzenes (1.3) into *cis*-3,5-cyclohexadiene-1,2-diols (1.4)[169,170]:

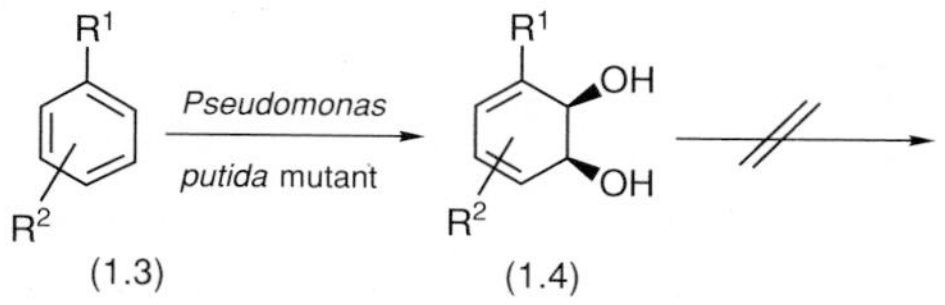

Recombinant DNA technology may revolutionize biotransformations [171,163,162]. With this technique, at least in principle, any DNA fragment originating from eucaryotes and procaryotes or even synthetic oligonucleotide fragments, can be incorporated into the genetic material of a suitable host organism (most often *Escherichia coli, Saccharomyces cerevisiae or Bacillus subtilis*).

Such recombinant DNA technology has been used to simplify the microbial transformation steps used in the conversion of D-glucose to vitamin C. 2-Keto-L-gluconic acid (1.7) can be prepared in one operation from D-glucose using a recombinant *Erwinia herbicola* strain, which contains the reductase gene of *Corynebacterium* to perform the reduction of 2,5-diketo-D-gluconic acid (1.6) [172]:

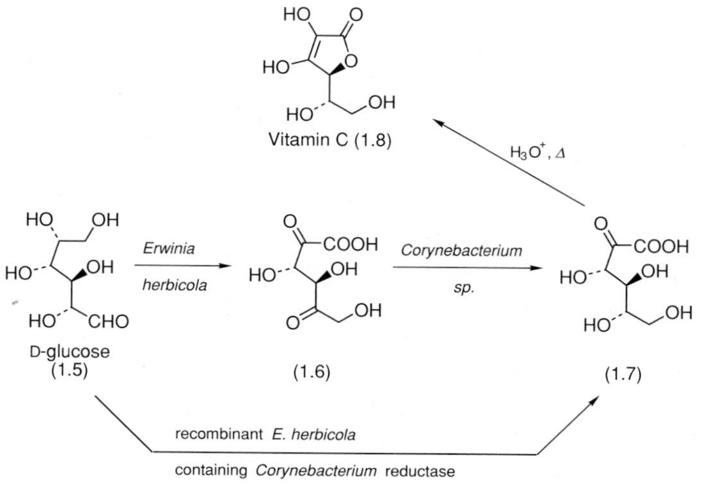

Vitamin C (1.8)

D-glucose (1.5)   (1.6)   (1.7)

recombinant *E. herbicola* containing *Corynebacterium* reductase

Chemical lactonization of 1.7 then yields vitamin C. Recombinant DNA techniques can be exploited to transfer an enzyme required for a certain biotransformation into an easy-to-handle microorganism [173].

1.3.4   Semisynthetic and Fully Synthetic Enzyme Models

Semisynthetic enzymes are novel protein based biocatalysts in which the catalytic site has been elaborated in an artificial way. For example, a semisynthetic esterase can be obtained from sperm whale Met-Myoglobin [174] by removing the heme group. In the deep hydrophobic pocket left behind an imidazole side chain exerting catalytic activity becomes exposed allowing the modified enzyme to hydrolyze p-nitrophenyl acetates. In a similar way, semisynthetic enzymes with oxidoreductase activity can be prepared. Thus, in flavohemoglobin a flavin coenzyme was bound covalently to globin subsequently allowing it, in a similar way to cytochrome systems, to hydroxylate aniline [175].

Enzyme models, sometimes called *synthetic enzymes* are catalytically active molecules which mimic the features recognized to be essential for activity in a given natural enzyme. They can bind substrates and possess groups exhibiting catalytic activity. Because such enzyme models are not proteins they have the advantage of being more stable and can thus be prepared by synthetic methods.

Hydrolysis [177,36], peptide synthesis [176], regio- and stereoselective reduction [178], transamination [179], and selective hydroxylation [180] exemplify the potentional of the synthetic enzymes.

1.3.5   Biocatalysis with Biomolecules other than Enzymes

i) Catalytic antibodies. Antibodies, i.e., proteins produced by the immune system in animals, are novel prospective biocatalysts [32,182,190]. Their application as biocatalysts was primarily facilitated by the advent of techniques to produce monoclonal, i.e., completely homogeneous, antibodies [183]. The importance of the so called abzymes (antibody-enzyme) lies in the fact that whilst the number of natural enzymes is limited to a few tens of thousands, immune systems are capable of generating millions of different antibodies as complimentary proteins to the stimulus of foreign antigenic material[183]. The potential of the immune system to produce selective antibodies to practically any molecule makes it a promising source of enzyme-like catalysts, with tailored selectivities and affinities.

So far, catalytic antibodies isolated have been reported to have catalyzed several types of reactions, including β-eliminations [184], redox reactions [185,195], Diels-Alder cycloadditions [186,196], cleavage of trityl protecting groups [197], or the stereoselective hydrolysis of alkyl esters [187]. Immobilized lipase-like catalytic antibodies were investigated in aqueous and organic media [198].

The primary approach for induction of catalytic antibodies has been to generate antibodies to haptens with a transition state analogue structure. The antibodies formed are the quasi-steric complements of the antigens. The procedure is explained in Fig. 1.6.

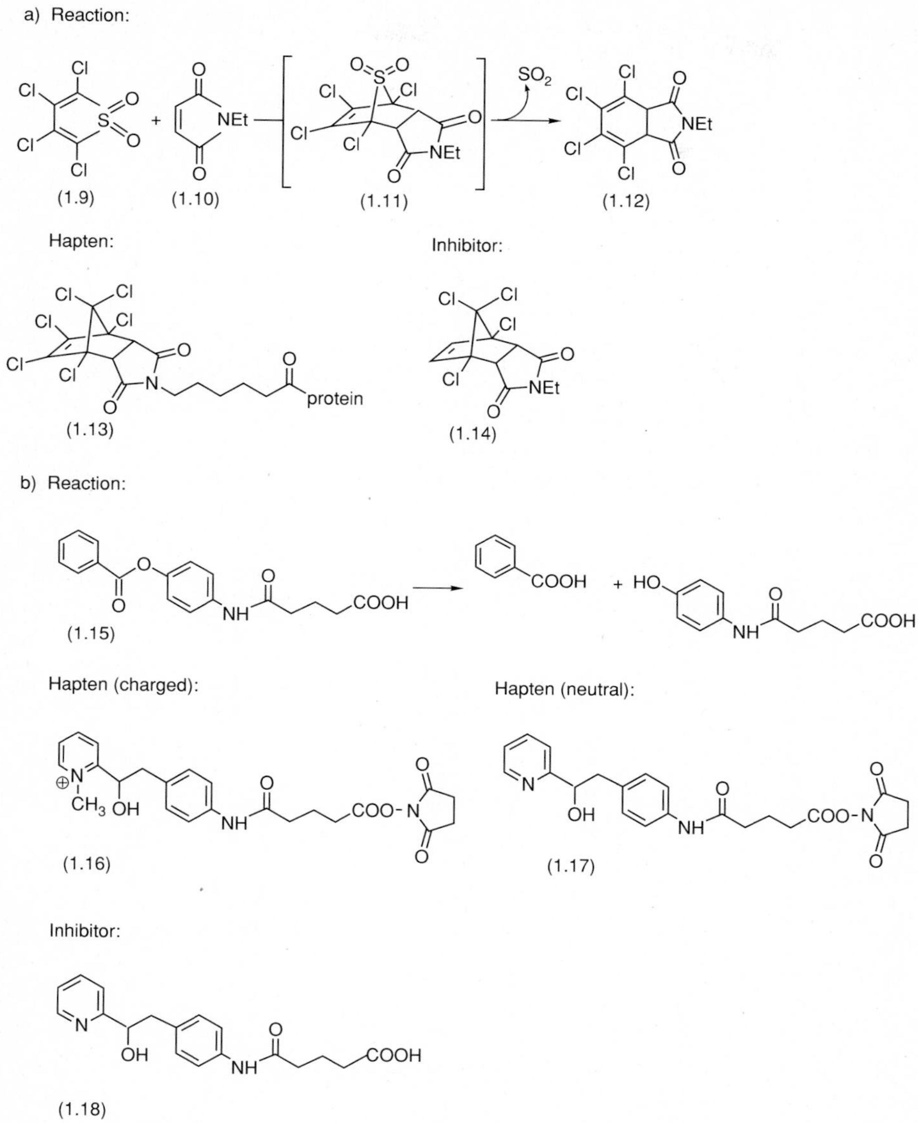

a) Reaction:

(1.9)   (1.10)   (1.11)   (1.12)

Hapten: Inhibitor:

(1.13)   (1.14)

b) Reaction:

(1.15)

Hapten (charged):   Hapten (neutral):

(1.16)   (1.17)

Inhibitor:

(1.18)

Fig. 1.6. Catalytic antibody generation against transition state analogs: a) abzyme generation for a Diels-Alder reaction; b) charge-controlled abzyme generation for hydrolytic reactions

In the Diels-Alder reaction of the educts 1.9 and 1.10, the stable final product (1.12) is formed by elimination of sulfur dioxide from the primary intermediate 1.11. The transition state of the reaction is product-like, therefore it was compound 1.13, an analogue of the primary product 1.11, which was used to induce antibodies [186]. One of the antibodies produced in this way was indeed found to catalyze cycloaddition. It was not surprising to find that compound 1.14, which is rather similar to the antigen (1.13), was a potent inhibitor.

In Fig. 1.6.b it is shown that in order to induce a catalytic antibody not only steric complementarity, but charge complementarity as well can be utilized [188]. In order to facilitate the hydrolysis of compound 1.15, apart from good binding, nucleophilic attack against the carbonyl group of the benzoyl ester moiety is also necessary. Thus steric complementarity is not sufficient for catalysis - clearly the presence of a strong nucleophilic negative charge in the vicinity of the carbonyl group is also a necessary presence. Further evidence can be construed from comparing the shape and bulk of hapten 1.16. It is similar to 1.15 and has a positive charge near the reacting group. Seven of the monoclonal antibodies induced by 1.16 proved in fact to be catalytically active, while not a single one of the antibodies induced by the hapten 1.17, sterically similar to 1.15 but neutral, was catalytically active.

The genetic code of antibodies with the required catalytic activity may subsequently be transferred into recombinant microorganisms whereby these tailor-made biocatalysts can then be produced in large quantities.

ii) Proteins with no intrinsic enzymic activity may assist as chiral additives in stereoselective chemical transformations. This is exemplified in the preparation of optically active sulfoxides by oxidation of sulfides in the presence of bovine serum albumin [181]. Bovine serum albumin can catalyze hydrolysis of racemic esters in an enantiomer-selective manner [194].

iii) Utilization of the catalytic activity of *biopolymers other than proteins*, e.g. that of RNA [189], is also being investigated.

2 Stereochemical Aspects of Biocatalysis

Biocatalytic processes can be selective in many ways. They may differentiate with respect to the type of the reaction catalyzed, to the constitution and stereochemistry of the substrate, or even with regard to the position and stereochemical features of the functional unit to be transformed. A large majority of biocatalytic transformations are highly stereoselective[1].

When offered their natural substrates under natural conditions enzymes generally exhibit total selectivity, i.e. the product is homogeneous within the error limits of existing analytical methods. This high selectivity is also often preserved towards artificial substrates. Biocatalytic reactions are usually stereoselective in a hidden way, i.e. one that can often only be demonstrated by special techniques, such as deuteration for example. A good instance of this is the reduction of cinnamic aldehyde by baker's yeast, where *(S)*-1-deutero-cinnamic alcohol is formed from the 1-deutero-aldehyde Fig. 2.1 [2].

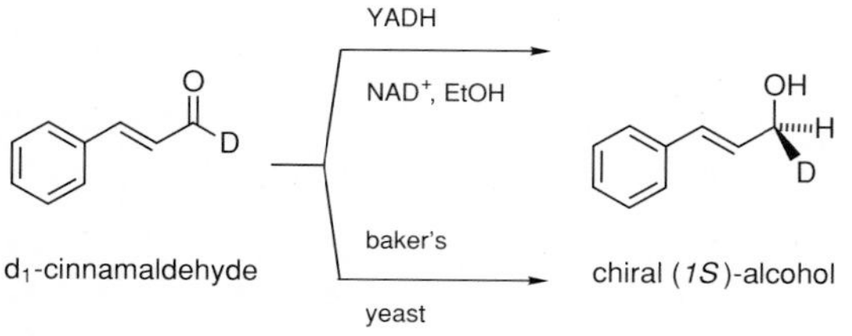

Fig. 2.1. Reduction of deuterocinnamaldehyde

Some knowledge of stereochemistry is needed in order to gain a full understanding of biotransformations. Evaluation of a stereoselective process for example, calls for

---

[1] According to Eliel [1] the term "stereospecific" is used for a reaction where substrates of different stereostructure are converted to products of different stereostructure, whereas the term "stereoselective" is intended to describe a reaction in which one or more of the possible stereoisomeric products predominates. However, since the terms stereo- (enantio- or diastereo-)specific are often used in a sense other than that defined by Eliel, we have avoided their use in the present work. We have taken "stereoselective" to encompass any transformation in which stereoheterotopic groups or faces in the same molecules are transformed to give a product ratio other than 1:1 or where stereoisomers are transformed at different rates.

the determination of the enantiomer composition of the products, whereas the outcome of selective biotransformations on a given substrate is determined by kinetic and thermodynamic parameters. All these problems will be discussed in the following sections.

2.1 Some Basic Concepts of Stereochemistry[2]

Constitution of a molecule is defined by the nature and number of the atoms it is built up from and by the order and nature of the bonds linking them (connectivity).

Configuration defines the mutual steric disposition of the atoms constituting the molecule. This, to a first approximation, disregards those arrangements which at room temperature differ only by an unhindered rotation around single bonds or bonds of sufficiently low order.

Different *conformations* of a molecule arise by rotations around single bonds.

Having characterized the make-up of individual molecules, let us now consider *the relationships between such molecules*. Molecules may differ in their composition and as such, they will also differ in their molecular formulae. Those molecules however, which have the same molecular formulae but are nevertheless distinguishable are called *isomers*. *Constitutional isomers* differ in their constitution, whilst *stereoisomers* have the same constitution but different configuration. Differences between constitutional isomers can be defined in terms of the degree of connectivity of their atoms (e.g. by connectivity matrices or systematic names), whilst description of stereoisomers requires additional identifiers, mostly those denoting directions or handedness.

Stereochemical characterization of molecules and their conformations (in general molecular models) is possible through observation of their symmetry: an *achiral* model is superimposable onto its mirror image, whilst a *chiral* model is not.

Stereoisomers are different molecules of the same constitution and their relationship can be twofold. Molecules which are different and related as object to mirror image are termed *enantiomers*, while stereoisomers which are not mirror images of each other are called *diastereomers*.

If a molecule is chiral, a full description of its stereoisomers requires a definition of *absolute configuration*. Molecules may contain more than one stereogenic element, this being the structural subunit around which a change in the arrangement of ligands under given energetic conditions and maintaining conservation of constitution, can give rise to stereoisomers. The configuration of each of these units must be defined.

---

[2]   For a more detailed treatment see refs.[3-8].

Absolute configuration of a chiral element is most generally defined by the Cahn-Ingold-Prelog (C.I.P) notation using the symbols *R* and *S* [9-10], configuration around double bonds is most often characterized by the symbols *E* and *Z* [11].

There is less agreement in the literature about the description of the *relative configuration* of two or more chiral centers in the same molecule. For chiral centers that are incorporated into rings, the prefixes *cis* and *trans* are used, whilst for open chain systems the prefixes *anti* and *syn* are preferred and used in the manner described below.[3]

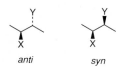

anti   syn

From the point of view of stereoselective reactions, relationships within a single molecule can be conveniently characterized by the equivalent relationships of its *homomorphic* and constitutionally equivalent groups and faces (Fig.2.2).[4]

Homotopic groups and faces are related by a rotational axis C_n, (where $\infty > n > 1$). These groups and faces are otherwise indistinguishable in terms of any physical, chemical, or biochemical transformations, or interactions they may undergo, this being inclusive of chiral systems as well. Transformation of homotopic groups or faces does not generate a new element of chirality.

Heterotopic groups or faces in contrast, can be differentiated through interaction with chiral systems. More specifically in fact, enantiotopic groups or faces can only be distinguished by chiral systems, whereas diastereotopic groups or faces may behave differently even when interacting with achiral systems.

Enantiotopic groups and faces can be exchanged by means of a reflection-rotation axis within the molecule, S_n, (in its simplest form a plane of symmetry, S_1). Reaction of such a species with an achiral reagent leads to a racemic product (i.e. two enantiomers in a 1:1 ratio).

Diastereotopic groups and faces cannot be exchanged by any symmetry element. The presence of such groups or faces within a molecule does not preclude the presence of symmetry elements in the same molecule, but these must not exchange the groups or faces in question. Transformation of diastereotopic groups or faces generally leads to diastereomers.

---

[3] According to IUPAC Rules *anti* and *syn* can also be used to characterize modes of approach in addition reactions. Because of its ambiguity the *erythro/threo* form of notation has been avoided since its meaning changes depending on whether the molecule is depicted in a Fischer projection or in a zig-zag conformation. Formally, the *erythro* compound was the isomer in which the vicinal substituents were at the same side when depicted in Fisher projection but the meaning of this notation has been changed in order to include the zig-zag conformation [12].

[4] According to Hirschmann [13] homomorphic species are those groups of atoms which become superimposable when detached from the rest of the molecule.

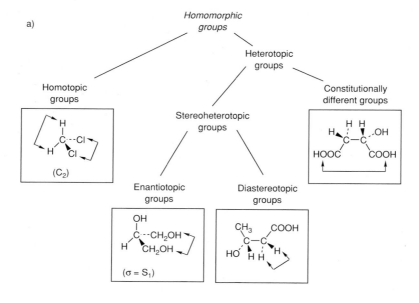

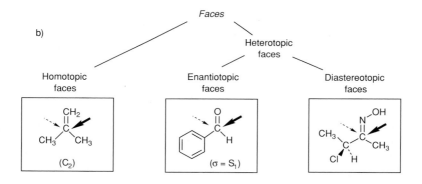

Fig. 2.2. Topology of tetragonal or trigonal centers: a) topicity of groups [17]; b) topicity of molecular faces

Groups of different constitution (the positions of which can be given by different numbers without referring to directions) but of identical structure differ, of course, in any of their interactions.

For an adequate description of biochemical processes, *intramolecular heterotopic (enantio- or diastereotopic) relationships* initially require more specific characterization since biocatalysts, as chiral agents, can readily distinguish between heterotopic groups or faces.

A common feature of enantiotopic and diastereotopic groups attached to tetragonal centers is *prochirality*. At prochiral centers an exchange of one of the groups in question for a new one which had not been a ligand of the center generates a new center of chirality. If an elevation favoring one of the heterotopic groups over its pair, whilst not surpassing the remaining ligands (all defined in terms of the C.I.P. convention), generates a center of *R* configuration, then that particular group of the pair is labelled *pro-R* (marked by *R* in lower index) whilst its partner is labelled *pro-S*. Note that the configuration of a product resulting from an actual substitution will be defined by the relative preferences of the newly introduced group in relation to the existing ones (cf. Fig. 2.3.a).

Faces of trigonal systems can be characterized in a similar way (Fig. 2.3.b) [14-16]. If in an Xabc arrangement the order of preference of the ligands regarded from one side is clockwise, that particular face is labelled *pro-R* (or *re*), whilst the opposite face can be assigned the label *pro-S (si)*. [5] Again, prochirality symbols assigned to faces do not predetermine product configurations. A chiral center arising from attack at the *re* face may be both *R* and *S*, depending on the relative preference of the entering ligand.

If both bridgehead atoms of a double bond are of the type Xabc, prochirality symbols should be assigned separately to the faces around both bridgehead atoms. (Fig.2.3.c).

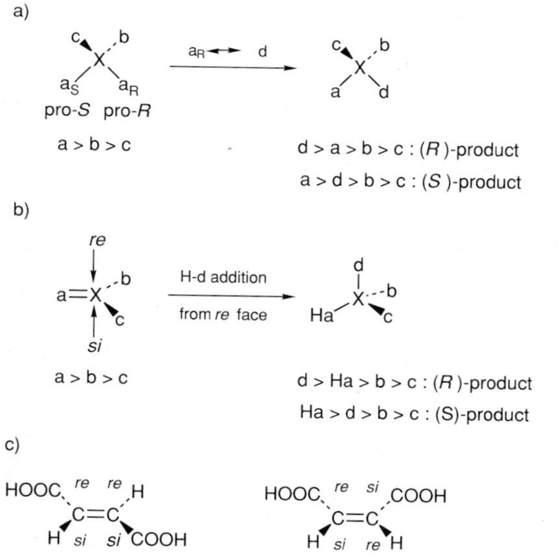

Fig. 2.3. Hypothetical transformations on prochiral centers

[5] Prelog suggested the use of the symbols *Re* and *Si* [15]. These have in almost every case the same meaning as *re* and *si*.

2.2 Determination of Enantiomeric Composition

As with all other stereoselective procedures, any meaningful evaluation of a biocatalytic transformation demands as prerequisite convenient and dependable methods for determining the structure and stereoisomeric composition of the products, including some measure of the enantiomeric purity. The most important up-to-date methods for the determination of the latter are briefly reviewed here.

Optical Purity and Enantiomeric Purity

Enantiomeric composition may be characterized by optical purity and/or enantiomeric purity [7,17]. Optical purity is defined as:

$$P = \frac{[\alpha]}{[\alpha]_{max}}$$

where $[\alpha]$ is the specific rotation of the sample and $[\alpha]_{max}$ the specific rotation of the pure enantiomer. It is often given as a percentage (the so called optical yield):

$$Op = \frac{[\alpha]}{[\alpha]_{max}} \times 100$$

Under ideal circumstances (i.e. high dilution), the numerical values of optical and enantiomeric purity, or enantiomeric excess (*ee*) are identical:

$$ee = \frac{E - E^*}{E + E^*}$$

where E is the mole fraction of the predominant enantiomer and E^* is the mole fraction of its enantiomer. Expressed as a percentage, *ee* gives the enantiomeric yield *(Ep)*:

$$Ep = ee \times 100 = (2E - 1) \times 100$$

Thus percentage distribution of the enantiomers can be calculated as follows:

$$\%(E - \text{enantiomer}) = \frac{E}{E + E^*} \times 100 = \frac{100 - Ep}{2} + Ep$$

Methods for the Determination of Enantiomeric Composition

Enantiomers can only be distinguished through observation of their interactions with chiral systems. The methods employed for their quantitative determination are based either on the differing chiroptical behavior of enantiomers, or on the diastereomeric interactions that may be induced in the course of the measurement or finally on the conversion of the enantiomers to a pair of diastereomeric derivatives.

In a classical review on the subject [7], it was envisaged that there were four possible approaches to the determination of enantiomeric composition, namely that:

a) determination is performed on the original enantiomeric mixture without separation;

b) determination is performed after the enantiomers have been separated;

c) the enantiomers (*R* and *S*) are first transformed to a pair of diastereomers (*R,R'* and *S,R'*) and determination is then performed without prior separation of the diastereomers;

d) the pair of diastereomers is generated as in (c), but these are first separated before determination is carried out on each individually.   A further subdivision is possible and that is whether the determination be carried out

1) without an auxiliary probe or

2) in the presence of a non-racemic chiral probe. With this in mind, the existing techniques can be divided into five classes [18,19]:

Class a.1

i) Polarimetry: Specific rotation of the enantiomeric mixture is measured in a polarimeter. If the specific rotation of the pure enantiomer is known, the enantiomeric composition can be calculated [20].

ii) Isotopic dilution: A labelled compound of known enantiomeric composition (conveniently a racemic mixture) is added to the sample to be analyzed. An aliquot of the mixture is isolated and its isotope content and optical rotation is measured. Enantiomeric composition can be calculated knowing the isotopic dilution factor and the change in optical rotation [21,22].

iii) Differential microcalorimetry: The phase diagram of the enantiomer is compared to that of the racemic mixture [23].

Class a.2

NMR spectroscopy: The method is based on the fact that in a chiral environment corresponding chemical shifts of the same nuclei may be different in different enantiomers (enantiotopic nuclei become diastereotopic by external comparison). The chiral environment is provided either by a chiral solvent [24,25] (Fig.2.4) or by a chiral paramagnetic shift reagent [30-33].

Amongst the latter, europium complexes of perfluoroacyl camphor derivatives are most commonly used (Fig. 2.5.a), which in moderately polar solvents establish weak bonds with the functional groups of the substrate, usually resulting in good resolution of the resonances of enantiomeric and enantiotopic nuclei.

Class b.2

Chromatography: The mixture of enantiomers is subjected to column chromatography [34], gas-liquid chromatography [35,36], or HPLC [37,38] on a chiral stationary phase,

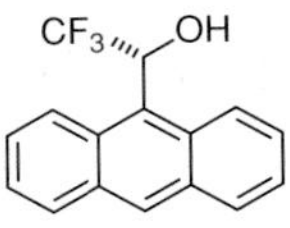

Fig. 2.4.   Some chiral solvating agents for NMR measurement of enantiomeric composition

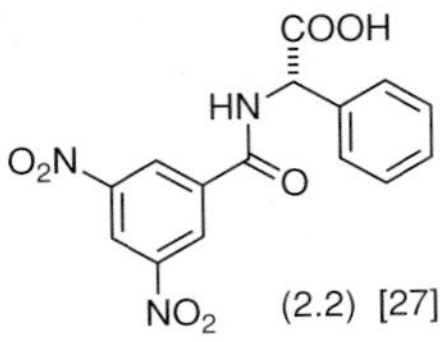

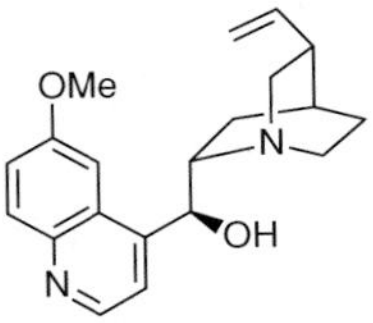

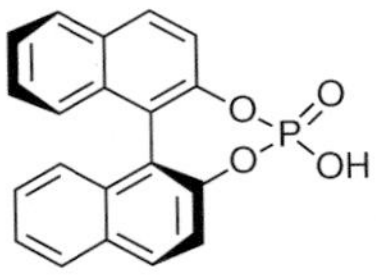

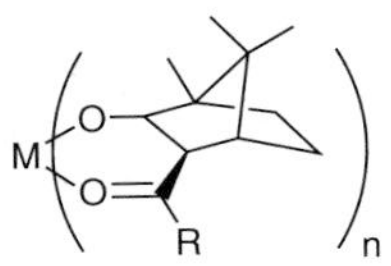

a) M = Eu; n = 3

b) M = Ni, Mn; n = 2

R = CF$_3$, C$_3$F$_7$

Fig. 2.5.   Perfluoroacyl camphor - metal complexes as chiral auxiliaries for determining enantiomeric composition: a) chiral shift reagents for NMR; b) chiral chelates for complexation GC

and the ratio of peak areas of the enantiomers is determined. The phases may be as follows: optically active polysiloxanes [39-43] as hydrogen bonding stationary phases for GC; chiral metal chelates [44-47] for complexation GC; chiral auxiliaries bonded to solid particles of ligand exchange [48] or donor-acceptor type [37,38] for HPLC; cyclodextrin derivatives for GC [49,50] or HPLC [51]; cellulose-triacetate for HPLC [52-54], etc.

A typical example would be polysiloxanes containing L- or D-valinamide groups as side chains for use in gas chromatography:

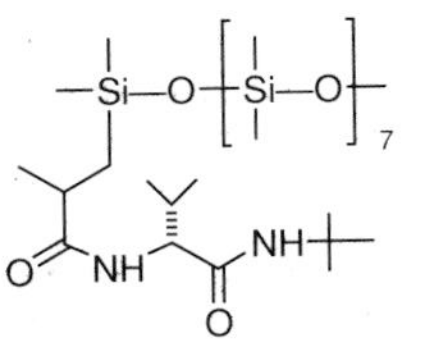

L-Chirasil-Val

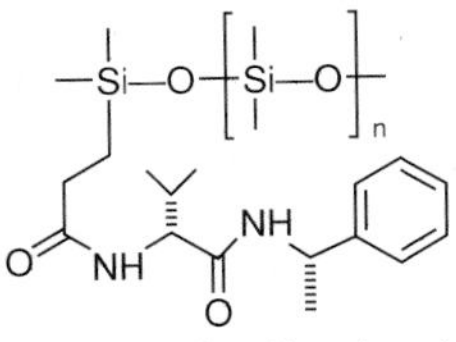

XE-60-L-valine-*S*-α-phenylethylamide

Capillary columns wetted with ChiraSil-Val® phase can be conveniently used for separating the enantiomers of amines, alcohols, glycols, carboxylic acids, hydoxy-carboxylic acids and aminoacids.

Similarly, complexation GC columns treated with chelates having similar structures to the lanthanide shift reagents can be used to separate alcohols, ethers, thioethers and carbonyl compounds (Fig.2.5.b).

The use of polysaccharides however, due to their limited thermal stability is mainly restricted to liquid chromatography where they are suitable for the analysis of hydroxy-acids, aminoacids, heterocyclic systems and aromatic (occasionally also aliphatic) compounds.

Class c.2

NMR spectroscopy: The enantiomers are transformed by an optically pure chiral reagent to a pair of diastereomers and the composition of the mixture is determined by normal ^{1}H, ^{13}C, ^{19}F, ^{31}P, etc. NMR spectroscopy [19,55]. As chiral reagents (Fig.2.6), α-methoxy-α-trifluoromethylphenylacetic acid (MTPA, *S*- or *R*-2.5) [56,57] can be used to derivatize alcohols and amines, α-methoxy-α-trifluoromethyl-benzyl isocyanate (*R*-2.6) [58] is used for primary and secondary amines, the benzoic acid derivative (2.7) [59] and tetrahydro-5-oxo-furan-2-carboxylic acid (2.8)[60] can be used for alcohols and amines, 2-chloro-1,3,2-oxazaphospholidine 2-sulfide and 2-oxide (2.9) [61] to derivatize alcohols, thiols, and amines, *(R,R)*-2,3-butanediol (2.10) [62] for ketones and bis[(trifluoromethyl)phenyl]-1,2-ethanediamine (2.11) [63] can be used for aldehydes.

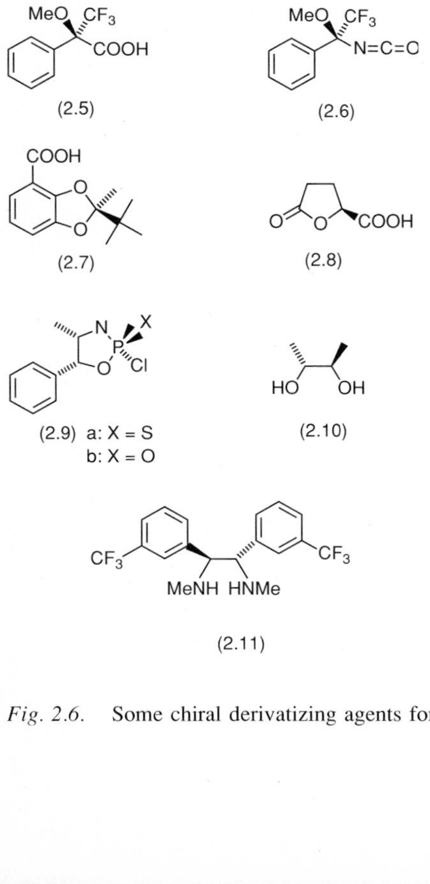

Fig. 2.6.   Some chiral derivatizing agents for NMR measurement of enantiomeric composition

Class d.2

Chromatography: After transformation of the enantiomers to a pair of diastereoisomers, the latter are separated on any suitable achiral stationary phase by GLC or HPLC; enantiomeric purity is calculated from the ratio of peak areas. As chiral derivatizing reagents, both those listed earlier and some additional ones (Table 2.1) [37,64-66] may be selected.

Table 2.1. Resolution of enantiomers by chromatography of diastereomeric derivatives (where no configuration is indicated both enantiomers of the reagent can be applied).

Analytes	Reagents	Method
Amino acids and	(*S*)-N-trifluoroacetylprolyl chloride	GC
amines	α-chloroisovaleroyl chloride	GC
	menthyl chloroformate	GC, HPLC
	(+)-10-camphorsulphonyl chloride	HPLC
	α-methoxy-α-methyl-1-	
	naphthaleneacetic acid	HPLC
	(-)-1,7-dimethyl-7-norbornyl	
	isothiocyanate	HPLC
	(+)-neomenthyl isothiocyanate	HPLC
	(*S*)-2-*tert*-butyl-2-methyl-1,3-	
	benzodioxole-4-carboxylic acid	HPLC
	(+)-*trans*-chrysantemoyl chloride	GC
Alcohols	menthyl chloroformate	GC, HPLC
	α-methoxy-α-(trifluoromethyl)-	
	phenylacetic acid	HPLC
	(-)-*trans*-chrysantemoyl chloride	GC, HPLC
	1-phenylethylisocyanate	GC, HPLC
	2-phenylpropionyl chloride	GC, HPLC
	(*S*)-tetrahydro-5-oxo-2-	
	furancarboxylic acid	GC, HPLC
Alcohols	1-(1-naphthyl)ethylisocyanate	HPLC
Ketones	2,2,2-trifluoro-1-	
	phenylethylhydrazine	GC
	1,2-octanediol	HPLC
Aliphatic and	menthol	GC
alicyclic acids	2-octanol	GC
	(1-phenyl)-ethylamine	GC, HPLC
	(1-naphthyl)-ethylamine	HPLC
Hydroxy fatty	(*R*)-2-butanol	GC
acids	(*R*)-3-methyl-2-butanol	GC
	menthol	GC
	menthyl chloroformate	GC
	2-phenylpropionyl chloride	GC

2.3 Classification of Selective Reactions

Selective transformations (Fig. 2.7) can be divided into two main groups. Selectivity may be manifested in the transformation of a mixture of two or more compounds (e.g. isomers or stereoisomers) (*substrate selectivity*) or between groups or faces in the same molecule in reactions giving more than one product from a single substrate (*product selectivity*)[6]. In biocatalytic transformations both types of selectivity are common. No general agreement has yet however been reached about the exact meaning of *chemo-selectivity*[7]. One interpretation is that it describes the intermolecular or intramolecular selective transformation of constitutionally different but chemically similar groups (e.g. of methoxycarbonyl and ethoxycarbonyl groups in hydrolysis).

Regioselectivity though, is generally defined as the preferred formation of one constitutional isomer in reactions which may lead to such isomers.

In the case of *stereoselective processes*[8] for sake of clarity, it seems necessary to draw a distinction between substrate and product selective transformations. Accordingly, depending on the symmetry of the substrate(s):

Diastereomer selective transformations are those transformations which involve two or more diastereomeric substrates and in which one of the substrates is transformed preferentially, whilst *enantiomer selective* processes are those in which transformation of one of an enantiomeric pair of substrates is preferred.

In *diastereotope selective* transformations, one of a diastereotopic pair of groups or faces in the same substrate is transformed preferentially, whilst in *enantiotope selective* transformations one of an enantiotopic pair of groups or faces is preferred. Diastereotope and enantiotope selectivities can be further classified depending on the nature of the prochiral units involved. When the prochiral units are trigonal centers these terms often become *diasteroface* and *enantioface selectivity* instead. In this book we have employed the more general terms only.

A biocatalytic transformation may be selective from more than one point of view. The following sections provide some examples of the various types of selectivities that occur and their applications.

---

[6] According to [67] in the course of a "product stereospecific" reaction a new chiral center is created, whilst in the case of "substrate specificity", selective transformation of enantiomers or diastereomers takes place.

   On the other hand, a reagent is termed [68] "substrate-selective" if it transforms different substrates under identical conditions at different rates to the products, whilst "product-selective" reactions occur when more than one product can be formed from a single substrate and the ratio of products differs from their statistical probabilities.

[7] The term "chemospecific" was introduced [69] "to define a reaction which is specific for a given structural unit even in the presence of other functionality that might have appeared to be as or more selective".

[8] For stereoselectivity and stereoselective reactions cf. refs.[68,70,71].

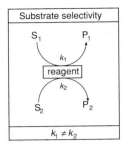
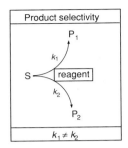

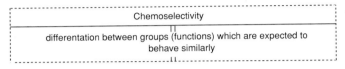

Regioselectivity	
differentation between constitutional isomers	differentation between identical groups which differ in constitution

Diastereomer selectivity	Diastereotope selectivity
diffferentiation between diastereomers	differentiation between diastereotopic groups or faces

Enantiomer selectivity	Enantiotope selectivity
differentiation between enantiomers	differentiation between enantiotopic groups or faces

Fig. 2.7.   Types of selectivities

2.3.1   Selective Biotransformations: Kinetics and Thermodynamics

To gain a better understanding of biotransformations we have considered it useful to outline briefly their most essential kinetic and thermodynamic features.

In the simplest case two substrates (A and B, substrate selectivity) or different structural units of the same molecule (A and A', product selectivity) compete for the binding site of a single enzyme.

Kinetic and thermodynamic description of the two types of selectivity are in many respects similar.

For substrate selective, homocompetitive processes the following model can be constructed:

$$A + Enz \underset{k_2}{\overset{k_1}{\rightleftharpoons}} Enz\text{-}A \underset{k_4}{\overset{k_3}{\rightleftharpoons}} Enz\text{-}P \underset{k_6}{\overset{k_5}{\rightleftharpoons}} P + Enz$$

$$B + Enz \underset{k_2'}{\overset{k_1'}{\rightleftharpoons}} Enz\text{-}B \underset{k_4'}{\overset{k_3'}{\rightleftharpoons}} Enz\text{-}Q \underset{k_6'}{\overset{k_5'}{\rightleftharpoons}} Q + Enz \tag{1}$$

where A and B represent the substrate molecules, Enz is the enzyme, and P and Q denote the products.

The corresponding model for a product selective transformation, provided that the products are not transformed any further, is as follows:

$$A \; \begin{cases} \underset{k_2}{\overset{k_1}{\rightleftharpoons}} Enz\text{-}A \underset{k_4}{\overset{k_3}{\rightleftharpoons}} Enz\text{-}P \underset{k_6}{\overset{k_5}{\rightleftharpoons}} P + Enz \\ \underset{k_2'}{\overset{k_1'}{\rightleftharpoons}} Enz\text{-}A` \underset{k_4'}{\overset{k_3'}{\rightleftharpoons}} Enz\text{-}Q \underset{k_6'}{\overset{k_5'}{\rightleftharpoons}} Q + Enz \end{cases} \tag{2}$$

Comparison of the two models side by side makes their close similarity readily apparent (in (2) B=A and Enz-B=Enz-A'), thus for a general discussion it is best just to consider model (1).

Hypothetical free energy relationships for various types of selective processes corresponding to this model are shown in Fig. 2.8. In Fig. 2.8.a, free energy relationships concerning substrate (A) are shown. Binding of substrate (A) to enzyme (Enz) to give the enzyme-substrate complex (Enz-A) involves a negative free energy of complexation (ΔG_{CA}). The enzymatic transformation can be envisaged as a process leading via the enzyme-substrate complex (Enz-A$^{\neq}$) to an enzyme-product complex (Enz-P). The free energy of activation requirement for the process Enz-A $\rightarrow$ Enz-A$^{\neq}$ is equal to $\Delta G_{TA}^{\neq}$.

Selective processes can also be distinguished regarding of whether they contain an irreversible step or not.

I) Irreversible Selective Processes [72]

If the transformation of substrate A shown in Fig 2.8.a is irreversible (i.e. it follows Michaelis-Menten kinetics) then only processes preceding the irreversible step and the irreversible step itself count for the outcome of the reaction. In this case the value of $\Delta G_{TA}^{\neq}(\Delta G_{TA}^{\neq} = \Delta G_A^{\neq} - \Delta G_{CA})$ is proportional (according to transition state theory [73]) to the equilibrium constant relating the transition complex (Enz-A$^{\neq}$) and the initial state (A + Enz) can be expressed by means of the kinetic constant $(k_{cat}/K_m)_A$

$$RT \ln \left(\frac{k_{cat}}{K_m} \right)_A = RT \ln \left(\frac{kT}{h} \right) - \Delta G_{TA}^{\neq} \tag{3}$$

where k is Boltzmann's constant and h denotes Planck's constant.

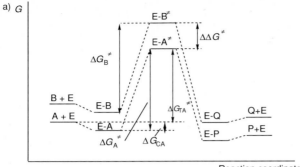

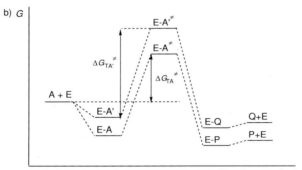

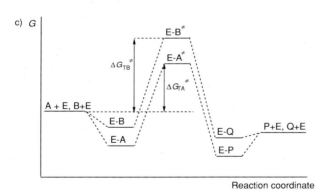

Fig. 2.8.   Hypothetical free energy relationships in selective enzymic reactions: a) chemo-, regio-, or diastereomer selective transformations; b) product type regio-, or diastereotope selective transformations; c) enantiomer selective transformations; d) enantiotope selective transformations (on page 41)

In Figs. 2.8.a-d, the hypothetical free energy relationships of various selective processes are depicted. In cases conforming to model (1) and satisfying the requirement of irreversibility, the correlation according to equation (3) can also be written up for substrate B (For Figures 2.8.b and d, B=A and $\Delta G_{TB}^{\neq} = \Delta G_{TA'}^{\neq}$), thus:

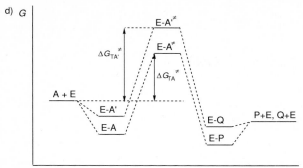

Fig. 2.8.(contd.)

$$RT \ln \left(\frac{k_{\text{cat}}}{K_{\text{m}}} \right)_{\text{B}} = RT \ln \left(\frac{kT}{h} \right) - \Delta G_{\text{TB}}^{\neq} \qquad (4)$$

Substracting equation (4) from equation (3) yields:

$$RT \ln \frac{(k_{\text{cat}}/K_{\text{m}})_{\text{A}}}{(k_{\text{cat}}/K_{\text{m}})_{\text{B}}} = -(\Delta G_{\text{TA}}^{\neq} - \Delta G_{\text{TB}}^{\neq}) = -\Delta\Delta G^{\neq} \qquad (5)$$

and

$$E = \frac{(k_{\text{cat}}/K_{\text{m}})_{\text{A}}}{(k_{\text{cat}}/K_{\text{m}})_{\text{B}}} = e^{-\frac{\Delta\Delta G^{\neq}}{RT}} \qquad (6)$$

From equation (6) it is apparent that for model (1) under irreversible conditions, selectivity (defined as E) depends only on the free energy difference ($\Delta\Delta G^{\neq}$) of the respective activated complexes (Enz-A$^{\neq}$ and Enz-B$^{\neq}$ or Enz-A'$^{\neq}$).

II) Reversible Selective Processes [74,75]

If each step of a selective process corresponding to model (1) is reversible, the general model is complex and needs some simplification. Such a simplified model is still useful for most practical purposes and reflects the real situation with sufficient accuracy to be considered seriously. For substrate selective processes this simplified model is as follows [74,75]:

$$\begin{array}{c} A + Enz \quad \underset{k_2}{\overset{k_1}{\rightleftharpoons}} \quad P + Enz \\[2ex] B + Enz \quad \underset{k_2'}{\overset{k_1'}{\rightleftharpoons}} \quad Q + Enz \end{array} \qquad (7)$$

where k_1, k_2, and further k_1, k_2, are the overall rate constants of the forward and reverse reactions respectively. For product selective processes the corresponding simplified model is:

$$A + \text{Enz} \xrightleftharpoons[k_2]{k_1} \begin{array}{l} P + \text{Enz} \\ \\ Q + \text{Enz} \end{array} \qquad \begin{array}{c} k_1{}' \\ k_2{}' \end{array} \tag{8}$$

The transformation in model (7) leads to an equilibrium characterized by the equilibrium constants K and K', such that

$$-\frac{\mathrm{d}A}{\mathrm{d}t} = k_1[\text{A}] - k_2[\text{P}] = (k_1 + k_2)[\text{A}] - k_2[\text{A}_0] \tag{9}$$

and

$$-\frac{\mathrm{d}B}{\mathrm{d}t} = k_1[\text{B}] - k_2[\text{P}] = (k_1 + k_2)[\text{B}] - k_2[\text{B}_0] \tag{9}$$

where $[\text{A}_0]=[\text{A}]+[\text{P}]$ and $[\text{B}_0]=[\text{B}]+[\text{Q}]$. From these two equations a combined expression can be obtained:

$$\frac{\mathrm{d}A}{\mathrm{d}B} = \frac{(k_1 + k_2)[\text{A}] - k_2[\text{A}_0]}{(k_1{}' + k_2{}')[\text{B}] + k_2{}'[\text{B}_0]} \tag{11}$$

which after integration yields:

$$\frac{\ln\left[1 - \left(1 + \dfrac{k_2}{k_1}\right)\left(1 - \dfrac{[\text{A}]}{[\text{A}_0]}\right)\right]}{\ln\left[1 - \left(1 + \dfrac{k_2{}'}{k_1{}'}\right)\left(1 - \dfrac{[\text{B}]}{[\text{B}_0]}\right)\right]} = \frac{k_1}{k_1{}'} \frac{\left[1 + \dfrac{k_2}{k_1}\right]}{\left[1 + \dfrac{k_2{}'}{k_1{}'}\right]} \tag{12}$$

Since $E = k_1/k_1{}'$, $K = k_2/k_1$ and $K' = k_2{}'/k_1$:

$$E = \left(\frac{1 + K}{1 + K'}\right) \frac{\ln\left[1 - (1 + K)\left(1 - \dfrac{[\text{A}]}{[\text{A}_0]}\right)\right]}{\ln\left[1 - (1 + K')\left(1 - \dfrac{[\text{B}]}{[\text{B}_0]}\right)\right]} \tag{13}$$

From this last equation it is evident that for reversible processes, both kinetic parameters (the degree of selectivity E depending on the nature of biocatalyst) and thermodynamic factors (equilibrium constants K and K', i.e. thermodynamic parameters independent of the catalyst) are crucial to the outcome of the transformation. It is also apparent that, even applying the simplified model, in the general case three parameters must be determined for the complete characterization of the process. When both k_2 and k_2' are equal to zero, equation (13) simplifies to that used to describe irreversible processes namely:

$$E = \frac{\ln([\text{A}]/[\text{A}_0])}{\ln([\text{B}]/[\text{B}_0])} \tag{14}$$

2.3.2   Applications of Biocatalysis and Utilization of Various Types of Selectivity

Biocatalysts show great versatility and have found applications in a wide range of synthetic transformations which take advantage both of the very mild conditions and of the diverse types of selectivities (eventually more than one at a time) (cf. Section 2.3.2.8) offered by enzymic systems. First and foremost though, it is high stereoselectivity which makes biocatalysis such an attractive alternative for the simple preparation of optically active compounds.

One of the main goals in optimization experiments is to improve selectivity. A list of available methods for doing this has been compiled in Table 2.2. These methods and combinations thereof can be used, in principle at least, for any kind of selectivity.

Table 2.2. Possibilities for improvement of selectivity in biotransformations.

I) *Single enzyme transformations*
  A) changing the enzyme
    1) search among commercial enzymes
    2) screening different organisms for a more suitable enzyme
    3) modifying the enzyme
      a) by immobilization
      b) by physical or conformational modification
      c) by chemical modification
      d) by genetic modification
  B) changing transformation conditions
    1) changing temperature or pressure
    2) changing the pH, ionic composition and/or ionic strength
    3) using additives so selectively influencing the transformation
      a) selective inhibitors
      b) selective activators
    4) changing solvent composition
      a) adding cosolvents to water
      b) using water-organic biphasic systems, microemulsions
      c) using organic solvents ("microaqueous conditions")
      d) changing solvent(s) to water ratio in a)-c)
  C) modifying the substrate(s)
    (structural modifications alter the course of the enzymic
    reaction sterically and/or electronically)
    1) permanent modifications
      (the modified part remains in the product)
    2) temporary modifications
      (the modified part is removed during the transformation or work up)

II) *Multienzyme transformations*
  A) all possibilities listed under I)
  B) selectively influencing different enzymes catalysing the same reaction e.g.
    1) selective enzyme induction at growing stage
    2) selective inhibition or activation of the different enzymes

---

In the following, a biocatalytic realization of each type of selectivity has been illustrated with a few selected examples.

2.3.2.1 Utilization of the Mild Conditions of Biocatalysis

Biocatalysis may be the method of choice for the transformation of small labile molecules. For example, hydrolysis of cyclopropyl acetate (2.12) to cyclopropanol (2.13) when it is carried out by base catalysis or by reductive cleavage, is accompanied by extensive ring opening. This can be avoided by using pig liver esterase (PLE) [76] as a biocatalyst.

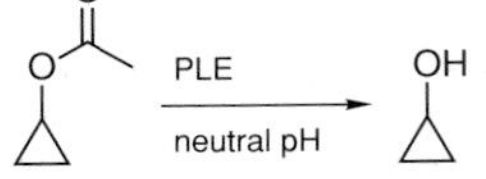
O
O
PLE
neutral pH
OH

(2.12) (2.13)

Similarly, esters (e.g. 2.14) of cyclopentadiene-1-carboxylic acid decompose during both acid and base catalyzed hydrolysis, whilst with PLE as the catalyst the free acid (2.15) can be obtained without difficulty [77].

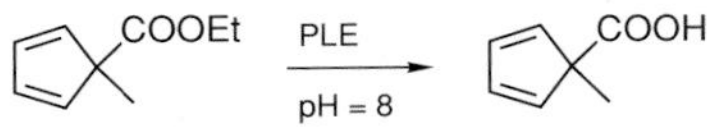
COOEt
PLE
pH = 8
COOH

(2.14) (2.15)

For a mild enzymic hydrolysis of a more complex molecule on an industrial scale, hydrolysis of penicillin G can be cited as a good example [78,79]. Similarly, biocatalysis has also been used for the clean hydrolysis of the prostanoid esters (2.16.a-d) to their corresponding free acids (2.17.a-d) (Fig.2.9), or for obtaining levuglandin E2 from its methyl ester [144].

Enzymic hydrolysis has been seen to work with quite complex substrates as well, for example in the transformation by transesterification and hydrolysis respectively [83] of chlorophylls (2.18) to chlorophyllide esters (2.19) and chlorophyllides (2.20):

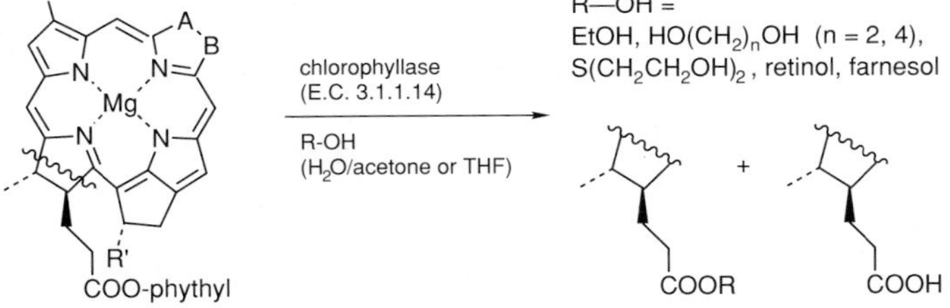
R''
A
B
N
N
Mg
N
N
R'
COO-phythyl
chlorophyllase
(E.C. 3.1.1.14)
R-OH
(H2O/acetone or THF)
R—OH =
EtOH, HO(CH2)nOH (n = 2, 4),
S(CH2CH2OH)2 , retinol, farnesol
+
COOR
COOH

(2.18) (2.19) (2.20)

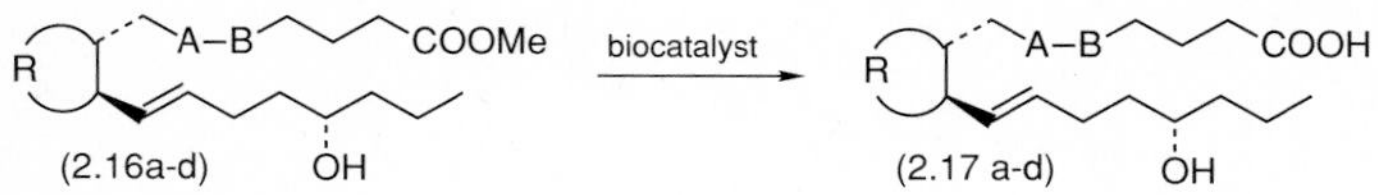

(2.16a-d) (2.17 a-d)

	R	A—B	Biocatalyst	Reference
a	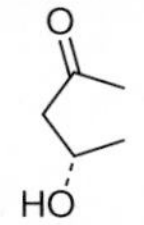	CH_2CH_2	BY	[80]
b	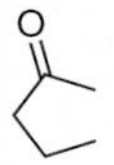	CH_2CH_2	BY	[81]
c	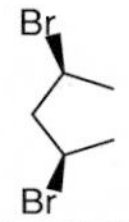	CH=CH	PPL	[82]
d	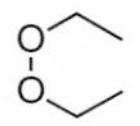	CH=CH	PPL	[83]

Fig. 2.9. Mild hydrolysis of prostanoid esters

The following example seems somewhat out of context in that a relatively stable molecule is converted under mild conditions to an unstable intermediate which subsequently undergoes further transformation spontaneously. Thus however, enzymic hydrolysis of the diepoxyester (2.21), a model in the study of the biosynthesis of monensin A, yields an intermediate in which a nucleophilic attack on the newly formed carboxylate triggers a cascade of spontaneous epoxy ring opening and closing events, leading ultimately to the tricyclic end product (2.22) [84a]:

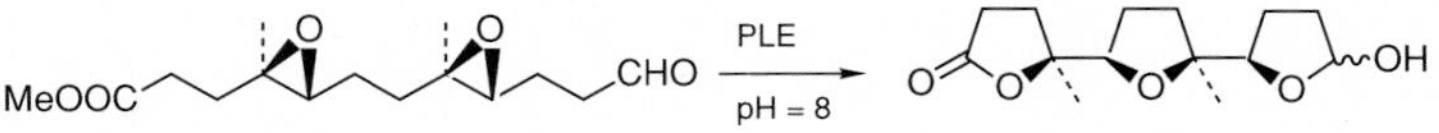

(2.21) (2.22)

Enzymic hydrolysis of compound (2.23) initiates an interesting process. Here, spontaneous decomposition of the product, a naphthol (2.24), is accompanied by the emission of light [84b]:

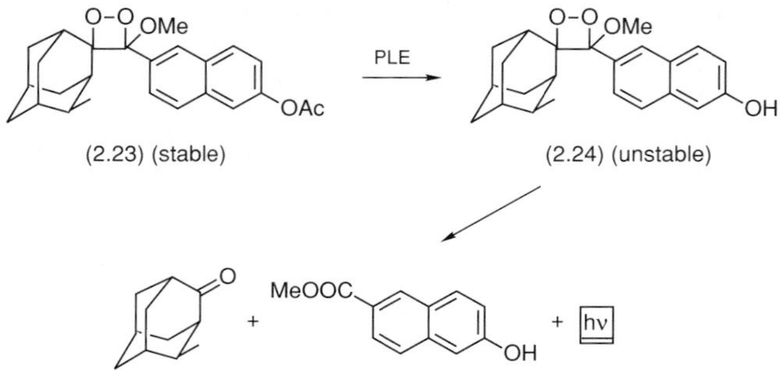

(2.23) (stable)   (2.24) (unstable)

2.3.2.2   Chemoselective Biotransformations

Biocatalysis permits differentiation between substrates which in chemical terms are close-ly similar. This is a rather common situation which will not be discussed here. Of more interest however, are selective transformations of similar groups in the same molecule (product chemoselectivity).

One manifestation of chemoselectivity in biocatalysis occurs when the selectivity of the catalyzed process differs from the non-catalyzed one. For example, with no enzyme catalysis the reaction of propargilic ester (2.25) with aromatic amines gives Michael type adducts (2.27), whilst in the presence of *Candida rugosa* lipase (CcL) [9] the amide (2.26) is obtained [85]:

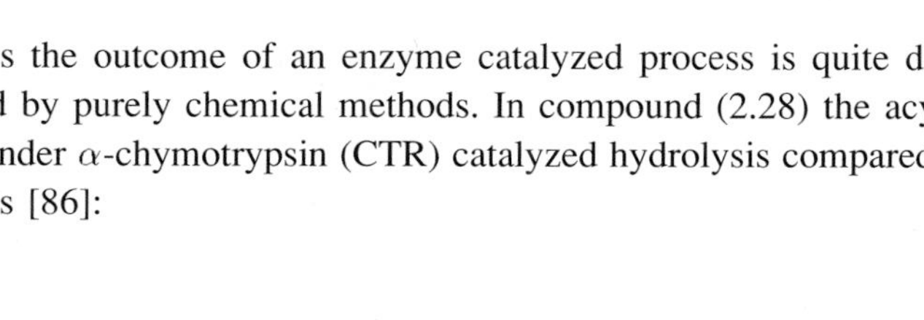

Sometimes the outcome of an enzyme catalyzed process is quite different from the one obtained by purely chemical methods. In compound (2.28) the acyl groups behave differently under α-chymotrypsin (CTR) catalyzed hydrolysis compared to base catalyz-ed hydrolysis [86]:

---
[9] We have retained this widely applied acronym which was derived from *Candida cylindracea*, an earlier name for the same organism. The more correct acronym would be CrL.

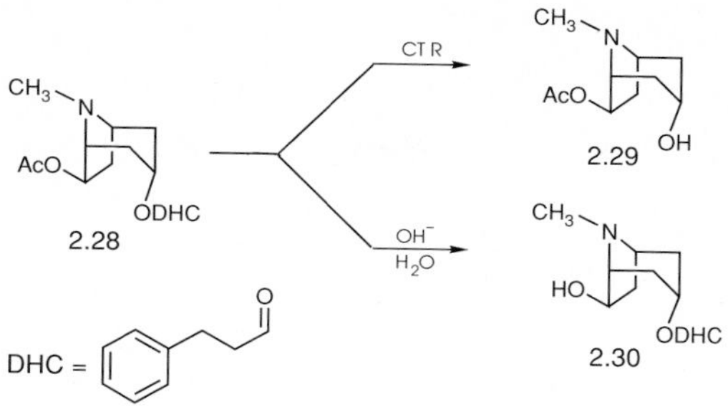

Under enzymic hydrolysis, the 3-phenylpropionyl group can be selectively removed in contrast to the alkaline hydrolysis situation where it is the acetyl group that is selectively removed.

Special groups which can be selectively removed with enzymes can be used to advantage as protective groups. E.g. the esters of the acids (2.31) and (2.32) can be selectively cleaved by CTR [87]:

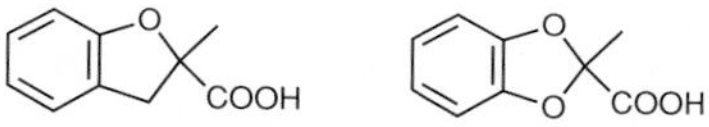

The chemoselectivity of a given biocatalyst can be reversed by creating a change of conditions or a careful selection of the group affected. Thus, in the transacylation of 6-aminohexanol (2.33) with various esters catalyzed by *Aspergillus niger* lipase (AnL), the ratio of *O*- to *N*-acylation depends on the nature of the acyl group being affected [88]:

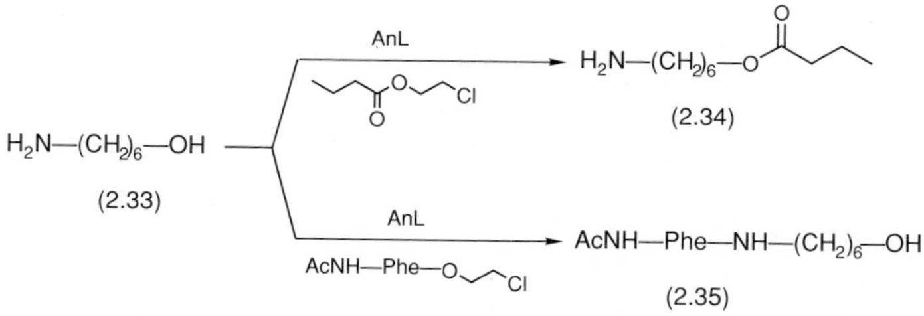

A similar change in chemoselectivity is experienced with *Pseudomonas* sp. lipase (PsL) and pig pancreatic lipase (PPL). With esters of aliphatic acids *O*-acylation is dominant, whereas with esters of aminoacids it is *N*-acylation that predominates.

2.3.2.3 Regioselective Biotransformations

In biotransformations, it is possible to effect regioselectivity both in terms of substrate selectivity (differentiation between constitutional isomers) and product selectivity (intramolecular differentiation between groups of the same structure but of different position).

Substrate regioselectivity
To allow the separation of α- and β-naphthol, enzymic hydrolysis of the corresponding sulfate esters (α, β-2.36) can be applied [89]. The less hindered ester (β-2.36) undergoes hydrolysis, whilst (α-2.36) remains unchanged.

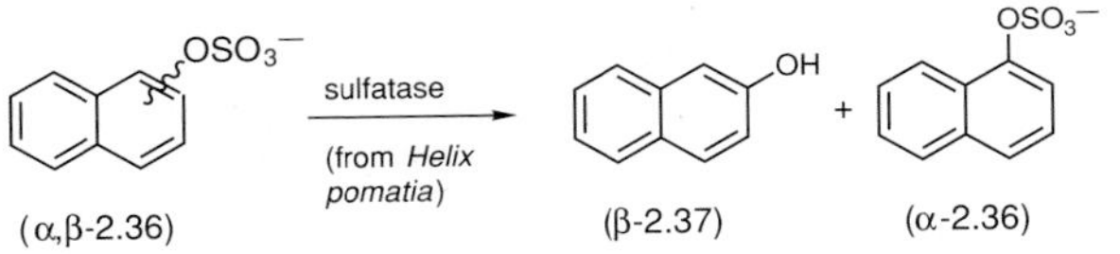

The same enzyme can also be employed for the regioselective hydrolysis of mixtures of the sulfate esters of *o*- and *m*- (and also of *o*- and *p*-) substituted phenols.

Product regioselectivity
Using microbial oxidation [90], the substrate ethylcyclohexane (2.38) was shown to be hydroxylated regioselectively (but also stereoselectively), inasmuch as only the 4-methylene group was attacked, the resulting alcohol (2.39) being mainly of *trans* configuration:

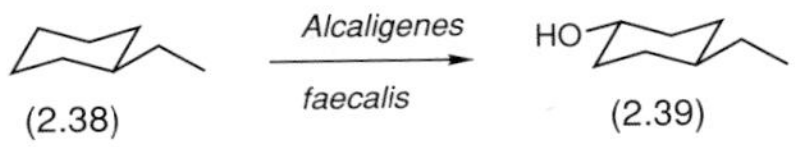

Interestingly, oxidation of butylcyclohexane (2.40) affected the side chain and not the ring yielding cyclohexylacetic acid (2.41) [91]:

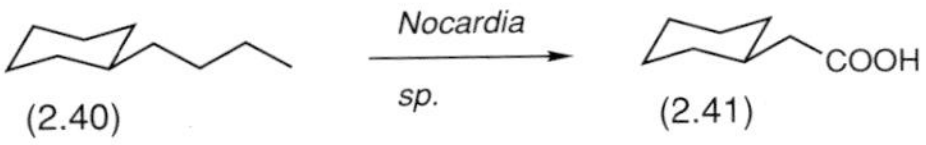

Through the use of enzymic methods, it is also feasible to effect the selective hydrolysis of constitutionally different ester groups. For example, selective hydrolysis of tartaric acid derivatives (2.42) in this way yields the corresponding semi esters (2.43) which are useful in the synthesis of β-lactams [92]:

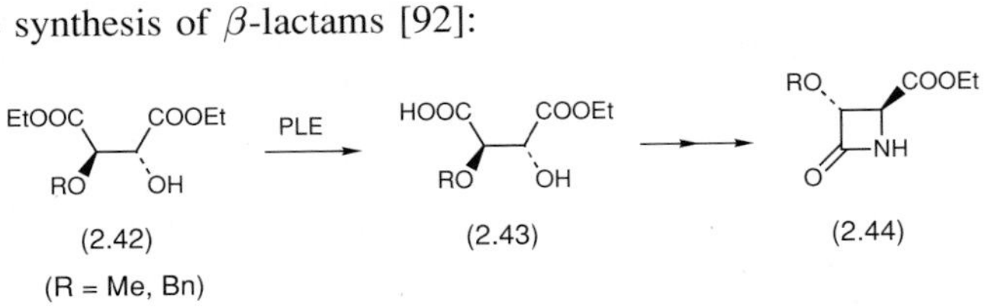

Regioselective enzymic hydrolysis has also been employed in the synthesis of the apor-phin alkaloid *N*-methyl-laurotetanine [93] where it was used to convert the diacetoxy compound (2.45) to the monoacetate (2.46):

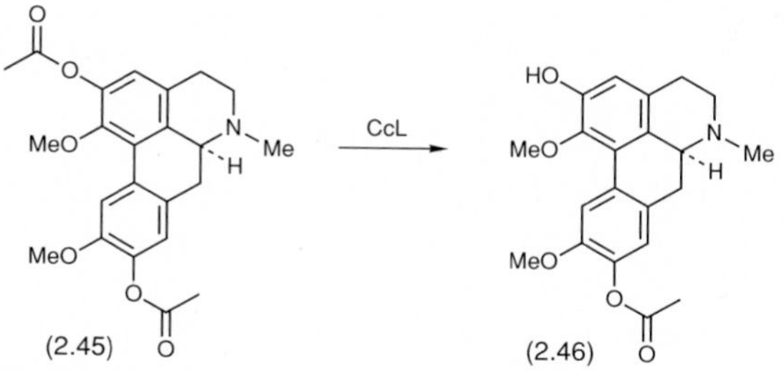

(2.45)   (2.46)

By applying another different enzyme, regioselectivity can often be inverted. The following examples are taken from the fields of steroid [94] and nucleoside [95] bioche-mistry respectively:

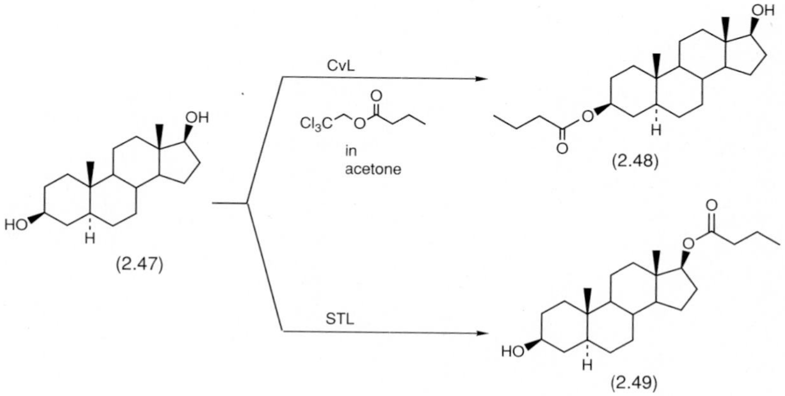

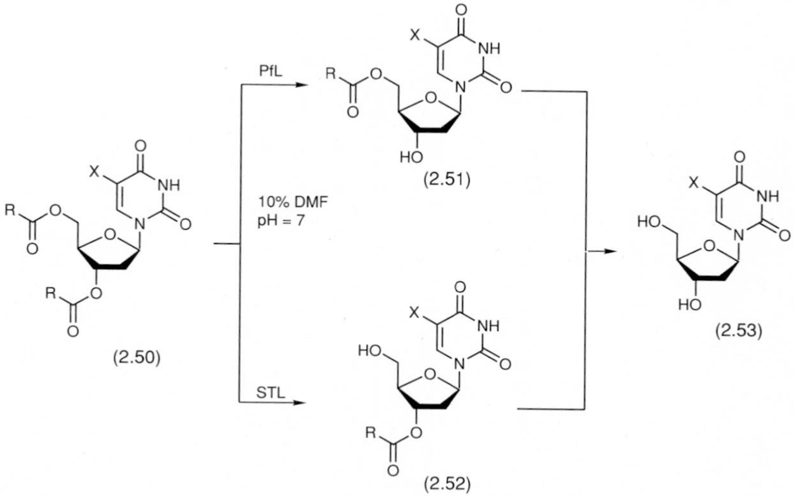

(2.47)

(2.48)

(2.49)

(2.50)

(2.51)

(2.52)

(2.53)

In both examples, one involving lipase and the other protease, each enzyme is catalyzing similar transformations on the same substrate but with opposite regioselectivity.

This regioselectivity of hydrolytic enzymes has more recently often been employed to manipulate protecting groups in carbohydrate, nucleoside and steroid syntheses.

2.3.2.4   Diastereomer Selectivity

A useful application for biocatalysis is the differentiation of diastereomers, the separation of which by other methods is often cumbersome. In this way also, it is possible to distinguish between diastereomers differing in double bond configuration *(E/Z)*. Thus in the hydrolysis of an *E/Z* mixture of (2.54) with immobilized PLE enzyme, it was the *E* isomer which hydrolyzed more rapidly [96]. The acidic product (*E*-2.55) could then be readily separated by extraction from the unchanged ester (Z-2.54):

Cis and *trans* isomers of cyclic compounds can be separated in a similar way. For example, in the *cis/trans* mixture of esters (2.56), only the *trans* component was hydrolyzed by PLE enzyme [97]. The product, (*trans*-2.57), was first resolved and then transformed to the optically active cyclopropanoid pheromone of brown algae:

In bridged and other polycyclic ring systems, *endo* and *exo* isomers can be differentiated by biocatalysis. Thus, in the hydrolysis of the mixture (*endo,exo*-2.58) with PLE enzyme the *exo* acids (*exo*-2.59) were formed more rapidly, whilst in the unchanged material from the transformation of the ester *trans*-(2.60) the *trans-endo*-ester (*endo*-2.60) remained in excess [98].

2.3.2.5   Diastereotope Selectivity

I) Selective transformation of diastereotopic groups

Most of the examples for the selective biocatalytic transformation of diastereotopic groups concern hydroxylations by microorganisms. Selective microbial hydroxylation of methylene groups with diastereotopic hydrogen atoms is well known and of prominent practical importance in the field of steroids. Practically any of the methylene groups of a steroid skeleton are amenable to microbial hydroxylation [99,100] and it is often possible to predict the regio- and stereoselectivity likely to occur in the reaction [101].

With suitably selected microorganisms, 5-α-pregnane (2.62) derivatives can be hydroxylated at several different positions in both a regio- and diastereotope selective manner [102]:

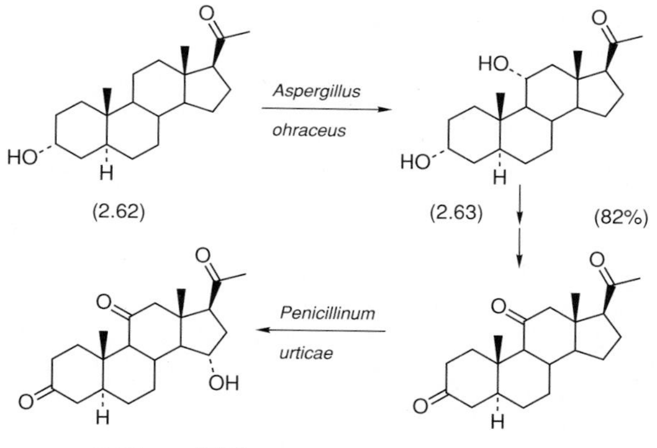

A rather exotic example of diastereotope selectivity is the oxidation of the ferrocene (2.66) in the 6-*exo* position by *Beauveria sulfurescens* [103]:

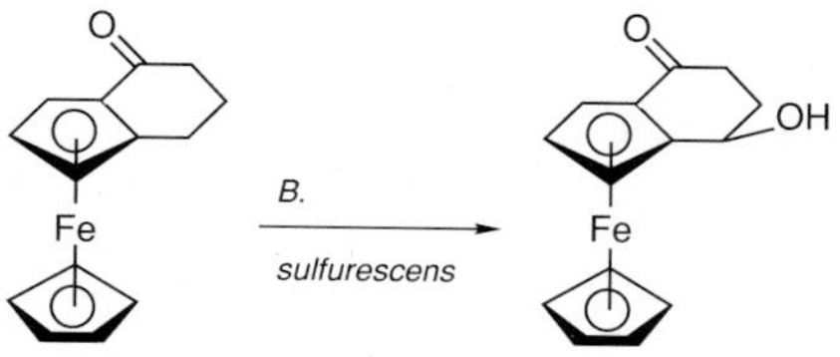

Not only diastereotopic hydrogens but also diastereotopic methyl groups can be selectively hydroxylated. An interesting example is the hydroxylation of compounds ((*1R*)-2.68) and ((*1S*)-2.68) once again with the aid of *B. sulfurescens* [104]:

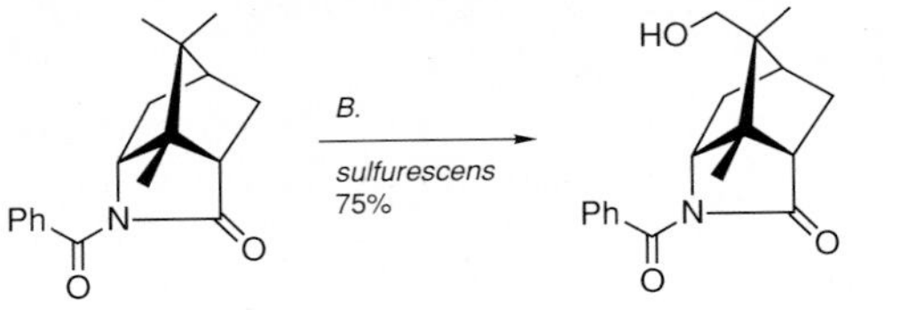

(1R)-(2.68)   (2.69)

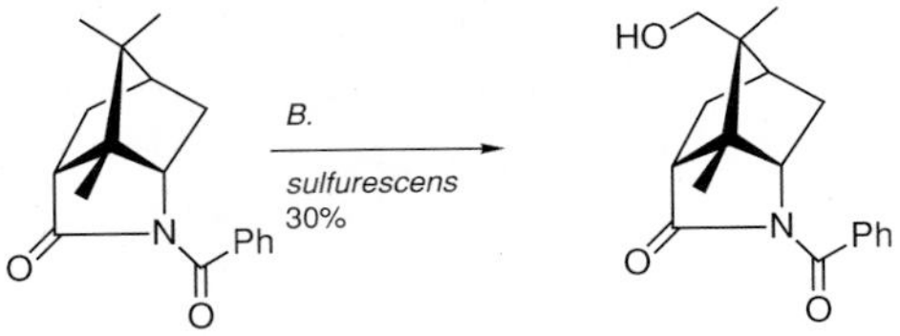

(1S)-(2.68)   (2.70)

For the two enantiomers, diastereotope selectivity is similar. In both substrates it is the *pro-R* methyl group which is oxidized yielding products (2.69 and 2.70) that are diastereomers.

Selective enzymatic transformation of diastereotopic groups is not restricted to those attached to prochiral carbon atoms. In the thiophosphate (2.71) used for the synthesis of thio-ATP (2.72) in the presence of adenylate kinase, selective phosphorylation of the *pro-R* oxygen atom attached to the tetragonal phosphorus atom takes place [105]:

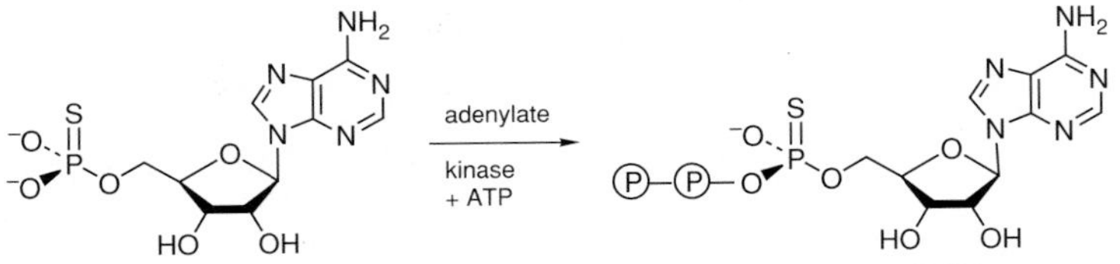

(2.71)   (2.72)

The most promising application of the selective transformation of diastereotopic groups is in the field of complex natural products, their analogues, and the advanced intermediates generated in their synthesis.

II) Differentiation of diastereotopic faces

In biocatalytic processes diastereotopic faces of various trigonal centers can be readily distinguished. Fermentation of the aldehyde (2.73) with baker's yeast in the presence of saccharose gives the α-hydroxyketone (2.74), the condensation product of the "acetaldehyde" formed from pyruvate. Diastereoselective reduction of the carbonyl groups in these compounds from the *re* side gives rise to the diols (2.75) [106]:

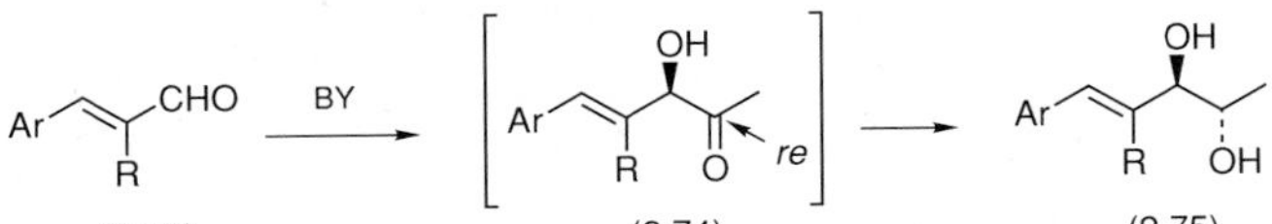

(2.73)   (2.74)   (2.75)

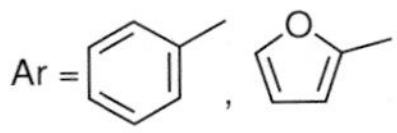

Ar =   R = H, CH₃

Aldolases may also catalyze diastereotope selective condensations as exemplified by the reaction of the aldehydes ((*R*)-) and ((*S*)-2.76) with dihydroxyacetone-phosphate (2.77) catalyzed by fructose-1,6-diphosphate aldolase isolated from rabbit muscle (FDPA) [107,108]

In this condensation, two new centers of asymmetry are generated, both created in a diastereotope selective manner. Note that, due to the chirality of the enzyme, the faces of the intermediate enamine (2.80) formed from (2.77) and a lysine side chain of the enzyme, are also diastereotopic [109]:

2.80

Biocatalytic reduction of chiral carbonyl compounds is also a diastereotope selective process, but since it is the reduction of racemic substrates that is of main practical interest and this involves two substrates, the problem will be addressed in Section 2.3.2.8 in dealing with the combination of different types of selectivities.

2.3.2.6   Enantiomer Selectivity

One of the most commonly exercised applications of biocatalysts is based on their ability to display enantiomer selectivity, that is to selectively transform one of a pair of enantiomeric substrates (Fig. 2.10).

If the process is completely selective the reaction stops with racemic mixtures in a state of 50 % conversion i.e. when the reactive enantiomer has been depleted. In this case both the product and the remaining substrate are enantiomerically pure and the yields for both, disregarding operational losses, are 50 % based on the racemic substrate (Fig 2.10.a).

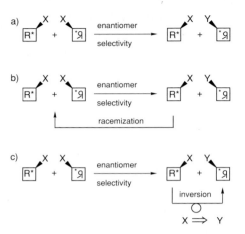

Fig. 2.10.   Enantiomer selectivity: a) simple kinetic resolution; b) enantiomer selective process involving racemization and recycling of remaining enantiomer; c) enantiomer selective process by conversion of the remaining enantiomer to product by inversion

Most often however, only one of the enantiomers is needed and if the racemization of the other one is possible, by recycling the racemized material (Fig.2.10.b) the 50 % limit can be exceeded. Whilst occasionally racemization/recycling can be performed in situ, generally it must be done after separation, as another step. The prospects for racemization are best when the molecule contains but a single center of chirality which can be readily inverted (Fig 2.10.c).

When the enantiomer selectivity of a biocatalytic process is not complete, enantiomeric purity is dependent, other than on the choice of the proper enzyme and conditions, on whether the process is reversible or not [72,74,75]. When it is irreversible - for example with enzyme catalyzed hydrolysis in water - the efficiency of resolution is governed solely by enzyme selectivity [74]. In this case, selectivity can be adequately described by the simple classical homocompetitive model:

$$
\text{Enz}
\quad
\begin{array}{l}
\xrightleftharpoons[k_2]{k_{1A}} \text{EnzA} \xrightarrow{k_3} \text{EnzA} \xrightarrow{k_4} \text{Enz} + \text{P} \\[2mm]
\xrightleftharpoons[k_2']{k_{1B}'} \text{EnzB} \xrightarrow{k_3'} \text{EnzB} \xrightarrow{k_4'} \text{Enz} + \text{Q}
\end{array}
\tag{1}
$$

where the selectivity of the enzyme (Enz) towards the two competing enantiomers (A and B) can be described by a single value (E), namely the ratio of the specificity constants (V_{max}/K_m) of the two competitive processes each following Michaelis-Menten kinetics.

$$
E = \frac{(V_{max}/K_m)_A}{(V_{max}/K_m)_B} = \frac{\ln(A/A_0)}{\ln(B/B_0)}
\tag{2}
$$

This equation can be converted to a form containing parameters which can be more conveniently determined [72a], thus:

$$E = \frac{\ln\left[1 - c(1 + ee(\mathrm{P}))\right]}{\ln\left[1 - c(1 - ee(\mathrm{P}))\right]} \tag{3}$$

$$E = \frac{\ln\left[1 - c(1 - ee(\mathrm{S}))\right]}{\ln\left[1 - c(1 + ee(\mathrm{S}))\right]} \tag{4}$$

where c is the conversion and $ee(\mathrm{P})$ and $ee(\mathrm{S})$ are the enantiomeric purities of product and substrate respectively. These quantities can be expressed by concentrations of substrates and products:

$$c = 1 - \frac{[\mathrm{A}] + [\mathrm{B}]}{[\mathrm{A_0}] + [\mathrm{B_0}]} \tag{5}$$

$$ee(\mathrm{P}) = \frac{[\mathrm{P}] - [\mathrm{Q}]}{[\mathrm{P}] + [\mathrm{Q}]} \tag{6}$$

$$ee(\mathrm{S}) = \frac{[\mathrm{B}] - [\mathrm{A}]}{[\mathrm{B}] + [\mathrm{A}]} \tag{7}$$

From all this one can conclude that in such practically irreversible homocompetitive processes, enantiomer selectivity is independent of substrate concentration and depends only on the specificity constants. It can also be seen that the enantiomeric purity of both product and unchanged substrate can be controlled to a certain measure by changing the degree of conversion. By determining the value of E using a few experimentally determined pairs of values for c and $ee(\mathrm{P})$ or c and $ee(\mathrm{S})$, enantiomeric purity for both product and substrate can be calculated for any degree of conversion [72a] (Fig. 2.11.a). Similar apparent E values (E_{app}) can be used as a measure of enantiomer selectivity in biotransformations with microorganisms. In such processes, however, several factors can cause serious deviations from the above model. Since the apparent E value is built up from individual E values of several enzymes catalyzing the same process simultaneously

$$E_{\mathrm{app}} = \frac{\displaystyle\sum_{i=1}^{n} \alpha_i (V_{\mathrm{max}}/K_{\mathrm{m}})_i}{\displaystyle\sum_{i=1}^{n} \alpha_i (V_{\mathrm{max}'}/K_{\mathrm{m}'})_i} \tag{8}$$

good correlations can only be expected when the environmental parameters such as temperature, pH value, substrate concentration remain constant throughout [72a].

Sometimes irreversible enantiomer-selective enzyme catalysis involves substrates having two chiral reaction centers. In such cases sequential biocatalytic kinetic resolution is possible and quantitative expressions are known [143].

If the competitive processes are reversible or other kinetic restrictions have to be considered, final enantiomer selectivity deviates significantly from the value predicted by the model assuming irreversibility [74,75]. The model for a reversible enantiomer selective process is the following:

$$\text{Enz} + \text{A} \underset{k_2}{\overset{k_1}{\rightleftharpoons}} \text{Enz} + \text{P}$$

$$\text{Enz} + \text{B} \underset{k_4}{\overset{k_3}{\rightleftharpoons}} \text{Enz} + \text{Q}$$

(9)

where A and B denote enantiomeric substrates, P and Q enantiomeric products, and k_1, k_2, k_3, and k_4 the apparent pseudo-first-order rate constants (V_{max}/K_m). Since the enzyme, being a catalyst, does not influence the equilibrium, the equilibrium constant (K) depends on the initial and final states alone. A feature of this model [74] is that the sense of chirality for the forward and reverse reaction is retained. If reaction conditions for the forward and reverse reactions are identical, then K values for both, i.e. k_1/k_3 and k_2/k_4 must be the same, therefore:

$$K = \frac{k_1}{k_3} = \frac{k_2}{k_4}$$

(10)

Thus for the irreversible case, selectivity is a function of both E and K:

$$E = \frac{\ln\left[1 - (1 + K)c(1 + ee(\text{P}))\right]}{\ln\left[1 - (1 + K)c(1 - ee(\text{P}))\right]}$$

(11)

where c and $ee(\text{P})$ are defined as in equations (5) and (6). Thus, as a result of reversibility, high enantiomeric purity cannot be realized even by carrying conversion well beyond 50 % (Fig.2.11).

The examples discussed below display some of the potential of enantiomer selective biocatalysis.

In the microbial Baeyer-Villiger type oxidation of the racemic ketone (2.81) [110], a reaction takes place which is enantiospecific [10]:

rac-(2.81)   *Acinetobacter sp.*   (2.82)   +   (2.83)

Both products can be obtained in good yield and enantiomerically pure. In the terminology adopted in this book, this is a manifestation of simultaneous enantiomer and regioselectivity, since regioselectivity for the individual enantiomers is different.

The most popular utilization of biocatalytic enantiomer selectivity however is for simple kinetic resolution. Thus, kinetic resolution of racemic *N*-acyl aminoacids (2.84) with acylase-I has attained real industrial importance [111-113]:

rac-(2.84)   acylase I   H⁺/Δ   racemization   D-(2.84)   +   L-(2.85)

---

[10] Using the term as defined by Eliel [1].

a)

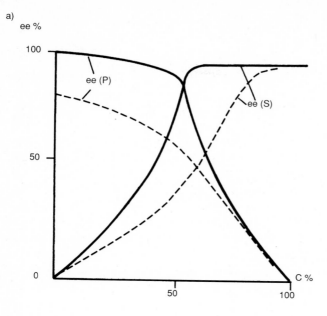

b)

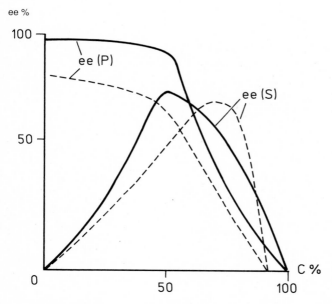

Fig. 2.11.   Enantiomeric excess of product [ee(P)] and remaining substrate [ee(S)] as a function of conversion *(c)* [74]: a) in the case of irreversible enantiomer selective reactions *(E*=100:—, *E*=10: - - -) b) in the case of reversible enantiomer selective reactions *(E*=100, *K*=0.1:—, *E*=10, *K*=0.1: - - -)

The yield of L-aminoacids (L-2.85) can be improved using acid catalyzed racemization followed by recycling of the unchanged D-acylamino acids (D-2.84).

The enantiomer selectivity of horse liver alcohol dehydrogenase (HLADH) was exploited in the preparation of the ketone (2.87) at the same time exemplifying a rare type of chirality (girochirality). The unchanged alcohol ((+)-2.86) was also enantiomerically pure [114]:

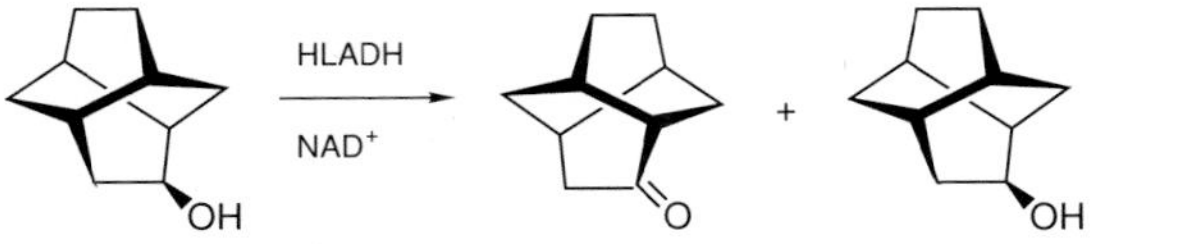

rac-(1.83)   (2.87) (83%ee)   (+)-(2.86) (~100%ee)

In purely practical terms, it is of course important to surpass the 50 % limit on the yield of simple kinetic resolutions. This is feasible if a highly enantiomer selective process can be combined with racemization of the substrate (but not that of the product)[11].

This strategy has been realized in the preparation of (S)-ketorolac (2.89), a potential anti-inflammatory and analgesic agent [115]:

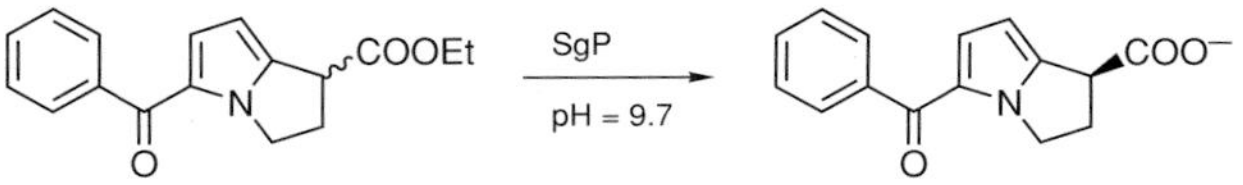

rac-2.88   (2.89) (92%, 85%ee)

Unfortunately, at such high pH not only is the rate of racemization of the substrate (rac-2.88) higher, but also both the rate of chemical hydrolysis and/or the rate of racemization of the product are higher. By conducting the same reaction at pH=8, whilst the process stops at 50 % conversion the enantiomeric purity is higher than 96 %.

The production of L-lysine from inexpensive α-amino-ε-caprolactame (rac-2.90) is based on this same principle [116,117]:

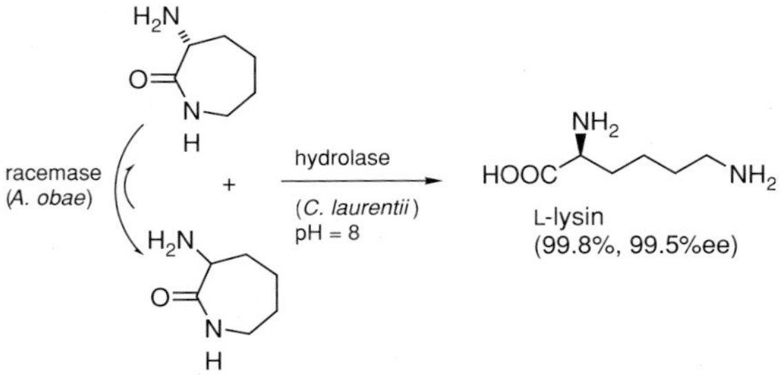

rac-(2.90)

L-lysin
(99.8%, 99.5%ee)

---

[11] In this case the enantiomeric purity of the product is independent of conversion. Enantiomeric purity is necessarily impaired if a non-enzymatic by-pass to the process exists or if the product is also susceptible to racemization.

In this transformation two enzymes are set to work simultaneously. Enantiomer selective hydrolysis of the lactame ring is performed by the hydrolase of *Cryptococcus laurentii*, whilst racemization of the unchanged substrate is looked after by a racemase produced by *Achromobacter obae*. Under the mild conditions of the transformation, neither chemical hydrolysis nor racemization of the product takes place and, owing to the high enantiomer selectivity of hydrolase [117], the racemic substrate can be totally converted to L-lysine.

An additional possibility to exploit this principle opens up if one of the products obtained by an enantiomer selective transformation can be converted to the other by configurational inversion. This approach was used in the production of the alcohol ((*S*)-2.92), an intermediate in the synthesis of a β-blocker (Fig. 2.12) [118]. It had been established that racemic oxazolidine esters of type (2.91) were substrates for several enzymes [119]. It was also established that for enantiomer selective hydrolysis of the racemic oxazolidinones (2.91) best selectivity could be achieved for a wide variety of R_1 and R_2 groups using the lipoprotein lipase of *Pseudomonas aeruginosa* (PaL)[120] and that the presence of the oxazolidinone ring is essential for high selectivity. In the absence of a carbonyl group or with 1,3-dioxolan-2-ones selectivity is poor [121]. Further chemical transformation of the products (2.92) is an attractive route to optically active β-blockers [122]. More recently the procedure has been further developed by running the hydrolysis in a column containing PaL enzyme fixed by adsorption [123]. Not only were the products of high enantiomeric purity, but also due to their differential adsorption, they could be eluted one after the other from the column. The ester ((*S*)-2.91) is eluted first using hexane or toluene, and then the alcohol ((*R*)-2.92) with a pH 7.5 phosphate

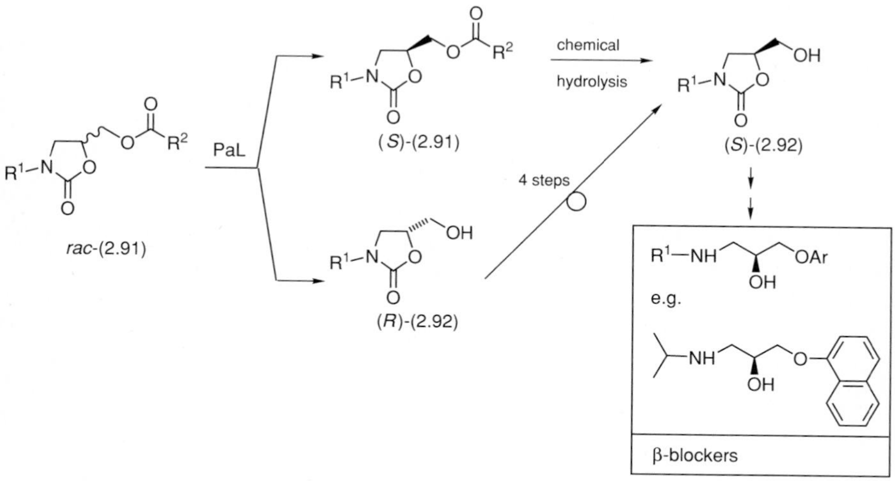

Fig. 2.12. Complete conversion of a racemic ester to a single enantiomerically pure alcohol by enantiomer selective hydrolysis and subsequent chemical inversion (The products are used in the synthesis of β-blockers.)

buffer and the column was shown to retain more than 95 % of its activity even after 20 cycles. In order then to utilize the unchanged substrate, a four step chemical inversion of ((R)-2.92) to enantiomerically pure ((S)-2.92) was worked out [118].

2.3.2.7   Enantiotope Selective Biotransformations

Enantiotope selectivity includes the selective transformation of both enantiotopic faces (Fig. 2.13.a) and enantiotopic groups (Fig2.13.b).In this process a prochiral structural unit in an achiral molecule is transformed to a center of chirality thus the achiral molecule becomes chiral.

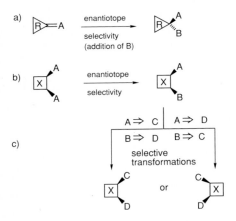

Fig. 2.13.   Enantiotope selective transformations: a) differentiation between enantiotopic faces; b) differentiation between enantiotopic groups; c) selective transformations of the product into both enantiomers of a target molecule

When the reaction is completely enantioselective, assuming there is a quantitative chemical yield, the total amount of the substrate can be converted to an enantiomerically pure product. (It is worth mentioning here that by virtue of a number of sophisticated schemes it is possible, in principle at least, to facilitate the transformation of a single enantiomer into both enantiomers of a target compound (cf. Fig. 2.13.c)).

In the simplest case, i.e. *when the products (P and Q) are not transformed further*, irreversible enantiotope selective processes can be described by the following model [72a]:

$$A \begin{array}{c} \overset{k_1}{\nearrow} P \\ \underset{k_2}{\searrow} Q \end{array}$$

(1)

where A is the substrate, P and Q are the enantiomeric products, and k_1 and k_2 the apparent first-order rate constants of the irreversible processes. An example of such a process would be the selective hydrolysis of prochiral dicarboxylic esters [72a] or occasionally the selective reduction of prochiral carbonyl compounds [72b]. In this case enantiotope selectivity is simply determined by the ratio of the two rate constants. Thus:

$$E = \frac{k_1}{k_2} = \frac{[P]}{[Q]} \tag{2}$$

from which it can be seen that the enantiomeric purity of the product (*ee*(P)) is independent of conversion.

$$ee(P) = \frac{E-1}{E+1} \tag{3}$$

This model becomes much more complicated, even if irreversibility holds, *when the products are transformed further*. For example, following the enantiotope selective hydrolysis of a prochiral diester the monoesters obtained as product may undergo further hydrolysis [124]. For this situation a different model (4) [124] shown below can be applied:

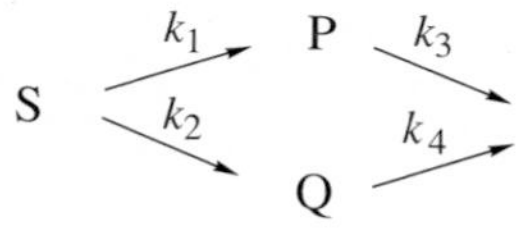

$$\begin{array}{ccccc} & & \xrightarrow{k_1} & P & \xrightarrow{k_3} & \\ S & & & & & R \\ & & \xrightarrow{k_2} & Q & \xrightarrow{k_4} & \end{array} \tag{4}$$

where k_1, k_2, k_3, and k_4 are the apparent first order rate constants. If, for the transformation of the products P and Q to R, stereochemical preference remains the same these constants follow the order $k_1 > k_2$ and $k_4 > k_3$. The process can now be described in terms of three parameters:

$\alpha = k_1/k_2$ (identical with parameter E in Eq. (2)),
$E_1 = k_3/(k_1 + k_2)$ and
$E_2 = k_4/(k_1 + k_2)$.

By inserting these parameters into the existing formulae, the following quantitative expressions can be derived:

$$[P] = \frac{\alpha[S_0]}{(\alpha+1)(1-E_1)} \left[\left(\frac{[S]}{[S_0]} \right)^{E_1} - \left(\frac{[S]}{[S_0]} \right) \right] \tag{5}$$

$$[Q] = \frac{\alpha[S_0]}{(\alpha+1)(1+E_2)} \left[\left(\frac{[S]}{[S_0]} \right)^{E_2} - \left(\frac{[S]}{[S_0]} \right) \right] \tag{6}$$

$$[R] = [S_0] - [S] - [P] - [Q] \tag{7}$$

$$ee(P) = \frac{[P] - [Q]}{[P] + [Q]} \tag{8}$$

Thus, by determining parameters α, E_1 and E_2 the quantity of the individual products and the value of $ee(\text{P})$ can be determined for any conversion with the aid of equations (5)-(8). Taking into account that $k_1 > k_2$ and $k_4 > k_3$, the enantiomeric purity of the product ($ee(\text{P})$) can be improved at the cost of its lower chemical yield by increasing the conversion percentage [125].

Since the kinetic treatment of *reversible enantiotope selective* processes is even more complex (cf. Section 2.3.1 about *reversible selective processes*) we have here made only some general comments. In a reversible enantiotope selective process, the enantiomer which was the preferred product in the forward reaction is reverted fastest in the reverse reaction. There are many examples of this phenomen e.g. hydrolysis vs. esterification with PPL enzyme [126], addition vs. elimination of ammonia with ammonia liase [127] or reduction vs. oxidation with GDH enzyme [128].

In enantiotope selective transformations performed with microbial systems a major problem is often that *the same reaction may be catalyzed by several competing enzymes* each with different kinetic parameters and different (or even opposite) selectivities [72b,130].

The following paragraphs present a few examples of selective biocatalytic transformations of enantiotopic groups or faces.

Enantiotopic groups are differentiated in the enzymic oxidation of prochiral glycerol to (S)-glyceraldehyde [131]:

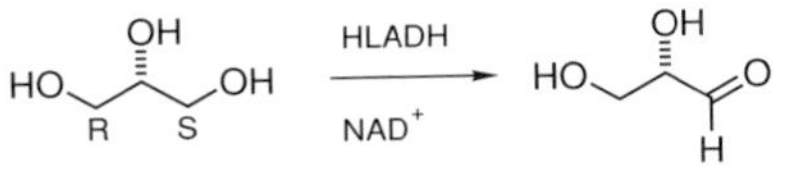

hydrolysis of the *pro-R* methoxycarbonyl group of the meso-diester (2.93) with PLE enzyme gives the chiral monoester (2.94) [132]:

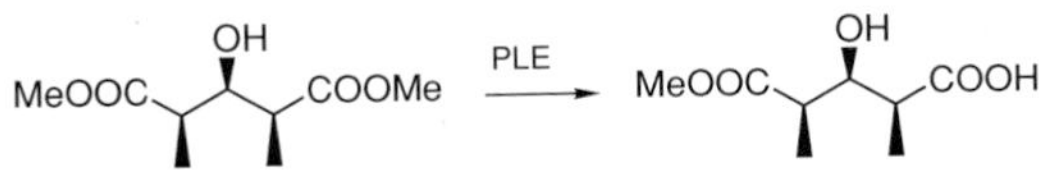

With a careful selection of the transforming microorganism, any of the R and the S configured carbon atoms in epoxymaleic acid (2.95) can be attacked to give the desired enantiomer of tartaric acid [133]:

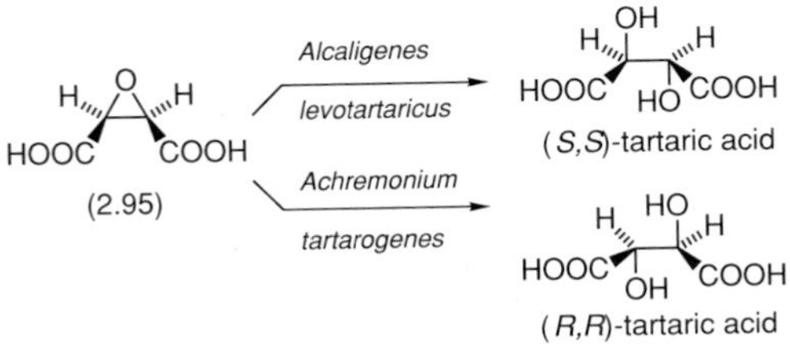

Non-bonding electron pairs of sulfides of type R_1SR_2 can be regarded as enantiotopic groups too and as such, microbial oxidation of sulfide (2.96) can be conducted to yield the *S* or the *R* sulfoxide (2.97) depending on the microorganism selected.

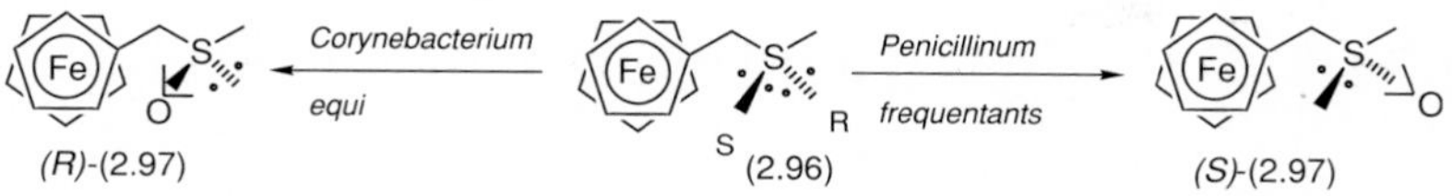

(*R*)-(2.97) (2.96) (*S*)-(2.97)

Differentiation of enantiotopic groups in biocatalysis is a phenomenon known for many years and its most common applications have been the enantioselective reduction of achiral carbonyl compounds. Thus reduction of ketone (2.98) from the *si* side performed with the aid of baker's yeast would be a classic example:

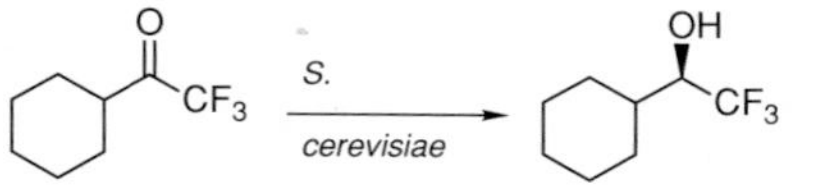

(2.98) (2.99)

Enantiotope selective transformations involving prochiral carbonyl compounds are not restricted to reductions. An example of such a non-reductive reaction is the formation of cyanohydrines from aldehydes catalyzed by oxynitrilase (E.C 4.1.2.10), yielding for example from benzaldehyde, the (*R*)-cyanohydrine ((*R*)-2.100) [136]:

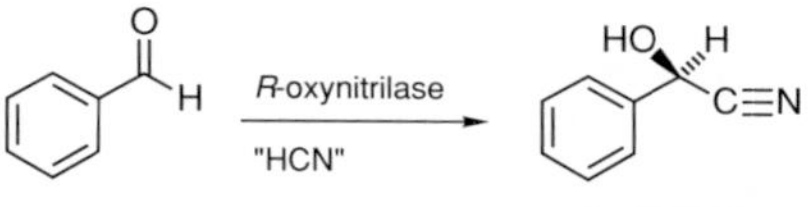

(*R*)-(2.100)

Bridgehead atoms of carbon-carbon double bonds may also participate in enantioselective processes as exemplified by the reduction of the α,β-unsaturated carbonylesters (*Z*-) and (*E*-2.101) by baker's yeast [137]:

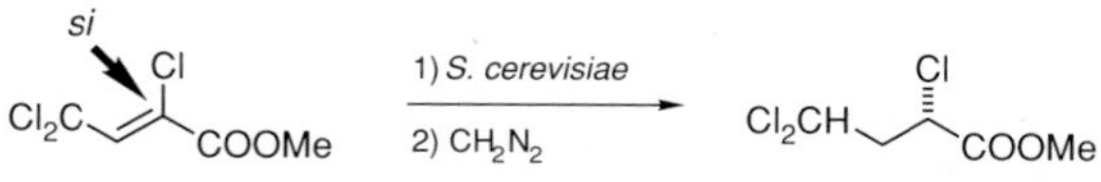

(*Z*)-(2.101) (*S*)-(2.102)

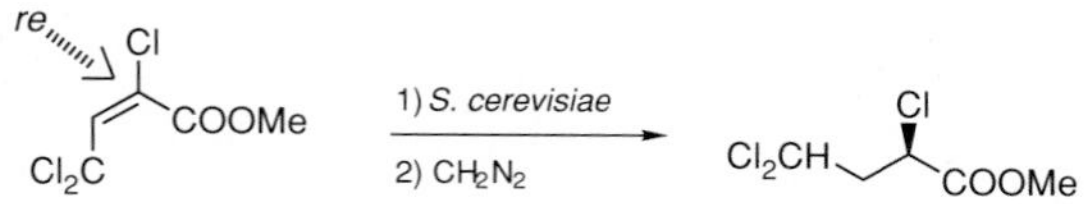

(*E*)-(2.101) (*R*)-(2.102)

This reduction is highly enantioselective at the C-2 atom for both geometrical isomers, the sense of selectivity being determined by the configuration of the double bond, i.e. it is *si* for the *Z* ester and *re* for the *E* ester. Reduction proceeds via the corresponding acids, as a premise demonstrated by the fact that the non-hydrolyzing ethyl esters resist reduction.

2.3.2.8   Biotransformations Involving More Than One Type of Selectivity

Diverse combinations of various types of selectivities can be encountered in biotransformations. Some of the many possibilities are considered here.

A relatively simple case is the coupling of chemo- or regioselectivity with stereoselectivity, although in this case it is usually only stereoselectivity that is emphasized[12].

Concurrent chemo- and enantiotope selectivity characterizes the hydrolysis of the prochiral dimethyl glutarate (2.103) in which α-chymotrypsin hydrolyzes predominantly the *pro-S* methoxycarbonyl group whilst both the *pro-R* methyl ester and the benzoyl ester groups remain intact [138]:

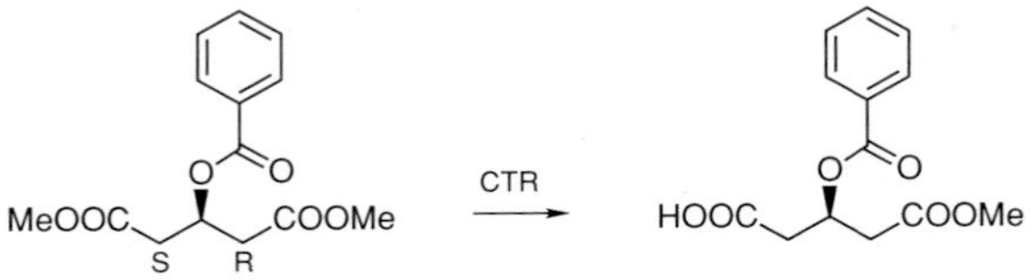

The oxidation of the racemic diol (2.105) in the presence of an alcohol dehydrogenase enzyme (HLADH) is both regio- and enantiomer selective. Here, dehydrogenation primarily affects the secondary hydroxyl of the (S,S)-diol yielding the corresponding ketone ((S)-2.106) whilst the (R,R)-diol can be recovered. Regioselectivity is not however complete, since about 7 % of a lactone is also formed via oxidation of the primary hydroxyl [139].

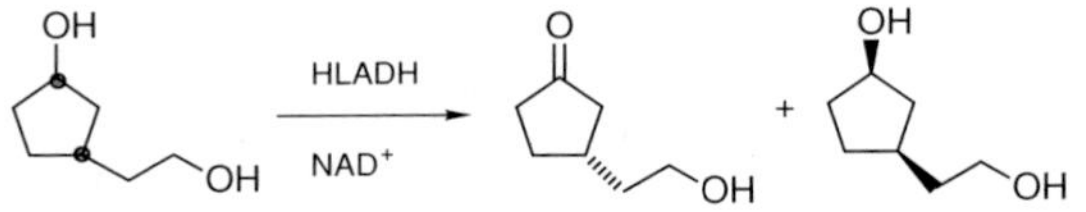

Stereochemical analysis is more complex when different types of stereoselectivities are in evidence concurrently. This is usually the case with the reduction of chiral (most often racemic) carbonyl compounds in which a new chiral center is created at a prochiral center of the racemic substrate. The results of such transformations are manifold (Fig. 2.14) [67,101]: (for the sake of simplicity, total selectivity has been assumed in the following examples).

a) Enantiomer and diastereotope selective reduction: Only one of the enantiomers is transformed and attack takes place at one of the diastereotopic faces. Both the product and the unchanged starting material are pure enantiomers (Fig. 2.14.a).

---

[12]   See, for example, the "enantiospecific" Baeyer-Villiger oxidation of (*rac*-2.81) in Section 2.3.2.6.

b) Enantiomer selective, diastereotope non-selective reduction: Again, only one of the enantiomers is transformed but both diastereotopic faces are equally attacked. The product is a mixture of diastereomers, while the remaining substrate is a single enantiomer (Fig. 2.14.b)

c) Enantiomer non-selective, diastereotope selective reduction: Both enantiomers are converted and diastereotopic faces in both are transformed with the same selectivity (e.g. both attacked from the *pro-R* face). The product is a 1:1 mixture of two optically active diastereomers (Fig. 2.14.c)

d) Enantiomer and diastereotope non-selective reduction: All four possible stereoisomers are formed as enantiomeric pairs in a 1:1 ratio (Fig. 2.14.d).

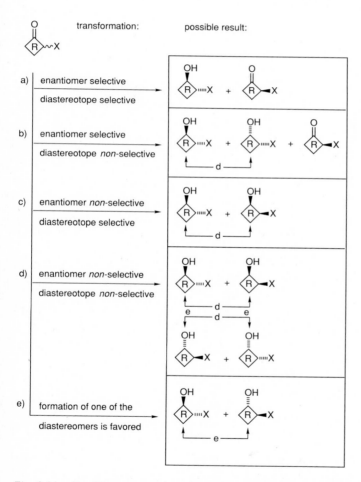

Fig. 2.14. Possible outcomes in reduction of a hypothetical racemic carbonyl compound (←d→: diastereomeric relationship, ←e→: enantiomeric relationship)

In biotransformations, types a) and c) prevail, type d) is rare [101] whilst types b) and e) are practically never encountered [although type e) (Fig. 2.14.e) is often met in chemical reductions].

It must be borne in mind however, that in the case of incomplete selectivities all possible stereoisomers are formed and product composition only approximates to the hypothetical limiting situations described above.

As an illustration, reduction of the racemic ketone (rac-2.107), corresponding to case c) is shown giving a separable mixture of diastereomers (2.108 and 2.109) [140]:

rac-(2.107)   (2.108)   (2.109)

After reoxidation (2.108) serves as an intermediate in the synthesis of cylobutanone analogues of penicillins.

Simultaneous enantiotope and diastereotope selectivity can be observed in the HLADH catalyzed reduction of the diketone (2.110). Of the two enantiotopic carbonyl groups of the symmetrical substrate only one is reduced, and this preferably from the *si* face, giving the alcohol (2.111) [141]:

(2.110)   (−)- endo-(2.111)

Interestingly, reduction of the same diketone (2.110) with baker's yeast occurred in a moderately enantiotope-selective but diastereorandom manner [145]. Examples also exist of the occurrence in a single biotransformation of more than two types of selectivity. Simultaneous regio-, enantiotope and diastereotope selectivity was experienced in the reduction of the triketone (2.112) with baker's yeast [142]:

(2.112)   (2.113)

While stereoselectivity was complete, i.e. the *pro-R* carbonyl remained unaffected and *re* selectivity was complete, the main product (2.113) was accompanied by 13 % of a diol arising from reduction of the exocyclic carbonyl.

3 Hydrolases: Preparative Biocatalysis

Hydrolases are extensively used for preparative biocatalytic purposes, often simply as reagents.

In order to help in the practical use of these biocatalysts we have briefly described here the most frequently used hydrolases, their preparation and some of their characteristics.

3.1 Hydrolases used for Preparative Purposes

Understandably, the first commercially available hydrolases exploited for preparative purposes were those that both possessed sufficient stability and did not require a cofactor. Among these, the most widely used are esterases, proteases and amidases [1-12]. Especially popular in synthetic operations are the carboxyl-esterases (EC 3.1.1.n), predominantly carboxylic ester hydrolases (EC.3.1.1.1) such as pig liver esterase (PLE) [13, 566] and horse liver esterase (HLE), and also lipases (EC.3.1.1.3).

The synthetic application of various lipases has recently become a dynamically developing field. The most extensively used is still the inexpensive pig pancreatic lipase (PPL), although more recently, several other lipases obtained from microorganisms (CcL, PfL, PsL, CvL, MmL, TvL, etc.) have also become commercially available and these are gaining ever increasing importance.

Esterases acting on substrates other than those mentioned above (EC.3.1.2.n), e.g. thiolester hydrolases, phosphomonoesterases, phosphodiesterases, and sulphate ester hydrolases, can also be utilized for synthetic purposes, although their scope is limited due to the rather strict structural requirements demanded of their substrates.

Glycosidases (EC.3.2), despite their considerable industrial importance, have only rarely been used in preparative biocatalysis, primarily because their substrates have a rather limited role as general synthetic intermediates.

Subclass EC.3.3 of the hydrolases comprises those enzymes that cleave ether bonds by hydrolysis. Amongst these, one mitochondrial epoxy hydrase (MEH) has been investigated in more detail.

Another often used group of hydrolases is that promote the cleavage of amide bonds. Amongst peptidases (i.e. amide bond cleaving enzymes specific for peptide bonds) it is the proteases (EC.3.4) that are of most practical use, whilst among the enzymes used for hydrolytic cleavage of non-peptidic amide bonds (EC.3.5) various acylases and amidases have been more widely applied.

In synthesis α-chymotrypsin (CTR) (EC.3.4.4.21.1), subtilisin (STC) (EC.3.3.3.16), trypsin (TR) (EC.3.4.21.4), papain (PP) (EC.3.4.22.2) and various microbial proteases have often been used.

Among acylases, the most frequently used is acylase I, which can be extracted from pig kidney or *Aspergillus* species (HKA and AA respectively) (EC.3.5.1.14), and also penicillin acylase (PA) (EC.3.5.2.6).

3.1.1   Preparation and Characteristics of Frequently Used Hydrolases

This section may offer some guidance towards the practical realizations of transformations catalyzed by hydrolases.

Hydrolases can be obtained both from microorganisms [14,16] and from various tissues of higher animals or plants [14,15].

Purity of enzymes

From an economic point of view, the degree of purity of an enzyme preparation intended for use in a particular transformation is not without importance. Often the crude microbial cell mass alone or a dried tissue preparation is both sufficiently active for use as it stands and highly selective. This situation is exemplified both in the preparation of L-lysine [17,18] and in the resolution of some racemic amino acid derivatives [19,20] where intact microbial cells can serve quite adequately as an enzyme source. Similarly, in the hydrolysis of prochiral dimethyl glutarates [21,22] or racemic lactones [23] dried pig or horse liver powder as the enzym source provides quite adequate activity and selectivity.

Sometimes, however, the purity of partially purified commercial enzymes is not sufficient and further purification aids in improving their selectivity [24]. Several methods are available for the purification of hydrolases and enzymes in general [25].

Characteristics of some frequently used hydrolases

Carboxylic-ester hydrolases (EC.3.1.1.1) [26] are widely distributed in the whole living universe. They have a broad specificity and can therefore be used for the hydrolysis of a wide selection of ester bonds. For synthetic purposes PLE enzyme is the primary enzyme of choice [13,27], although application of HLE enzyme is also known. At low concentrations of benzene or acetonitrile, activation of the serine hyrolase PLE can be observed [28] and substrate activation has occasionally been reported [29,30]. PLE is a mixture of isoenzymes each comprising a complex of three subunits. At pH 5 the subunits dissociate, whilst at neutral pH they are seen to reassociate [30,31] and, since the isolated isoenzymes show similar activity and specificity, essentially their mixture can be treated as being functionally homogeneous [32]. Various active site models predicting substrate specificity and selectivity have been forwarded and these are detailed in references [33-35, 565]. For synthetic purposes PLE can be used either as the crude acetone powder [21,37] or as a purified soluble or immobilized preparation [38].

Lipases (EC 3.1.1.3) [39,40] have a central role in lipid digestion and perform the cleavage of ester bonds in triglycerides, exerting their activity at the lipid-water interface [40,41]. Certain lipases such as RaL, AnL, and PPL, are specific for the cleavage of the primary ester bonds in triglycerides (i.e. 1,3-specific), whilst others such as CcL [42] attack both types of ester bond. Individual lipases show different specificities regarding the chain length of the acyl group; esters with a longer carbon chain are hydrolyzed faster. Some lipases that show high activity for soluble short chain glycerol esters would be better classified as esterases. Information concerning commercially available lipases is summarized in Table 3.1 although it should be borne in mind that commercial enzymes are not homogeneous and may exhibit more than one kind of enzyme activity (e.g. PPL [24] or CcL [61,62]).

Proteases (EC.3.4.4) are amongst the enzymes produced in largest quantity. They are extensively used in industrial processes, for example as detergents and food additives and additionally utilized both in the manufacturing of aspartame and for pharmaceutical purposes too.

The salient characteristics of proteases applicable in preparative biocatalysis have been compiled in Table 3.2. They can be classified into four main groups. *Serine proteases* (EC.3.4.21), e.g. α-chymotrypsin (CTR), trypsin (TR), and subtilisin Carlsberg (STC) contain in their active center a serine residue participating in the catalytic process. Serine proteases are active in the basic to neutral pH range. In animals they participate in digestion. *Cysteine or thiol proteases* (EC.3.4.22) perform their catalytic role by means of the SH group of a cysteine side chain. The range of their pH optimum is broad. Commercial thiol proteases are mainly extracted from higher plants, but a demand has also developed for enzymes of microbial origin too.

In *carboxyl proteases* (EC.3.4.23) which are active at acidic pH, it is a carboxyl group (usually of an aspartic acid) which plays the key role in catalysis. They can be found in the stomach of higher animals but are also common in fungi.

In *metalloproteases* (EC.3.4.24), a metal atom bound to the active center participates in the catalytic process. A good diagnostic indicator for such enzymes is inhibition by chelating agents such as EDTA. They are frequently found in bacteria and molds and function mainly at neutral pH.

The specificity of the various proteases differs widely [68]. Pronase, a protease prepared from *Streptomyces griseus,* for example, cleaves practically any kind of peptide bond, whilst others such as α-chymotrypsin or papain show distinct specificity (cf. Table 3.2).

Table 3.1. Sources and abbreviations of commercial lipase preparations.

Source [ref.]	Abbreviation[a]	Suppliers[b]
BACTERIA		
Alcaligenes sp [43]	AlcL	A, N
Achromobacter sp [44]	AchL	MS
Bacillus subtilis [45]	BsL	TK
Chromobacterium viscosum [46]	CvL	S, T, F
Pseudomonas aeruginosa [47]	PaL	A
Pseudomonas fluorescens [48]	PfL	A
Pseudomonas sp	PsL	S, A, BM, F
YEASTS		
Candida cylindracea[c] [49]	CcL	S, A, MS, BM, F
Candida lipolytica [50]	ClL	A, F
MOLDS		
Aspergillus niger [51]	AnL	A, F
Geotrichum candidum [52]	GcL	S, A
Humicola lanuginosa [53,54]	HlL	A
Mucor miehei	MmL	A, N
Mucor javanicus	MjL	F
Penicillinum camamberti	PcL	RP
Phycomyces nitens [55]	PnL	TY
Rhisopus arrhisus [56,57]	RaL	S, BM, F
Rhisopus delemar [58]	RdL	S, A, F
Rhisopus japonicus	RjL	A, NS
Rhisopus niveus	RnL	F
Rhisopus oryzae	RoL	A
Rhisopus sp	RhL	A, SE
PLANT		
wheat germ [59]	WGL	S
MAMMALIAN		
pig pancreas [60]	PPL	S, A, BM, F

[a]Abbreviations used throughout this book. Commercial abbreviations may differ from these.
[b]Abbreviations for suppliers: A: Amano, BM: Boehringer Mannheim, F: Fluka, MS: Meito Sangyo, N: NOVO, NS: Nagase Sangyo, RP: Rhone-Poulenc, S: Sigma, SE: Serva, T: Toyo Yozo, TK: Towa Koso, TY, Takeda Yakuhin. [c]*Candida rugosa.*

Amidases (EC.3.5.1)

Penicillin amidases (EC.3.5.1.11) [69] are enzymes, often called penicillin acylases, that have been used since the 1960's for the manufacture of penicillic acid from penicillins. Amidases specific for penicillin G (benzylpenicillin) are widely distributed in bacteria, whilst molds generally produce amidases specific for penicillin V (phenoxymethylpenicillin). The hydrolysis of substrates other than penicillins requires a penicillin acylase (PA, specific for penicillin G) produced by *Escherichia coli,* an enzyme which with suitable substrates may also show esterase activity. *Aminoacylases* (EC.3.5.1.14) are enzymes that are enantiomer selective in the face of *N*-acylamino acids, inasmuch as they only hydrolyze the derivatives of L-amino acids. Acylases have been used for almost 40 years for the resolution of synthetic and racemic acylamino acids [70-72]. Enzymes with acylase activity can be isolated from hog kidney (HKA) or from *Aspergillus oryzae* (AA) [72,73]

Table 3.2. Some commercial proteinases.

Enzyme	Source	Abbr.[a]	Preference (bond[b]/ref.)	Supplier[c]
Serine Proteinases				
α-Chymotrypsin	animal pancreas	CTR	Tyr,Trp,Phe, Leu [63]	S
Trypsin	animal pancreas	TR	Arg,Lys [64]	S
Subtilisins	alkalophilic bacilli, e.g.			
	Bacillus subtilis		broad [65]	N
Subtilisin Carlsberg	*Bacillus lichienformis*	STC	broad [66]	S, F
Pronase	*Streptomyces griseus*	SgP	broad	
Cystein Proteinases				
Papain	papaya latex	PP	Arg,Lys, (Phe)X[d]	S
Bromelain	pineapple		Lys,Ala,Tyr, Gly	S
Carboxyl Proteinases				
Pepsin	pig stomach		broad (aromatic)	S
Rennin	calf stomach		broad (aromatic)	S
Microbial	*Aspergillus* sp			T
	Mucor sp		rennin-like	
Metalloproteinases				
Thermolysin	*Bacillus thermoproteolyticus*		[67]	S
Other microbial	*Aspergillus* sp *Bacillus subtilis*			

[a]Abbreviations used throughout this book. [b]Enzymes act preferentially at the carboxyl group of the listed amino acid residues. [c]Abbreviations of suppliers: F: Fluka, N: NOVO, S: Sigma, T: Taka Ind. [d]Enzyme acts at the carboxyl group of the amino acid next to Phe.

although for practical purposes the latter is better suited [74,75]. A recent development is an even more stable, heat resistant acylase obtained from *Bacillus thermoglucosidus* [76] which may well soon acquire a degree of importance in industrial syntheses.

3.2 Biotransformations with Hydrolases: General Considerations

The use of hydrolases (EC.3) is probably the fastest growing area in the field of biocatalysis. The principal reason for this is their ease of application since they do not require either special equipment or specialist know-how. In addition, hydrolases function without coenzymes.

In aqueous solution hydrolases catalyze the following general reaction:

$$X\text{-}Y + H_2O \xrightleftharpoons{\text{Enz}} X\text{-}H + HO\text{-}Y \tag{1}$$

More specifically, hydrolases are most often used to catalyze the transformation

$$R^1\text{--}CO\text{--}Z\text{--}R^2 + H_2O \xrightleftharpoons{\text{Enz}} R^1\text{--}CO\text{--}OH + H\text{--}Z\text{--}R^2 \tag{2}$$

where, depending on the enzyme and the structural features of the substrate, Z may be O, NR^3, or S. In aqueous medium the equilibrium is shifted towards hydrolysis although under special conditions, primarily by changing the solvent, the equilibrium can be reverted.

Processes characterized by equation (2) comprise

a) if the dominant process is the one from left to right: Ester and anhydride hydrolysis (Z=O), amide or peptide hydrolysis (Z=NR^3), thioester hydrolysis (Z=S) etc., and

b) if the dominant process is that from right to left: Ester or anhydride formation (Z=O), amide or peptide formation (Z=NR^3), thioester formation (Z=S), etc.

Certain transformations pertinent to other enzyme classes may also be carried out using hydrolases. Thus hydrolases may function as transferases in the strict sense, if the substrate R^1-CO-Z-R^2 in equation (2) instead of reacting with water reacts with another partner which is a nucleophile:

$$R^1\text{--}CO\text{--}Z\text{--}R^2 + H\text{--}Nu \xrightleftharpoons{\text{Enz}} R^1\text{--}CO\text{--}Nu + H\text{--}Z\text{--}R^2 \tag{3}$$

Various transesterifications (Z=O, Nu=alcohol), transacylations, transamidations, etc. are among the transformations represented by equation (3).

It should be mentioned here that equations (2) and (3) correlate with the concepts of thermodynamically (i.e. equilibrium) controlled condensation reactions and kinetically controlled condensations by hydrolytic enzymes respectively that are often referred to in the literature [77,78].

In thermodynamically (equilibrium) controlled condensations hydrolases catalyze the process described by equation (2) where the desired situation is a process shifted towards R^1-CO-Z-R^2. In this case the enzyme only accelerates conversion whilst product ratio is determined by the equilibrium constant. Thus equilibrium concentrations are independent of the nature of the enzyme since they can only be influenced by changing reaction conditions and thereby the equilibrium constant.

In kinetically controlled condensations, in turn, hydrolases catalyze a process corresponding to equation (3). If the target compound is R^1-CO-Nu, then in an aqueous medium H-Nu and H-OH are competing nucleophiles (i.e. hydrolysis occurs as a side reaction). Similarly, the product H-Z-R^2 is also a competing nucleophile. The outcome of the process is controlled both by initial concentrations and reaction conditions.

The considerations above apply primarily to hydrolases which act via acyl-enzyme intermediates [79-84] (Fig. 3.1). Such enzymes are the serine hydrolases (certain esterases (PLE), lipases (PPL, CcL, etc.)), proteases (CTR, TR) and the cysteine (thiol) hydrolases (e.g. papain).

With hydrolases which do not form covalent intermediates, such as metalloproteases or carboxyproteases, the nucleophile, provided that it can interact with the active center, will act as a competitive inhibitor in kinetically controlled synthesis. In this case, these hydrolases can only be used in equilibrium controlled processes and not as transferases [77].

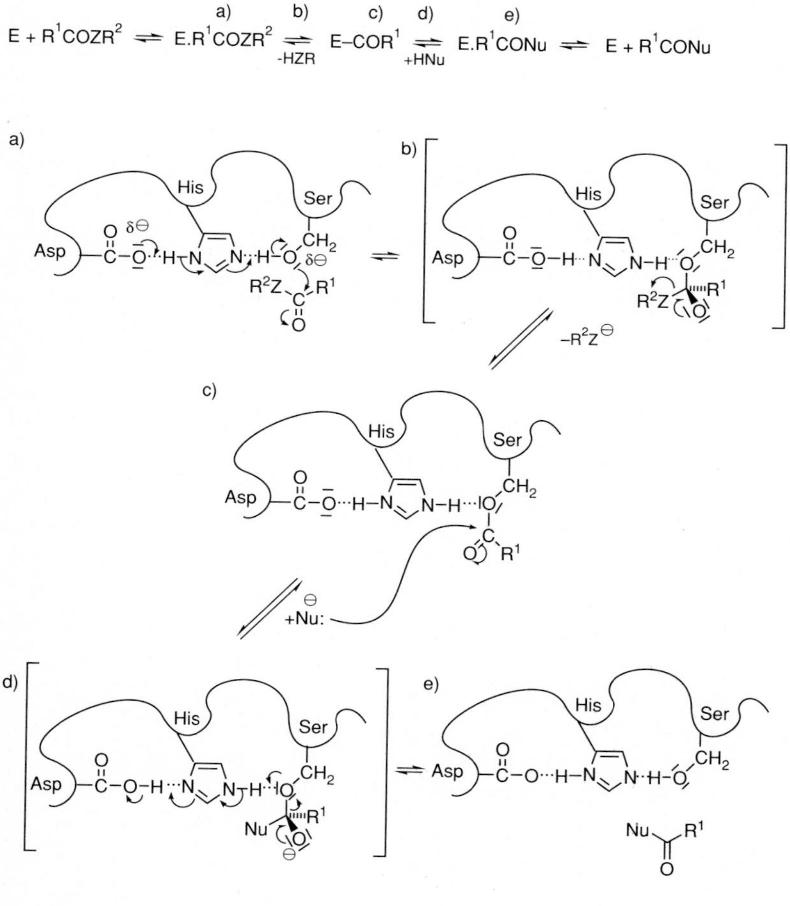

$$E + R^1COZR^2 \overset{a)}{\rightleftharpoons} E.R^1COZR^2 \overset{b)}{\underset{-HZR}{\rightleftharpoons}} \overset{c)}{E-COR^1} \overset{d)}{\underset{+HNu}{\rightleftharpoons}} \overset{e)}{E.R^1CONu} \rightleftharpoons E + R^1CONu$$

Fig. 3.1. A simplified mechanism for hydrolysis (transacylation) by serine proteases (hydrolases) (after [84]): a) enzyme-substrate complex; b) acylation of the enzyme via tetrahedral intermediate; c) acyl-enzyme; d) deacylation (transacylation) step; e) enzyme-product complex

In special cases, hydrolases may even act as lyases, e.g. in additions onto the C=C bond of α, β-unsaturated esters (Michael additions) [85]:

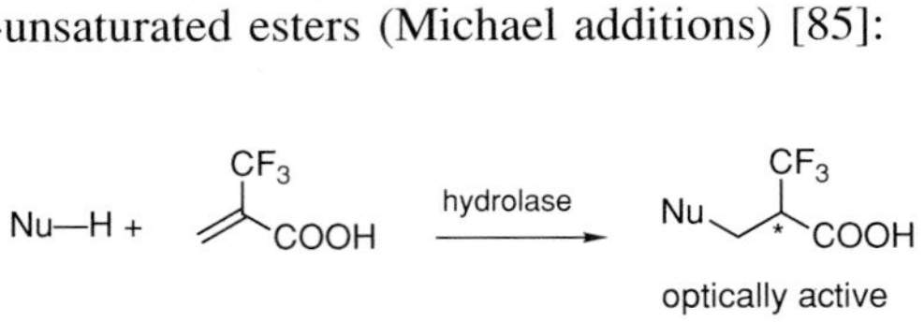

Here the formation of an acyl-enzyme intermediate can be assumed (Fig. 3.1.c), but the nucleophile, instead of attacking at the carbonyl group of the ester function, attacks at the β-carbon of the double bond.

3.2.1   Hydrolysis

In enzymic hydrolysis, predominantly aqueous solutions or mixtures of a water miscible solvent and water are used. Occasionally hydrolysis is conducted in a solvent that is immiscible but saturated with water [86].

In a pure aqueous medium it is primarily pH, ionic strength, ionic composition, and temperature that are the parameters by which the course of a given hydrolytic process can be effectively influenced. Sometimes various additives and changes of concentration may help.

Aqueous phases containing cosolvents, (i.e. water soluble organic solvents) are used in order to enhance the solubility of substrates that are poorly soluble in water [87]. The most common cosolvents are alcohols (methanol, ethanol, *t*-butanol), acetonitrile, acetone, tetrahydrofuran (THF), dimethylformamide (DMF), and dimethylsulphoxide (DMSO). Whilst at low concentration cosolvents usually do not impair the activity of enzymes, at higher concentration they can often cause denaturation of the protein. In an appropriate concentration however, cosolvents may even increase reaction rate [29,30].

In the presence of cosolvents, stereoselectivity of hydrolases often improves. In the selective hydrolysis of the *pro-S* ester group of the dimethyl 3-methylglutarate by PLE for example, a 79% ee was achieved in pure water at 20 °C, whilst at -20 °C in the presence of 20% methanol this rose to 97% [88]. A striking example of solvent effects on selectivity is the hydrolysis of dimethyl (2-phenyl-2-methyl) malonate by PLE, giving *pro-R* preference (52% ee) in water and *pro-S* preference (49% ee) in 50% aqueous DMSO [89].

The effect of organic solvents is much dependent on the constitution of the substrate. For example, whilst a 10% addition of *t*-butanol much improved *pro-R* selectivity in the PLE catalyzed hydrolysis of 3-cyclohexene-1,2-dimethanol diacetate (ee increased from 55% to 96%), in the case of the saturated analog the same measure of cosolvent caused the selectivity to drop from 12% ee to nil [90].

3.2.2   Non-Hydrolytic Processes in Organic Solvents

In a *one phase system* hydrolysis can be described by the following equation:

$$A + B \quad \overset{Enz}{\rightleftharpoons} \quad C + H_2O \tag{1}$$

The equilibrium constant for this can be defined in terms of the activities of the components $a_A \ldots a_{H_2O})$ [91] as:

$$K = \frac{a_C \cdot a_{H_2O}}{a_A \cdot a_B} \tag{2}$$

For non-ideal solutions of component i, activity and concentration are correlated as:

$$a_i = \gamma_i [i] \quad (\text{for water } a_{H_2O} = \gamma'_{H_2O} [H_2O]) \tag{3}$$

Thus the equilibrium constant expressed in terms of concentrations is:

$$K = \frac{[C][H_2O]}{[A][B]} \frac{\gamma_C \gamma'_{H_2O}}{\gamma_A \gamma_B} \tag{4}$$

For ideal solutions ($[i] \to 0, \gamma_i \to 1$) thus:

$$K_{aq} = \frac{[C]}{[A][B]} \tag{5}$$

Equation (4) suggests that there are two ways in which the equilibrium can be shifted in favor of product C (equilibrium controlled condensation):
a) diminishing the concentration of water and
b) making the solution non-ideal in such a way that $\gamma_C \gamma'_{H_2O} / \gamma_A \gamma_B$ should be much less than one [91].

Thus in one-phase systems it is by a change of the activity of water [92] as compared to the other components of the systems through which a shift of the equilibrium concentrations can be realized [91]. The simplest way to diminish the concentration of water is by the addition of a water miscible organic solvent. Thus, for example, aminoacid derivatives can be directly esterified with CTR enzyme in ethanol [93,94] or with papain in methanol [95] both containing less than 10% water.

Removal of water formed in the reaction of course shifts the equilibrium in favor of condensation. For instance, in the direct esterification of glycerol and various fatty acids, water can be eliminated from the system under reduced pressure [96].

In *two-phase systems* the equilibrium conditions of the process according to equation (1) (equilibrium controlled condensation) are thoroughly changed due to distribution between the organic and the aqueous phase [91,97]. A judicious choice of the solvent means the system can be dramatically shifted towards condensation. This approach was thoroughly studied in connection with the formation of amino acid esters and peptides

[77,78,98] as exemplified by the esterification of aminoacid derivatives in two phase systems with CTR [99,100] or papain [95,101]. A sophisticated way to shift the equilibrium of condensation reactions is to work in two phases moving in opposite directions [102].

A special utilization of two-phase systems are the various micellar microemulsions. Reverse micelles [103-105] are spontaneously formed aggregates of amphiphilic molecules in organic solvents. Water containing enzymes can be solubilized inside these reverse micelles by the hydrophilic end groups of the amphiphilic molecules. These head groups associate to form structures having polar cores with hydrophobic tails extending outwards into the bulk organic solvent leading to optically transparent, stable water-in-oil microemulsions. Amphiphiles are usually ionic molecules, but some non-ionic compounds have also been used successfully for such purposes [106,107].

Transferase activity of hydrolases (see Section 3.2, Eq. (3) and its discussion) can be exploited for transesterification, transacylation, transamidation, etc. in a pure or cosolvent containing aqueous phase, in a two phase system, or in an organic solvent alone. Water of course, promotes competing hydrolysis, but there are exceptions to every rule, for example the selective transpeptidation of alanine methyl ester by muramoyl pentapeptide carboxypeptidase is, even in water, the dominant process [108].

The developments of the last 4 or 5 years have shown that several hydrolytic enzymes, mainly lipases, remain active in *solvents of very low water content or even in seemingly anhydrous solvents* [109,116,117,119]. The advantages of performing biotransformations in organic solvents are listed in Table 1.4, and so only certain special features will be discussed here.

With few exceptions *enzymes are insoluble in organic solvents* [115]. For this reason, enzymes in organic solvents form suspensions of protein aggregates. Thus it is not surprising that under mechanical dispersion [109] or sonication [115] enzyme activity (or more precisely the accessible active surface) increases.

Hydrolases (and other enzymes) can be made more soluble in organic solvents by attaching hydrophobic side chains to the protein core. Modified enzymes solubilized in organic solvents by coupling with polyethyleneglycol (PEG) chains are amenable to biotransformations in apolar media [120-122]. PEG modified enzymes can be used in immobilized form; e.g. such a modified CTR enzyme in copolymeric gel was used for peptide synthesis [567].

In apolar media enzymes become more heat-resistant. For example, in an apolar solvent PPL enzyme may remain active even at $100\,^{\circ}\text{C}$ [111]. In addition, the stability of hydrolases in organic solvents may also be enhanced by the introduction of antioxidants [123].

In organic solvent system enzymes are only active in the presence of some water, the so-called *essential water content,* that serves to maintain the native conformation of the peptide chain [115,124,125]. This water content is present in two relatively well defined forms, namely as adsorbed or bound water and as water dissolved in the organic solvent [125,126]. It is the bound water component which is essential for optimal functioning. Essential water content varies between wide limits. Thus, for example, α-chymotrypsin remains active in the presence of as little as 50 moles of water per mole enzyme [127], whilst other enzymes may require much more [119]. In the case of CTR the source of essential water can be salt hydrates co-suspended with the enzyme in organic solvents [568]. Since hydrolases, relatively speaking, need a lot of water the process intended to proceed in the organic solvent may be accompanied by hydrolysis. Thus, thermolysin used for peptide synthesis by segment condensation is inactive in anhydrous amyl alcohol, it is slightly active when 1% of water is added, and whilst acceptable activity is achieved with a 4% water content, at this level of hydration hydrolysis becomes significant [128]. Part of the water component can be substituted by so-called *water mimics* [128,569], which are polar, water soluble cosolvents such as formamide, ethylene glycol and others. In a system containing 1% of water and 9% of formamide, the same level of activity was achieved as with 4% of water, but hydrolysis was almost completely suppressed. Note that it is not possible to substitute the entire water content with water mimics [128].

In media with low water content, although pH cannot practically be measured, the ionizable groups of the enzyme are almost in the same state as they would be at a pH characteristic to the essential water content. *The pH in organic solvents* refers to the pH of the solution from which the enzyme was lyophilized [114,129]. Indeed, the results of an NMR study on α-lytic protease showed that there was no difference between the ionization state of protein in the solid state and in the solution from which it was isolated [129].

There is some information available about the state of *the active center in organic solvents*. In the case of subtilisin, a Hammet analysis (linear free energy correlation model) was invoked for the study of the hydrolysis of *p*-substituted phenyl acetates in water and for their transesterification with hexanol in organic solvents [130]. Since it is the acyl-enzyme formation which is the rate determining step (cf. Fig. 3.1), kinetic data are not affected by the nature of the nucleophile. Hammet constants agreed, within experimental error limits for water and five different organic solvents indicating that in the transition state the active center was shielded from the solvents and remained in its native conformation. A solid phase NMR study on α-lytic protease [129] led to similar conclusions. The state of the active center was the same in water, acetone and octane. However, in DMSO, which dissolves the enzyme, the original ordering of the catalytic triad in the active center was destroyed and the enzyme thereby inactivated. Lyophilization from DMSO had the same disruptive effect.

It is a commonly observed phenomenon that in enzyme catalyzed condensations (or transacylations and transesterifications) carried out in an organic solvent *stereoselectivity is higher* than in the corresponding hydrolysis [113,131]. This is primarily a characteristic of those enzymes such as lipases, which are capable of induced fit [132], i.e. their protein skeleton is sufficiently flexible such that in interaction with the substrate, they can adopt a conformation that is optimal for substrate binding. It has recently been discovered [133] that CcL enzyme undergoes a *permanent conformational change* in the presence of organic solvents and deoxycholate. As a consequence of this non-covalent protein modification, enantiomer selectivity improved remarkably. Such substrate-induced conformational change has also been observed in the case of CTR acting in organic solvents [134]. The conformationally modified enzyme can catalyze esterification of the inducing substrate D-aminoacid derivative, whilst untreated natural enzyme can not. In organic solvents, enzymes assume a frozen conformation which may explain their characteristically enhanced selectivity [113].

The situation is different for non-induced fit enzymes which have a more rigid conformation. Here the binding of substrate to the active center induces much less change in the structure of the active center. As such, the structure of the active center remains unchanged in an organic solvent and hence no increase in selectivity can be expected when using these enzymes in organic solvents. In fact, a study on the hydrolysis of L- and D-amino acid derivatives in water and their transesterification in organic solvents by several proteases revealed a solvent dependent drop in selectivity [135]. This may be explained by the assumption that binding of the correct enantiomer displaces more water from the binding site than that of the wrong one and thus, in a hydrophobic solvent, the prevailing enantiomer is at a relative disadvantage as compared to the wrong one.

The equilibrium conditions of both the hydrolytic and also the transferase activity of hydrolases can be influenced. In transformations characterized by equation (3) (Section 3.2) both R^1-CO-Nu (e.g. peptide synthesis, transesterification, etc.) and H-Z-R^2 (e.g. alcoholysis) can be the desired product.

The methods available to shift equilibria in the required direction may be distinguished on the grounds of whether they influence the equilibrium in a "reversible" or "irreversible" manner. A reversible shift can be induced by applying one of the substrates in large excess, e.g. as solvent. An example of this approach is transesterification with various alcohols in tributyrin (glyceryl tributyrate) as solvent [111] (where the alcohol is representative of H-Nu, solvent is R^1-CO-Z-R^2 and product is R^1-CO-Nu).

In the *"reversible"* case, reaction rate and/or selectivity may be increased if the R^2 group of the transesterifying partner (R^1-CO-Z-R^2) contains electron attracting substituents (e.g. using trichloroethyl butyrate [136]). Heteroatoms in the alcohol component (H-Nu) (e.g. $HOCH_2CH_2X$ where X=Hal, dialkylamino- or alkoxyalcohols) may show a similar effect [137].

"Irreversible" methods are more efficient in shifting equilibria and increasing reaction rate. Some typical and often used "irreversible" transesterification agents are the vinylesters [138,139] whose irreversibility originates from the tautomerism of the liberated alcohol.

acetaldehyde (bp: 22 °C)

Oxime-esters may also serve as "irreversible" transesterification partners [140]. In the acylation of various alcohols catalyzed by PPL enzyme, diacetyl-monooxime acetate proved to be faster than even vinyl acetates. The liberated oxime is thus a very weak nucleophile.

The process becomes similarly "irreversible" when the R^1-CO-Z-R^2 partner is an anhydride. When using anhydrides in enzymatic acylation, the removal of the acid formed may be crucial [570]. The application of succinic anhydride for this purpose is also interesting from the point of view of product separation, inasmuch as the half ester formed and any unreacted alcohol that might remain can be separated by extraction [141]:

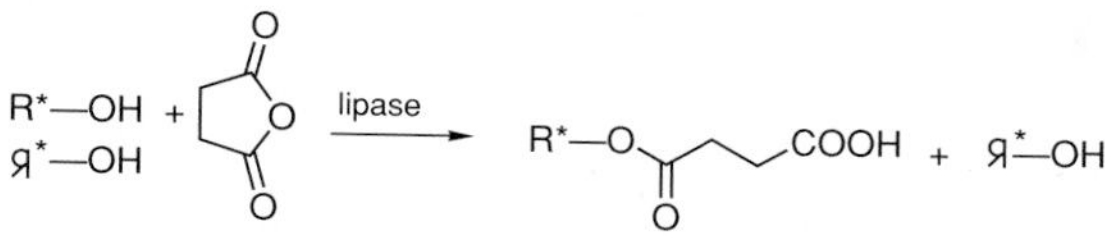

In hydrolase catalyzed processes conducted in organic solvents, enzyme immobilization may affect both rate and selectivity [142,143]. Additionally, immobilization can also enhance enzyme stability, for example multipoint attachment of CTR to a solid matrix which fixes the intrinsic conformation of the protein protects the enzyme from inactivation by alcohols [144].

Selection of the right solvent is crucial. If the solvent is not one of the partners it is important that it should not participate in any reaction catalyzed by the enzyme. Equally important is that the solvent should not denature the enzyme. The native conformation of most hydrolases for instance, is lost in DMSO or DMF. Solubility of the substrate should also be considered; sugars for example, are soluble only in hydrophilic, water miscible solvents thus enzymic acylation of sugars can only be carried out in such solvents as pyridine [145] or DMF [147]. In hydrophobic solvents there would be no interaction between substrate and enzyme since both are insoluble therein. Occasionally additives can be applied to improve solubility which may enhance rates even in apolar solvents, an example of which is the transesterification of *N*-acetylphenylalanine ethyl ester in octane. In this CTR enzyme catalyzed transformation, the reaction rate increased on addition of the crown ether 18-crown-6 which serves to solubilize the substrate [148].

3.3 Biotransformations with Hydrolases: Utilizable Selectivities

From the wide range of selectivities encountered with biocatalysts, not all but many can also be realized with hydrolases.

Thus hydrolases are often used to transform sensitive compounds under mild conditions; they are also suitable agents to effect chemo-, regio- and diastereomer selective transformations, although examples of transformations involving enantiotopic or diastereotopic faces are rare. The most popular fields of application of hydrolases are, however, enantiomer and enantiotopic group selective reactions and it is in these particular two areas that fastest progress can be expected. Of course, simultaneous manifestation of more than one kind of selectivity also occurs in hydrolase catalyzed processes, most often in the case of enantiomer selective transformations. These applications will be described in detail in the following sections.

3.3.1   Hydrolases: Non-Selective Transformations under Mild Conditions

Lipase catalyzed hydrolysis of lipids has great industrial significance. Besides the hydrolysis of triglycerides (e.g. in a hollow fiber membrane bioreactor with CcL enzyme [110]) the transformation of oils and fats by transesterification is of prime economic importance [110,149,150]. From a nutritional point of view, it is significant that triglycerides rich in eicosapentaenoic and docosahexaenoic acid can be produced by incubation of triglycerides of saturated fatty acids with a concentrate of polyunsaturated fatty acids [151] or by selective hydrolysis of fish oil with lipase [571]. Direct enzymic esterification may also be utilized, e.g. in the manufacturing of monoglycerides (for use as non-ionic surfactants) by the condensation of glycerol-acetonide and free fatty acids in the presence of MmL enzyme [96].

Similarly, so-called flavor esters can be prepared by esterification and transesterification of terpenes [153-155].

In a broader sense, enzymic peptide synthesis [77,78,98,156,157] also belongs to this field, since in this mild reaction catalytic condensation without racemization is a central issue. In contrast to purely chemical methods, as a result of the substrate specificity of the enzymes used for this purpose, less or even no protection at all of the substrates is necessary. Only a few special points are discussed here and for more details you are referred to the literature.

In peptide synthesis it is primarily proteases that are used. This includes serine proteases [157] (e.g. CTR [158-162,164] or subtilisin [163]), thiol proteases (e.g. papain [165-170] or thiol-subtilisin [171] a thiol protease obtained by transformation of a serine protease) or metalloproteases (e.g. thermolysin [128,172-174]).

In general, proteases are selective for the formation of peptides containing only L-amino acids. In an enantiomer selective reaction, if the enzyme is offered a racemic amino acid, only the L-enantiomer is transformed [168,173].

Peptide syntheses can also be catalyzed by hydrolases other than proteases, e.g. by lipases [175-177] and PLE [176]. With lipases, lack of amylase activity is an advantage and they can also catalyze the incorporation of D-amino acids and other more unusual residues. This is of note since unnatural peptides containing D-amino acids are generally not accessible by recombinant DNA technology.

Recently it has been observed that at high concentrations of organic solvents, proteases may also be capable of incorporating D-amino acids into peptides [161,165,169,178,179] and moreover, proteases specific for D-amino acids have also been found [108,180,181].

In enzymic peptide synthesis, usually only a single peptide bond is established in one operation, although it is not unprecedented that under special conditions a tripeptide can be formed in one operation from three amino acid derivatives [169].

With the use of different enzymes, more than one fragment can be coupled, whereby oligopeptides containing up to 15-20 amino acid units can be prepared. Examples of this are the preparation of dynorphin-(1-8) and of an encephalin elongated at the C-terminal (Leu-encephalin) with the aid of papain, CTR, and TR [182], or the synthesis of the cholecystokinin octapeptide with minimal side chain protection using papain and thermolysin [183].

Some peptide syntheses have been developed up to an industrial scale. Of particular economic importance is the production of aspartame, an artificial sweetener, [168,172,173] and the transformation of porcine insulin to human insulin based on trypsin catalyzed fragment condensation and transpeptidation [78,98,184]. Neuroactive oligopeptides [78,98,185], peptide antibiotics [186] or the tripeptide precursor of penicillin [178] are also potentially useful candidates for similar industrial scale enzymic syntheses.

3.3.2   Hydrolases: Chemoselectivity – Substrate Specificity

The chemoselectivity of hydrolases is well known and some examples of this are shown in Fig. 3.2.

A classical case demonstrating chemo- and regioselectivity is that of the action of phospholipases on phospholipids (3.1) [187].

Hydrolases specific for various functional groups or structural units are especially suited to the realization of chemospecific transformations. Suitable examples of this are the hydrolysis of the acetoxy-dimethylester (3.2) with an acetyl-specific carboxyl esterase [190], hydrolysis of the peracylated sugar derivative (3.5) with a penicillin acylase specific for phenacetyl groups [193] or the preparation of the anti-ulcer agent cetraxtate.HCl from its benzyl ester (3.4) with a benzyl ester specific hydrolase [192]. The hydrolysis of the α-methyl ester of BOC-Asn(NAcGlc)-OMe was chemoselectively achieved by papain [573].

Different hydrolases may be selective for different structural units. It was established that lipases (e.g. PPL) preferentially hydrolyze esters of primary alcohols, whilst carboxyesterases (e.g. PLE, HLE) cleave esters of both primary and secondary alcohols [196-198]. Chemo- and regioselectivity of the two types of enzymes is often opposite [198].

The above mentioned features of hydrolases may be exploited to *introduce or remove protecting groups* enzymatically. Thus the phenacetyl group which is removable by penicillin acylase can be used both in peptide chemistry for *N*-terminal protection [167,193,199,200-202], but also in sugar synthesis [193,194]. Papain catalyzed amidation by phenylhydrazine is suitable for carboxyl protection in peptides [203] as is the preparation of allylic esters [204,205] or n-heptyl esters [572], which can later be hydrolyzed enzymatically without racemization.

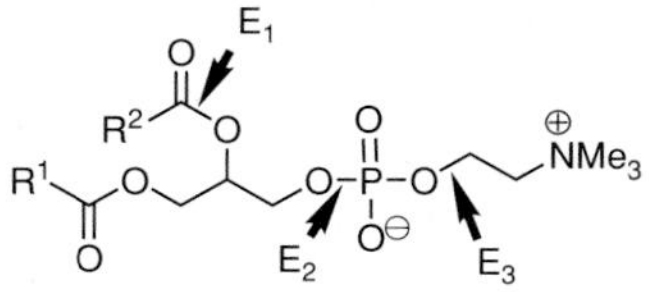

Fig. 3.2.   Chemoselective attack of hydrolases at selected substrates E_1=phospholipase A [187,188]; E_2=phospholipase C [187]; E_3=phospholipase D [189]; E_4=acetyl specific esterase from *Nocardia mediterranei* [190]; E_5= subtilisin [191]; E_6=cetraxate benzyl ester hydrolase [192]; E_7=penicillin acylase [193]; E_8=lipase (CvL,PsL,PPL,CcL) [195]

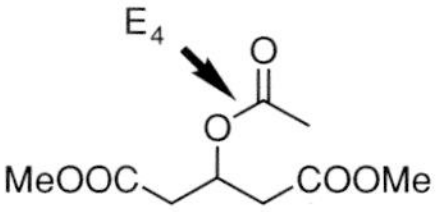

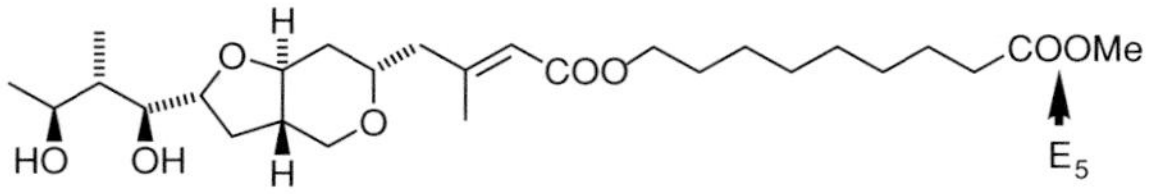

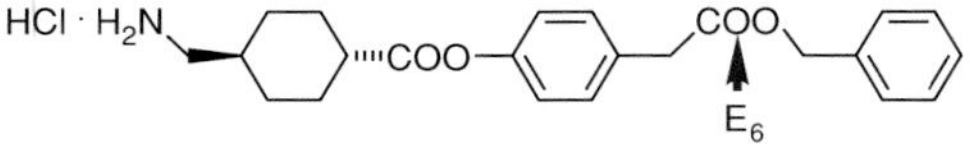

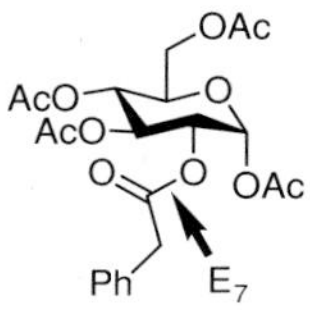

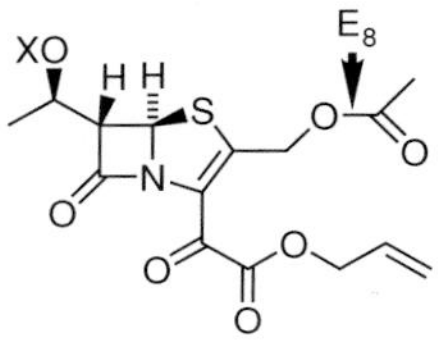

3.3.3   Hydrolases: Regioselectivity

Regioselectivity (in the product selectivity sense) is possible with substrates which contain at least two constitutionally different homomorphic groups (cf. Section 2.1). Most of the regioselective processes realized with hydrolases so far concerned three, relatively well distinguished groups of compounds: (i) amino acid derivatives with carboxyl or amino functions in the side chain, (ii) steroids, and (iii) sugars and congeners.

i) Bifunctional amino acid derivatives

Derivatives of monoamino-dicarboxylic acids (Asp, Glu) are generally attacked by hydrolases at the α-carboxylic acid end. Good examples of this are the hydrolysis of dibenzyl esters by Alcalase R (from *B. lichienformis*) [206], their esterification with papain [95,101,207], and also their peptidation with papain [166]. This α-selectivity of papain was also conserved in the hydrolysis of (Z)-α-dehydroglutamate esters [208]. On the contrary, CTR hydrolyzes the same substrate with opposite regioselectivity [574].

The sense of regioselectivity can sometimes be influenced by structural modification of the substrate. For instance, the cycloalkyl esters of arginin hydrolyze preferentially at the β position [209], whilst in the subtilisin or PsL catalyzed condensation of the bulky *tert*-butyl ester of L-lysine, it is rather the α-amino group and not the usual ε-amino function which is affected [210].

ii) Steroids

With polyhydroxysteroids, regioselectivity (also in combination with other selectivities) is primarily useful in transesterifications (transesterification of hydroxysteroids [211,212] or alcoholysis of steroid esters [213] in organic solvents). Selectivities recorded in such transformations are shown in Fig. 3.3. It can be seen that lipases (CcL, Figs. 3.3a-c) attack at positions 3 and 17β [213] and furthermore they have a preference for 3β configuration [213], although occasionally 3α derivatives are also transformed [212]. Groups attached to carbons 7, 12β, and 17α remain unchanged [212]. Subtilisin, in turn, is specific for 17α oriented functions however [211].

iii) Sugars and sugar derivatives

Regioselectivity is also manifested in the enzymic esterification of sugars and sugar derivatives (Tables 3.3 and Fig. 3.4), as well as in the hydrolysis and alcoholysis of sugar esters (Fig. 3.5).

Acylation of sugar derivatives (Fig. 3.4 and Table 3.3)

Regioselective acylation of unprotected monosaccharides [145,146, 579, 580] and disaccharides [147,222] and further sugar derivatives [581, 582] can be carried out with the aid of hydrolases. For such purposes, only organic solvents in which the sugars are soluble (pyridine, DMF) and enzymes which tolerate such solvents can be used.

Acylation of primary hydroxyls is generally preferred (3.7-3.11, 3.13-3.23), but if these are protected, under certain conditions secondary hydroxyls can also be regioselectively acylated (3.12, 3.19). The sense of regioselectivity may differ for different enzymes. For example, with the furanose derivative (3.19), HrL and RjL enzymes markedly differ in regioselectivity. Other examples of regioselectivity in such transformations are the enzyme catalyzed acylations (transesterifications) of the pyranose derivatives (3.7-3.17), the furanose derivatives (3.18), (3.19), the disaccharide (3.20), the nucleosides and analogs (3.22), (3.23) and riboflavin (3.21).

Acylation can also be carried out by direct esterification [214] using irreversible transesterification reagents [139,218], or with anhydrides [221], although most often haloalkyl esters are used for this purpose [145,147,215,219,220,222]. The enzymes used have generally been lipases, but subtilisin was also found to be suitable [147,222].

Hydrolysis of peracylated sugars (Fig. 3.5)

With peracylated sugars, again the esters of primary alcohols were preferentially hydrolyzed (3.24, 3.29). It was reported however, that the anhydrosugar ester (3.25) could

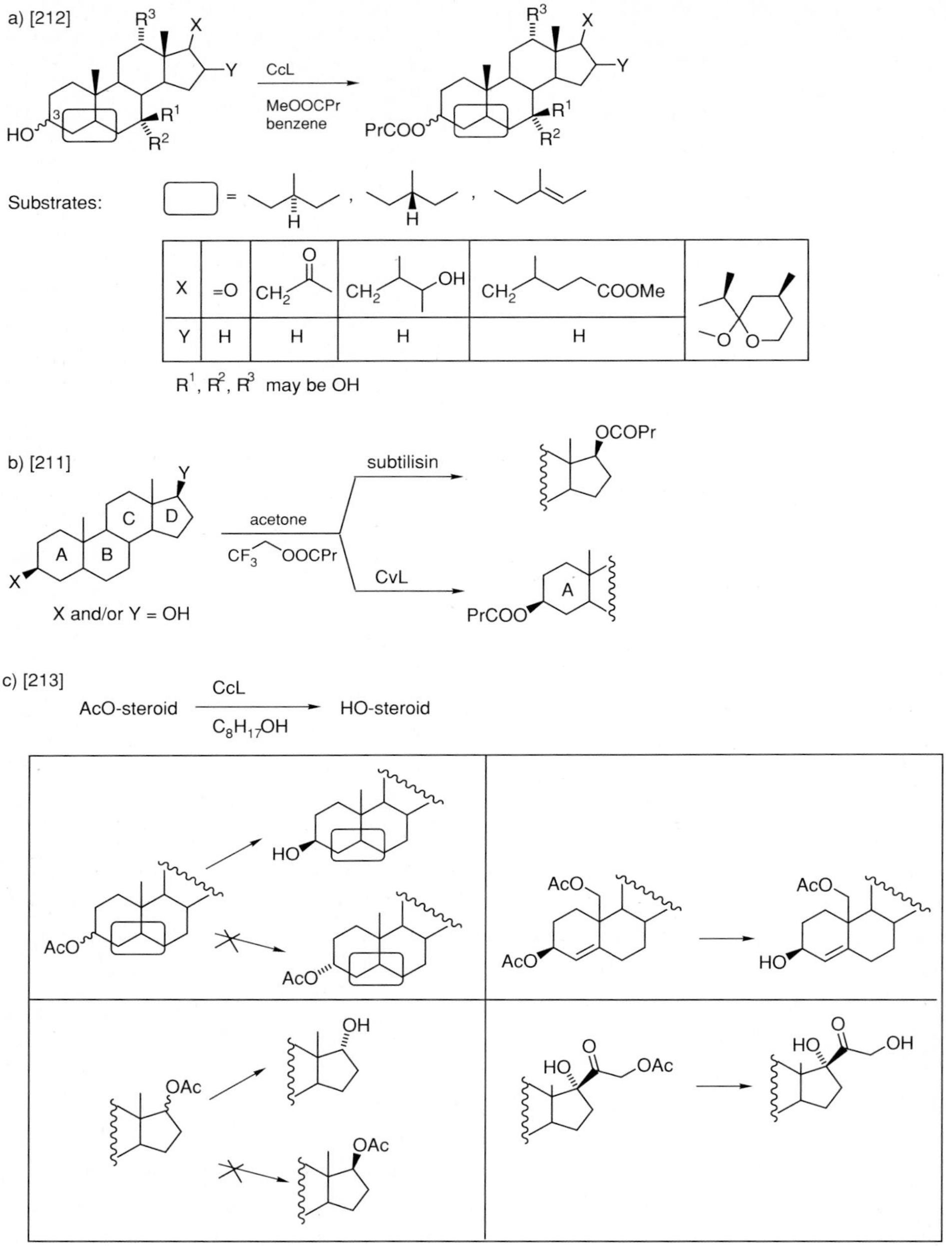

Fig. 3.3.   Selectivities in hydrolase catalyzed steroid transesterifications

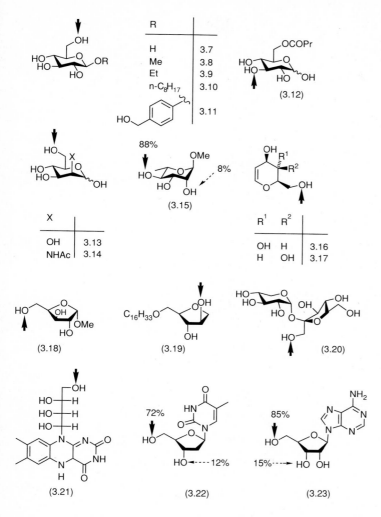

Fig. 3.4. Selected examples of regioselective acylations of sugars and related compounds by hydrolases (details in Table 3.3)

be selectively hydrolyzed by a proper selection of the enzyme affecting ester groups in any position [224].

Synthetic application of the regioselective protection and deprotection of sugar derivatives is exemplified by the hydrolysis of tetrabutyryl-α-methyl-glucopyranoside (3.24). The product is then converted to an antitumor agent 6-fluoro-sugar [223].

Naturally, hydrolases can be used for regioselective biotransformations of substrates other than the three types of compounds discussed above. For example, the regioselective

Table 3.3. Selected examples of regioselective acylation of sugars and related compounds by hydrolases.

No.	Enzyme	Acylating agent	Ref.
3.7	PPL	$Cl_3CCH_2OCOC_{11}H_{23}$	[145]
3.8	CcL	$CH_2=CHOAc$	[139]
3.9	CaL	HOOC-alkyl	[214]
3.10	AnL	Cl_3CCH_2OCOPr	[215]
3.11	STC	Cl_3CCH_2OCOPr	[217]
3.12	CvL	Cl_3CCH_2OCOPr	[215]
3.13	CcL	$CH_2=CHOAc$	[139]
3.14	CcL	$CH_2=CHOAc$	[139]
3.15	PPL	CF_3CH_2OCOPr	[216]
3.16	CcL	$CH_2=CHOAc$	[218]
3.17	PfL	$CH_2=CHOCOPh$	[218]
3.18	PPL	CF_3CH_2OAc	[219]
3.19	RjL,HlL	Cl_3CCH_2OCOPr	[220]
3.20	STC	$CF_3CH_2OCOC_5H_{11}$	[147]
3.21	STC	Cl_3CCH_2OCOPr	[219]
3.22	PfL	$(C_5H_{11}CO)_2O$	[221]
3.23	STC	Cl_3CCH_2OCOPr	[222]

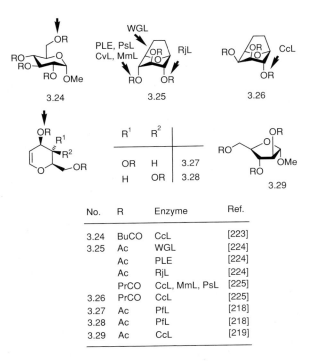

No.	R	Enzyme	Ref.
3.24	BuCO	CcL	[223]
3.25	Ac	WGL	[224]
	Ac	PLE	[224]
	Ac	RjL	[224]
	PrCO	CcL, MmL, PsL	[225]
3.26	PrCO	CcL	[225]
3.27	Ac	PfL	[218]
3.28	Ac	PfL	[218]
3.29	Ac	CcL	[219]

Fig. 3.5.   Selected examples of regioselective deacylations of sugar esters and related compounds

hydrolyses or acylations of the polyhydroxy alkaloid castanospermine can be conveniently carried out with the aid of hydrolases [226]:

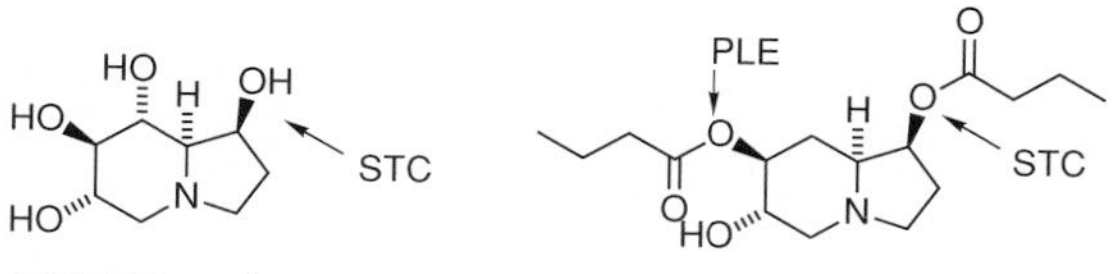

castanospermine

And indeed the hydrolase catalyzed regioselective esterification of chloramphenicol [227] or the regioselective hydrolysis of diethyl (S)-malate by PLE used in the synthesis of tulipalin-B [228] are further examples.

3.3.4   Hydrolases: Diastereomer and Diastereotope Selectivity

Purely diastereomer selective transformations that are realizable with hydrolases have already been discussed in Section 2.3.2.3. Relatively few examples of this type of selectivity, e.g. diastereomer selective acylation of (E,Z)-terpenic allylic alcohols [583] or enzymatic recognition of diastereomeric esters [584], have been reported and thus there is ample room for further studies.

In hydrolase catalyzed transformations, diastereomer selectivity can be combined with other selectivities too (Fig. 3.6). Depending on the conversion and initial isomeric composition, hydrolysis of a stereoisomeric mixture of pyrethroic esters with PLE enzyme (Fig. 3.6.a) may be solely diastereomer selective, both diastereomer and enantiomer selective, or finally purely enantiomer selective. Hydrolyses have been carried out both with the purified enzyme [229] (X=Me, Cl, R=Me) and with a crude acetone powder extract [230] (X=Me, Cl, Br, R=Me, Et). Diastereomer and enantiomer selectivity are coupled in the lipase catalyzed hydrolysis of (chloro-fluoro)methyl ar(alk)yl carbinol esters (e.g. 3.32) (Fig. 3.6.b) and in the PPL catalyzed hydrolysis of the chlorinated cyclopentenol acetate (3.34) (Fig.3.6.c). The product of this latter reaction (3.35) was used for the synthesis of punaglandin-4, a chlorine containing prostanoid isolated from a coral [232].

Of the selective hydrolytic transformations of diastereotopic groups, an interesting example is that of the following dimethylester, the chirality of which arises from the hindered inversion of an sp^3 nitrogen atom:

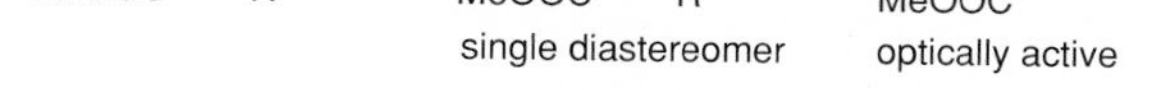

This hydrolysis is diastereotope and enantiomer selective both with aziridines (X=CH, R=Cl, enzyme: lipases)[233], and oxaziridines (X=O, R=i-Pr, t-Bu, enzyme: PPL)[234].

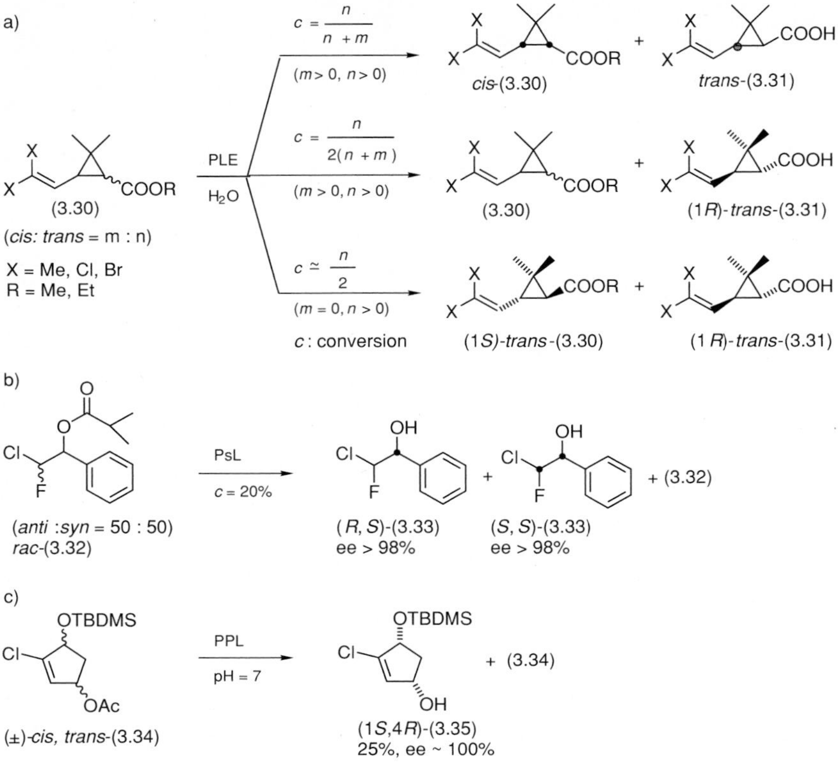

Fig. 3.6.   Diastereomer selectivity (with enantiomer selectivity) in enzymic hydrolysis: a) hydrolysis of pyrethroid esters [229,230]; b) moderate diastereomer selectivity parallel with enantiomer selectivity in hydrolysis of chloro-fluoro-methyl substituted carbinol esters [231]; c) diastereomer and enantiomer selective hydrolysis of a chlorocyclopentenyl acetate [232]

The half esters isolated were pure diastereomers and the unreacted substrates optically active.

3.3.5   Hydrolases: Enantiomer Selectivity

The potential for hydrolases in enantiomer selective hydrolyzes was recognized relatively early and was first applied in the kinetic resolution of amino acid derivatives [71,235].

Nowadays this method is used predominantly for the transformation and kinetic resolution of racemic esters and alcohols (Fig.3.7). with the esters of chiral carboxylic acids (Fig. 3.7.a), the intended enantiomer selective transformation is usually hydrolysis, but by making use of the transferase activity of hydrolases, nucleophiles other than water (Nu = OR', NHR', SR', etc.) can also be partners. The same applies to the transformation of esters of chiral alcohols (Fig.3.7.b).

In the enantiomer selective acylation of chiral alcohols there are several ways open (Fig. 3.7.c), namely esterification, transesterification, and acylation with anhydrides.

Enantiomer selective, hydrolase catalyzed transformations are amenable, in principle, to many other types of compounds, such as chiral carboxylic acids, epoxides, amines, thiols, sulfonic acids, phosphoric acid derivatives, etc. but for various reasons only a few examples, and for some types no examples, have been reported.

For some general aspects of the processes discussed in this chapter cf. Sections 2.3 and 3.2.

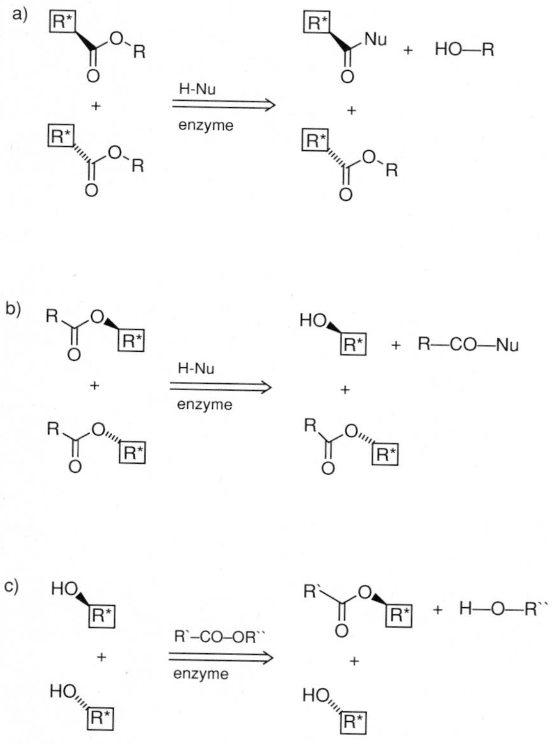

Fig. 3.7. Possibilities for enantiomer selective transformations by hydrolases: a) differentiation between enantiomeric esters of a chiral carboxylic acid (e.g by hydrolysis: Nu=OH, alcoholysis: Nu=OR' ...); b) differentiation between enantiomeric esters of chiral alcohols (e.g. by hydrolysis: Nu=OH, alcoholysis: Nu=OR' ...); c) differentiation between enantiomeric alcohols (e.g. by direct esterification: R''=H, transesterification: R''=alkyl, aralkyl..., anhydride acylation: R''=COOR³, ...)

3.3.5.1 Transformations of Amino Acid Derivatives

The enantiomer selectivity of hydrolases was first applied to the resolution of amino acids and their derivatives [71,84].

It was predominantly proteases and amidases that were used for this, and amongst these most often α-chymotrypsin (CTR) and acylase I (HKA, AA).

Suitable substrate candidates for the kinetic resolution of amino acids by hydrolysis are *N*-acylamino acids, amino acid esters and hydantoin derivatives.

i) Hydrolysis of N-acylamino acid derivatives

It is primarily acylase I (from hog kidney (HKA) and *Aspergillus sp.* (AA)) that is used for the hydrolysis of these compounds, although some other enzymes (e.g. carboxypeptidase A, cf. *entries l and m* in Table 3.4) may also be effective [236].

Acylase I (predominantly AA) has also found industrial application [72] in the resolution of synthetic amino acids. Substrate tolerance and a number of other features of the two acylase I enzymes mentioned have been thoroughly studied [74,75]. From the point of view of synthesis it is important that the high L selectivity of HKA enzyme is also retained in the hydrolysis of α-methyl amino acids (Table 3.4), which incidentally, are not transformed by AA enzymes. It is of interest that the selectivity of HKA and AA enzymes is high when the substrates are the chloroacetyl rather than the acetyl derivatives. Their substrate tolerance was investigated in detail [75]. Both enzymes were shown to contain Zn^{2+} although their activity is not much influenced by other divalent cations (AA however is activated by Co^{2+}). HKA enzyme is sensitive to oxidation whilst AA enzyme is less so. More recently however, a heat resistant acylase isolated from *Bacillus*

Table 3.4. Selected examples for hydrolase catalyzed enantiomer selective hydrolysis of unusual *N*-acyl-amino acids.

Entry	R^1	R^2	R^3	Enzyme	L-amine (3.37) Y(%)	L-amine (3.37) ee(%)	D-amide (3.38) Y(%)	D-amide (3.38) ee(%)	Ref.
a	C_2H_5	H	Ac	HKA	40	>95	35	>95	[237]
b	C_2H_5	Me	$COCH_2Cl$	HKA					[238]
c	$n\text{-}C_8H_{17}$	H	$COCH_2Cl$	AA	45	94	41	96	[239,240]
d	$n\text{-}C_{11}H_{23}$	H	$COCH_2Cl$	AA	37	91	50	92	[241]
	$CF_3(CH_2)_n$								
e	n=1	H	Ac	HKA	48	>95	46	>95	[242]
f	n=3	H	Ac	HKA	49	>95	48	>95	[242]
g	CF_3	Me	$COCF_3$	HKA	53	98		97	[243]
h	$C_3H_7(CH_3)CH^c$	H	Ac	AA	28	98	70	a	[244]
i	$HOOC(F)CH$	H	Ac	HKA		b		b	[245]
j	$HOOC(F)CH^d$	H	Ac	HKA		b		b	[245]
k	$(CH_3)_2C=CH$	H	$COCH_2Cl$	HKA		b		a	[246]
l	$PhCH_2$	Me	$COCF_3$	$CP\text{-}A^e$		b		a	[236]
m	$(CH_3)_2CH$	Me	$COCF_3$	$CP\text{-}A^e$		b		a	[236]

a: Data not given; b: Optically pure; c: *syn* Isomer; d: *anti* Isomer; e: CP–A=carboxypeptidase-A.

thermoglucosidus [76] has come to light that may well be of industrial importance in the future.

Table 3.4 contains data relating to the selective hydrolysis of some unusual *N*-acylamino acids realized on a preparative scale. The α-aminobutyric acid derivatives *(entry a)* were used to prepare optically active ethyloxirane [237], while the L-amino acid (3.37.b) *(entry b)* served to facilitate the assignment of absolute configuration [238]. Compounds obtained in transformation under *entries c, d, and h* were converted to insect pheromones and analogs [240,241,244], while the compound under *entry k* (3.36.k) was an intermediate in the synthesis of penicillin [246]. With the assistance of acylase enzymes, various fluorinated amino acid derivatives can be obtained in high enantiomeric purity *(entries e, f, g, i, j)* [242,243,245].

ii) Hydrolysis of amino acid esters

For the enantiomer selective hydrolysis of amino acid ester derivatives, a number of proteases have been most frequently used (Table 3.5), although several applications of lipases for the same purpose have also been reported *(entry j* in Table 3.5)[1].

α-Chymotrypsin (CTR) can be used for the hydrolysis of a wide range of amino acid ester derivatives [84]. Based on kinetic data obtained in such hydrolyses and on data from X-ray diffraction studies, the topology of the active site of the CTR enzyme could be elucidated. A model of the active centre and a rationalization of L selectivity is shown in Fig. 3.8 [84,258]. The surface marked *ar* is hydrophobic and capable of receiving aromatic structural units whilst parts marked *am* are able by hydrogen bonding to bind *N*-acyl or OH groups, or to accommodate without hydrogen bonding other groups like CH_3, Cl, $OCOCH_3$, or $CH_2COOC_2H_5$[2]. Nucleophilic attack takes place at the site marked *n,* where both ester and amide groups can be attacked. The pocket marked *h* is small and can therefore accommodate only small groups, such as H, Cl, OH, whilst CH_3 is too bulky to fit in. By showing parts *am* and *n* as being open we wish to indicate that these sites are capable of accommodating longer chains (even a tripeptide). A good substrate for this enzyme is an aromatic L-amino acid derivative for example (Fig. 3.8.a), whereas the binding of its D-enantiomer is inhibited due to the small volume of site *h*.

The enzymic hydrolysis of the esters of various non-natural amino acids is also feasible *(entries c-m* in Table 3.5) and interestingly the L selective character of the process remains unchanged with a wide variety of enzymes and substrates.

For examples of the utilization of optically active unusual amino acid derivatives one can turn to the transformation of the L-acids obtained in *entries c and d* (Table 3.5) to (+)-chloramphenicol [250,251] or of the L-acid prepared according to *entry i* for the synthesis of the alkaloid (+)-desoxoprosopinine [254].

---

[1] With commercial PPL enzyme it is uncertain whether we are dealing with genuine lipase catalysis because this preparation is known to exhibit significant protease activity too.

[2] For a summary of CTR catalyzed enantiomer selective hydrolysis see ref. [84].

Table 3.5. Selected examples for hydrolase-catalyzed enantiomer selective hydrolysis of amino acid esters.

$$\underset{\substack{(3.39)\\ \text{(D,L-ester)}}}{R^1\!-\!\overset{\substack{R^2\;NHR^3}}{\underset{}{C}}\!-\!COOR^4} \;\xrightarrow[\;H_2O\;]{enzyme}\; \underset{\substack{(3.40)\\ \text{(L-acid)}}}{R^1\!-\!\overset{\substack{R^2\;NHR^3}}{C}\!-\!COOH} \;+\; \underset{\substack{(3.41)\\ \text{(D-ester)}}}{R^1\!-\!\overset{\substack{R^2\;NHR^3}}{C}\!-\!COOR^4}$$

Entry	R¹	R²	R³	R⁴	Enzyme[a]	Conv. (%)	L-Acid Y(%)	L-Acid ee(%)	D-Ester Y(%)	D-Ester ee(%)	Ref.
a	CH_3	H	Ac	Et	CTR	50		96		88	[248]
b	$EtOOCCH_2CH_2$	H	Ac	Et	CTR	50		97		97	[249]
c	O_2N	H	Ac	Me	CRT	–		>95		>95	[250]
d	O_2N	H	$COCHCl_2$	Me	STC	50	48	>95	50	>95	[251]
e	(4-Cl-C₆H₄-CH₂)	H	BnOCO	Me	TH[c]	~50		d		g	[252]
f	(naphthyl-CH₂)	H	BnOCO	Me	TH[c]	~50		d		g	[252]
g	(C₆H₅CH₂-S-CH₂)	H	BnOCO	Me	TH[c]	~50		d		g	[252]
h	(furan-2-yl-CH₂)	H	BnOCO	Me	PP[e]	50		>95		>97	[253]
i	(furan-2-yl-CH₂)	H	EtCO	Me	PP[e]	50		>95		>95	[254]

Table 3.5. (contd.)

Entry	R¹	R²	R³	R⁴	Enzyme[a]	Conv. (%)	L-Acid Y(%)	L-Acid ee(%)	D-Ester Y(%)	D-Ester ee(%)	Ref.
j	X[f]	H	BnOCO	CH₂CH₂Cl	AnL	15-40		85-95		g	[255]
					PsL	8-50		15-87		g	[255]
					PPL	20-40		20-99		g	[255]
					CTR			g		g	[256]
k	(indol-3-ylmethyl)	Me	H	Me	CTR			g		g	[256]
l	PhCH₂	Me	H	Me	CTR					g	[256]
m	(4-fluorophenyl)CH₂	Me	H	Me	CTR			g		g	[256]
n	MeOOC(CH₂)ₙ n=1,2	H	Ac	Me	PPL	50	45	100		g	[257]

a: See list of abbreviations. b: *anti*-isomer. The *syn*-isomer reacts with neither CTR nor PLE [250]; c: thermitase (from *Thermoactinomyces* sp.) 25-30% DMF; d: Nearly optically pure; e: Papain, 20% DMF; f: X=alkyl, aralkyl, allyl, 4-thiazolylmethyl; g: Data not given.

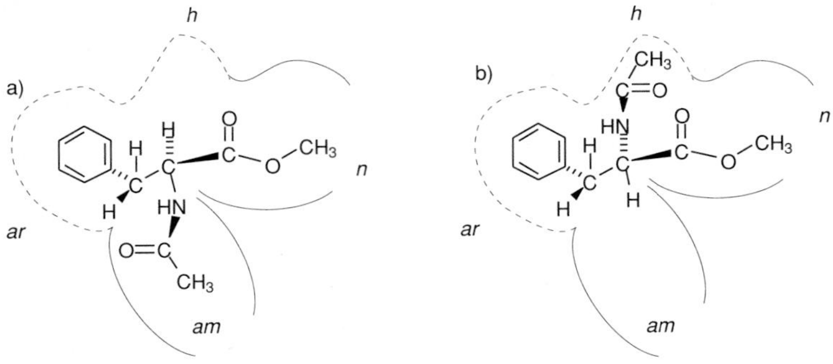

Fig. 3.8.   Active center model for CTR: a) close fitting of an L substrate; b) poor fitting of a D (unnatural) substrate

Optically active phosphorus-containing amino acid derivatives can also be efficiently prepared by enzymic methods (Fig. 3.9). The compounds shown in the figure are then used, with the aid of additional enzymes (primarily phosphodiesterase and alkaline phosphatase[3]) for chemoenzymatic tripeptide syntheses. The phosphorus containing peptide antibiotic bialaphos (3.43, R = Ala-Ala-OH) and its cyclic analog (3.44, R= Ala-Ala-OH) [263-266], and further the tripeptide analogs of the antibiotic plumbemycin A (3.45, 3.46, R = Ala-Asp-OH, Asp-Ala) [261] were all prepared by this method.

An interesting case is the synthesis of alaphosphalin, where enantiomer selectivity appeared in the papain catalyzed peptide condensation of the phosphoric ester analog of alanine [267]:

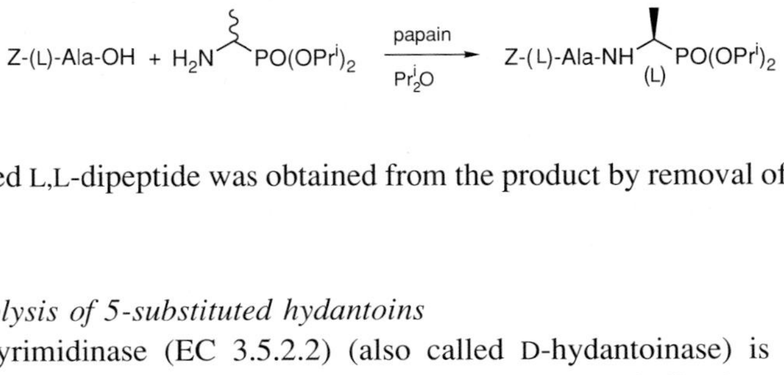

$$\text{Z-(L)-Ala-OH} + \text{H}_2\text{N}\overset{\displaystyle\frown}{}\text{PO(OPr}^\text{i})_2 \xrightarrow[\text{Pr}^\text{i}_2\text{O}]{\text{papain}} \text{Z-(L)-Ala-NH}\overset{\displaystyle\downarrow}{\underset{\text{(L)}}{}}\text{PO(OPr}^\text{i})_2$$

The desired L,L-dipeptide was obtained from the product by removal of the Z protecting group.

iii) Hydrolysis of 5-substituted hydantoins

Dihydropyrimidinase (EC 3.5.2.2) (also called D-hydantoinase) is an enzyme that is rather common in nature. It catalyzes the hydrolysis of dihydropyrimidines to *N*-carbamoyl-β-amino-acids and of racemic 5-substituted hydantoins to *N*-carbamoyl amino acids. This latter transformation can be manipulated so as to yield a D-*N*-carbamoyl amino acid in almost quantitative yield, since the transformation is D specific and at pH 8-10 the remaining L substrate is rapidly racemized (Fig. 3.10) [8,235,268,269]. The carbamoyl derivatives can then be converted either chemically [8,235] or enzymatically

---

[3]   For synthetic application of these enzymes see ref. [262].

[268] into D-amino acids which are valuable intermediates for the drug industry. D-phenylglycine and D-4-hydroxyphenyl glycine for example are incorporated into side chains of semisynthetic penicillins and cephalosporins. Various D-hydantoinase enzymes hydrolyze with high D specificity substrates having a wide selection of 5-substituents (alkyl, aralkyl, aryl).

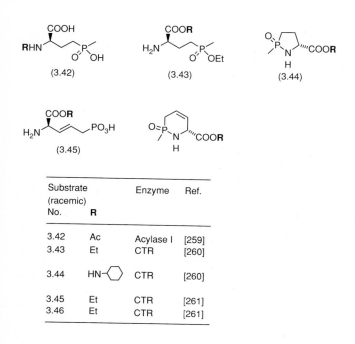

Substrate (racemic)		Enzyme	Ref.
No.	R		
3.42	Ac	Acylase I	[259]
3.43	Et	CTR	[260]
3.44	HN–⬡	CTR	[260]
3.45	Et	CTR	[261]
3.46	Et	CTR	[261]

Fig. 3.9. Enantiomer selective hydrolysis of phosphor containing unusual α-amino carboxylic acids. [The product L-amino acids (3.X a, **R**=H) are shown in the figure, remaining fractions of substrate (3.X b, **R** is the same as for the racemic substrate) have D-configuration.]

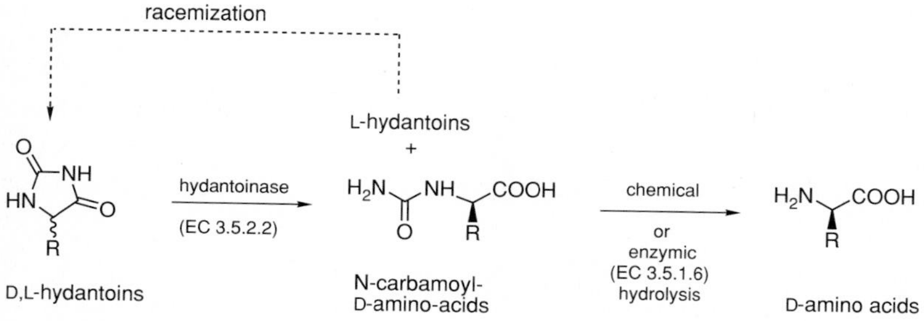

Fig. 3.10. D-Hydantoinase catalyzed ring cleavage as a key step for D-amino acid production

3.3.5.2   Derivatives of Racemic Acids: Ester Hydrolysis, Alcoholysis, Esterification

Enantiomer selectivity in the enzymic hydrolysis and alcoholysis of esters of racemic carboxylic acids can be conveniently used for resolution on a preparative scale. Most often it is ester hydrolysis that is employed for this purpose, although there are scattered examples of both alcoholysis of racemic esters and esterification of racemic acids too. Information concerning enzymic hydrolysis of racemic esters is compiled in Figs. 3.11-3.14 and Tables 3.6-3.9 respectively[4].

Available data suggest that for the hydrolysis of the esters of racemic acids (primarily if the ester group is attached to a tertiary, quaternary or sp^2 carbon) carboxyl esterases (PLE, HLE) or proteases (CTR, SgP) are the enzymes of choice. Lipases have a more limited scope and can best be applied in the hydrolysis of ester groups linked to secondary (or perhaps even tertiary) carbon atoms. This becomes apparent when comparing the β-acids (3.59-3.62.a) (Fig. 3.11 and Table 3.6.a, R=H) obtained by lipase catalysis of 2-substituted succinic diesters with the α-acids (3.63.a) and (3.64.a) (Fig. 3.12 and Table 3.7, R=H). In lipase catalysis both hydrolysis and alcoholysis take place solely at the β carbon [313], whilst catalysis with esterases (PLE, HLE) affects both the α- and the β-esters [278]. Finally, chymotrypsin only hydrolyzes the α-ester. In addition, enantiomer preference in lipase and CTR catalyzed transformations are opposite, thus it is advisable to test several enzymes for a given hydrolysis, since their selectivities may be complementary. Another feature which can be extracted from the data is that sense of selectivity with a given enzyme is generally invariant within a family of compounds. Thus in the series of oxiraneacetic acids (3.52-3.55) it is always the $2S$ enantiomers which hydrolyze faster.

i) Hydrolysis of ester groups attached to a secondary carbon atom (Fig. 3.11 and Table 3.6)

Optically active β-hydroxy-β-methyl carboxylic esters (3.47-3.49.b) are important building blocks in the synthesis of the anticholesteremic agents compactin and its analogs [270]. Substituent X was shown to have an important role in PLE enzyme catalyzed hydrolysis; enantiomer selectivity dramatically dropped when X contained polar functional groups (e.g. a hydroxyl). PLE enzyme gives good enantiomer selectivity with the esters of oxiraneacetic acids (3.52-3.55) [273,274] too. Interestingly, when trichlorethyl ester of oxiraneacetic acid was transesterified with polytheleneglycol by PPL, the product solidified on cooling and was easily removed by filtration [585]. It can be shown that a sulphur atom may also serve as the centre of chirality as exemplified by lipase catalyzed hydrolysis of the racemic sulphoxide esters (3.56).

---

[4] Only the results of experiments on a preparative scale in which at least one of the enantiomers was isolated and the best reported selectivities (generally over 70-80% ee) were included. For more details see the original literature.

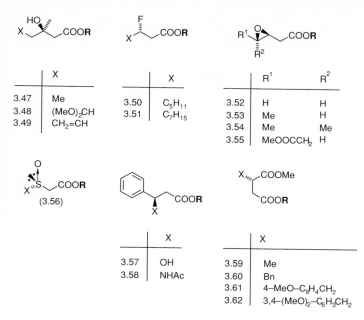

Fig. 3.11. Enantiomer selective hydrolysis of racemic α-unsubstituted carboxylic acid esters [Structures of the faster hydrolyzing enantiomers (3.X a, **R**=H) are shown. Remaining esters (3.X b, **R** $\neq$H) have enantiomeric structure. For details and references cf. Table 3.6]

ii) Hydrolysis of ester groups attached to a tertiary carbon atom (Fig. 3.12 and Table 3.7)

Several of the products arising from the transformations to be discussed here are important synthetic intermediates. (Very recently, such hydrolyses were used for preparing intermediates in the synthesis of FK506, a potent immunosuppressive macrolide [589, 590].) Thus the acid (3.68.a) obtained by PLE catalysis is a synthon in the synthesis of monensin A. Note that enantiomer preference is the opposite in a process catalyzed by *Gliocadolium roseum* [282,314]. The fluoro compounds (3.71.a,b) were used for the synthesis of 16-fluoroprostanoids [283], the benzoylmercaptanes (3.75.a,b) for the preparation of novel antihypertensive agents [287]. In the latter process, among the lipases tested AnL enzyme proved to be highly selective. Several of the 2-arylpropionic acids (3.74 and 3.79-3.82) are potent anti-inflammatory drugs, the 2-aryloxypropionic acid derivatives (3.76-3.78) in turn however are herbicides.

Optimization of the enzyme catalyzed transformation of 2-aryloxypropionic acid derivatives, particularly those of 2-(4-chlorophenoxy)-propionic acid (3.77), has been the subject of several studies [136,288-291,315]. The latter can be regarded as a test case and as such will be discussed here in detail. Firstly, the optimal enzyme was sought and CcL enzyme was found to be the best [288]. Other investigative approaches employed a single enzyme and varied conditions. E.g. with PPL enzyme and the methyl ester (3.77, R=Me)

Table 3.6. Enantiomer selective enzymic hydrolysis of racemic α-unsubstituted carboxylic acid esters.

Substrate[a] (racemic) R	Enzyme	E value	Conversion (%)	Acid product[a] (a: R=H) Yield(%)	ee(%)	Remaining ester[b] (b: R≠H) Yield(%)	ee(%)	Ref.
3.47 Me	PLE		88		c	12	>98	[270]
3.48 Me	PLE		67		c	26	94	[270]
3.49 Me	PLE		84		c	11	94	[270]
3.50 Et	CcL		37		+4.54[od]			[271]
			48				-2.44[od]	
3.51 Et	CcL		34		+3.78[od]			[271]
			50				-3.13[od]	
3.52 C$_8$H$_{17}$	PPL		60		e		>95	[272]
Me	PLE	11	50	30	74	40	82	[273]
3.53 Me	PLE	16			e		e	[273,274]
3.54 Me	PLE	17			e		e	[273]
3.55 Me	PLE	21.5	50	40	>97	40	>97	[273]
3.56 Me[f]	PsL[g]		50	20-30	80-100	30-50	>98	[275]
3.57 Et	CTR			50	h	42	h	[276]
3.58 Et	CTR			14	84	47	h	[248,276]
3.59 Me	PPL		50	47	95		>96	[257,277]
3.60 Me	PPL		50	45	98			[257]
3.61 CH$_2$CN	PsL		42	40	>95	55	65	[278]
3.62 CH$_2$CN	PsL		47	35	82	42	68	[278]

a: Faster reacting enantiomer shown in Fig. 3.11; b: **R** is the same as in the substrate. Configuration is enantiomeric to that shown in Fig. 3.11.; c: Low e.e.; d: $[\alpha]_D^{22}$ value (c=1,EtOH); e: Data not given; f: X=Ph, p-NO$_2$-Ph, p-Cl-Ph, p-MeO-Ph, Bn, cyclohexyl); g: Amano K-10 lipase, 10% toluene; h: High optical purity.

for which selectivity under the usual conditions was poor (E=1.9 [288]), it was found that changing the ionic composition of the medium quite markedly improved selectivity [290]. A borate buffer containing both potassium and sodium chloride was found to be optimal (Table 3.7) whilst omission of any of the salts was detrimental. In other studies, the original selectivity of the reaction (E=17, with 3.77, R=Me [133]) was shown to be much enhanced by enantioselective inhibition elicited by the addition of dextromorphan (to $E > 100$ [291]) or by an irreversible modification of enzyme conformation provoked by deoxycholate treatment (to $E > 100$ [133]).

High selectivity could be achieved in the alcoholysis of the racemic methyl ester (3.77, R=Me) with CcL enzyme [289], whilst esterification with methanol in an organic solvent was less selective [136]. With a proper choice of alcohol component (e.g. with cyclohexanol) high enantiomeric purity could be secured even with transesterification reactions [289].

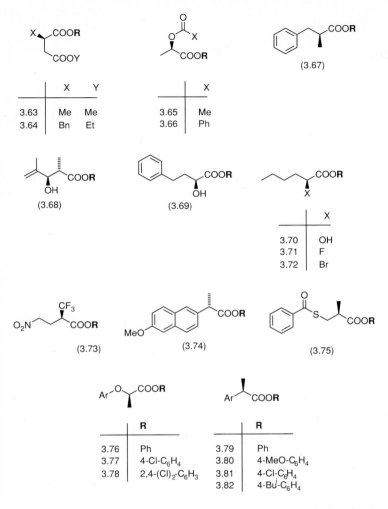

Fig. 3.12. Enantiomer selective hydrolysis of racemic α-monosubstituted carboxylic acid esters [Structures of the faster hydrolyzing enantiomers (3.X a, **R**=H) are shown. Remaining esters (3.X b, **R** ≠ H) have enantiomeric structure. For details and references cf. Table 3.7]

2-Arylpropionic acid derivatives (3.74, 3.79-3.82) also proved to be favorite substrates for ester hydrolysis studies. Inexpensive, crude HLE enzyme (in the form of the acetone powder) was found to be sufficiently selective [284] and hydrolyzed (*R*)-esters faster. CcL enzyme was also highly selective but preferred the *S* enantiomers [133,286] and its selectivity could be further improved by enantiomer selective inhibition [291] or deoxycholate treatment [133].

iii) Hydrolysis of ester groups attached to quaternary or sp2 carbon atoms (Fig. 3.13 and Table 3.8; for additional very recent results, see [586-588]).

Table 3.7. Enantiomer selective enzymic hydrolysis of racemic α-monosubstituted carboxylic acid esters.

Substrate[a] (racemic)	R	Enzyme	E value	Conversion (%)	Acid product[a] (a: **R**=H)		Rem. ester[b] (b: **R**≠H)		Ref.
					Y(%)	ee(%)	Y(%)	ee(%)	
3.63	Me	CTR		50		76		70	[241]
		CTR		50	31	85	36	88	[84]
3.64	Et	CTR		50		c		d	[279]
3.65	Et	CTR		50		46		46	[280]
3.66	Et	CTR		50	19	d	44	82	[281]
3.67	Et	CTR		50		94		87	[279]
3.68	Me	PLE		50		64		63	[282]
3.69	Et	PfL		35	33	92			[283]
				55			43	99	[283]
3.70	Et	PfL		50	48	79	43	95	[283]
3.71	Et	PfL		60	53	69	37	>99	[283]
3.72	Et	PfL		50	48	69	40	73	[284]
3.73	Bn	PsL		49		>98		>98	[285]
3.74	CH₂CH₂Cl	CcL	>100	39		>98		63	[286]
3.75	Me	AnL	>100	32		>98		45	[287]
3.76	Me	CTR	6.6						[288]
3.77	Me	CcL[e]	10.5						[288]
	Me	CcL		50	43	96	48	94	[289]
	Me	PPL[f]		50	49	89	49	>99	[290]
3.78	Me	CcL[e]	1.9						[288]
	Me	CcL[g]	20						[291]
3.79	Me	HLE		56	45	53	38	>96	[284]
	CH₂CH₂Cl	CcL[g,h]	>100	28		98[h]		39[h]	[291]
3.80	i-Pr	HLE		52	35	91	42	>96	[284]
3.81	i-Pr	HLE		52	35	91	42	>96	[284]
	CH₂CH₂Cl	CcL[h,i]	>100						[133]
3.82	Me	HLE		58	36	66	31	>96	[284]
	CH₂CH₂Cl	CcL[h]	50	42		95[h]		70[h]	[133]

a: Faster reacting enantiomer shown in Fig 3.12; b: **R** is the same as in the substrate. Configuration is enantiomeric to that visualized in Fig. 3.12; c: Nearly optically pure; d: High optical purity; e: 20% DMSO; f,g: 0.03M dextrometorphan; h: Opposite enantiomeric preference as with HLE; i: CcL enzyme pure A form treated with deoxycholate.

Such transformations are interesting in several ways. In remarkable contrast to chemical hydrolysis, with a careful choice of enzyme (proteases, carboxyl esterases) facile hydrolysis of sterically hindered esters is feasible. With racemic nitroesters (3.84), although resolution can be realized with high selectivity by chymotrypsin catalyzed hydrolysis to (3.84.b), only the unchanged ester can be utilized due to the occurrence of decarboxylation in the nitro compound (3.85).

Enantiomer selective hydrolysis of axially and planarly chiral compounds is also possible, as shown by the PLE catalyzed transformation of the allenes (3.86-3.88) and of the iron-carbonyl complex (3.89).

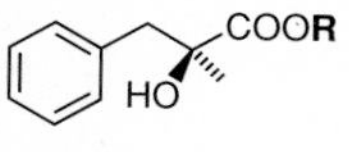

(3.83)

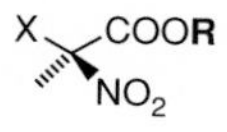

(3.84)

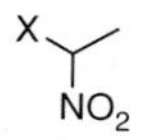

(3.85)

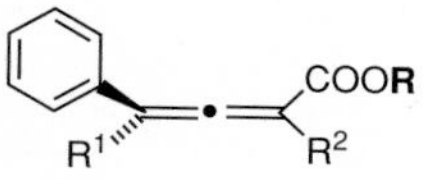

	R^1	R^2
3.86	Me	Me
3.87	Me	Et
3.88	Et	Et

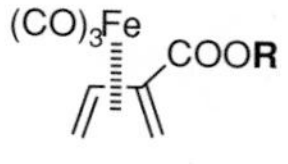

(3.89)

Fig. 3.13. Enantiomer selective hydrolysis of racemic carboxylic acid esters, with ester groups attached to a quaternary or sp^2 carbon [Structures of the faster hydrolyzing enantiomers (3.X a, **R**=H) are shown. Remaining esters (3.X b, **R** ≠ H) have enantiomeric structure. For details and references cf. Table 3.8]

Table 3.8. Enantiomer selective enzymic hydrolysis of racemic carboxylic acid esters with ester groups attached to quaternary or sp^2-carbon.

Substrate[a] (racemic)	**R**	Enzyme	Conv. (%)	Acid product[a] (a: **R**=H) Y(%)	ee(%)	Rem. ester[b] (b: **R**≠H) Y(%)	ee(%)	Ref.
3.83	Et	CTR		15	>70	46	75	[292]
3.84[c]	Bu	CTR	60	(3.85)[c]	d		>95	[293]
3.86	Me	PLE		33	90	50	61	[294]
3.87	Me	PLE		52	63	43	73	[294]
3.88	Me	PLE		17	93	79	22	[294]
3.89		PLE	40		85			[295]
			60				85	[295]

a: Faster reacting enantiomer shown in Fig. 3.13; b: **R** is the same as in the substrate. Configuration is enantiomeric to that visualized in Fig. 3.13; c: X=allyl, phenyl, 2-indolylmethyl; d: The acid products (3.84) undergo rapid chemical decarboxylation to yield nitro compounds (3.85)[c].

iv) Hydrolysis of ester groups attached directly to ring carbons (Fig. 3.14 and Table 3.9)

Similarly in this category, it is primarily proteases and esterases that are the enzymes of choice, although occasionally ester groups attached to tertiary ring carbons can be hydrolyzed by lipases too.

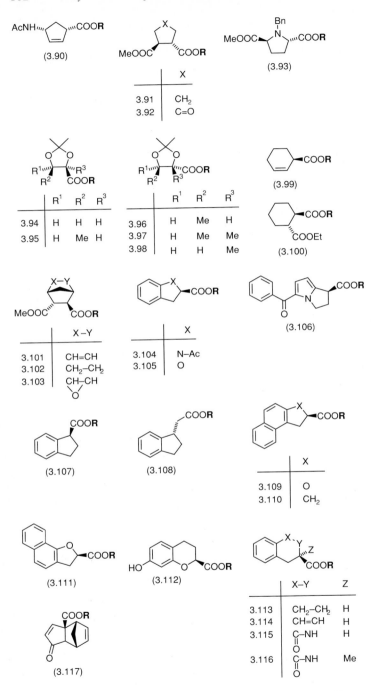

Fig. 3.14.   Enantiomer selective hydrolysis of racemic carboxylic acid esters, with ester groups attached directly to rings (3.108 is an exception) [Structures of the faster hydrolyzing enantiomers (3.X a, **R**=H) are shown. Remaining esters (3.X b, **R** ≠ H) have enantiomeric structure. For details and references cf. Table 3.9]

Table 3.9. Enantiomer selective enzymic hydrolysis of carboxylic esters with ester groups attached directly to rings.

Substrate[a] (racemic)	R	Enzyme	E value	Conv. (%)	Acid product[a] (a: R=H)		Rem. ester[b] (b: R≠H)		Ref.
					Y(%)	ee(%)	Y(%)	ee(%)	
3.90	Me	PLE[c]			47	97	43	87	[296]
3.91	Me	PLE			43	59	54	49	[297]
3.92	Me	PLE		50	34	95	45	95	[298]
3.93	Me	PLE			54	23	39	28	[297]
3.94	Bu	PPL[d]		57	53	71	40	>95	[299]
3.95	Bu	CcL		50	35	93	47	94	[299]
3.96	Bu	CcL		69	38	42	19	95	[299]
3.97	Bu	CcL		45	41	95	35	77	[299]
3.98	Bu	CcL		50	40	33	45	32	[299]
3.99	Me	CTR			15	>70	19	e	[300]
3.100	Et	PLE		67		45		83	[301]
3.101	Me	PLE			45	73	45	70	[302]
	Me	PLE[f]	40	64	55	53	32	>95	[303]
3.102	Me	PLE			45	82	40	90	[302]
3.103	Me	PLE			45	82	40	95	[302]
3.104	Me	g			25	98			[304]
3.105	Me	CTR[h]		50	20	93	18	96	[305]
3.106	Me	SgP	>100	50		>96		>96	[306]
3.107	Me	CTR		50	45	88	41	62	[84,307]
3.108	Me	CTR		50	48	78	48	76	[84,307]
3.109	Me	CTR[i]		50	26	94	20	j	[308]
3.110	Me	CTR[i]		50	27	j	28	j	[309]
3.111	Me	CTR[i]		50	19	j	25	j	[308]
3.112	Et	PfL		55	50	75	45	>99	[283]
3.113	Me	CTR			47	73	43	k	[283]
3.114	Me	CTR			44	98	45	k	[310]
3.115	Me	CTR			20	e	32	e	[311]
3.116	Me	CTR			42	e	24	e	[292]
3.117	Et	PLE			40	>99	48	83	[312]

a: Faster reacting enantiomer shown in Fig. 3.14; b: **R** is the same as in the substrate; c: 10% MeOH; d: Adsorbed to Celite; e: High optical purity; f: 10% acetone; g: Esterase from a soil isolate; h: 8% acetonitrile; i: 50% DMSO; j: Optically pure; k: Data not given.

Amongst the products obtained in this way is the cyclopentene carboxylic acid (3.90.a), an intermediate for the carbocyclic analogs of ribonucleosides [296]. Comparing the *trans*-dicarboxylic methyl esters (3.91) and (3.92) it is conspicuous that the substitution of a methylene group by a carbonyl group significantly increased selectivity in PLE catalyzed hydrolysis. In the hydrolysis of the dioxolane carboxylic acid butyl esters, it was observed that both the substrate tolerance and selectivity of CcL enzyme were better than that of PPL enzyme. Note also that with CcL enzyme, it was possible to hydrolyze the sterically severely hindered esters (3.97) and (3.98) in which the ester group was linked to a quaternary carbon atom. Hydrolysis of the racemic bicyclic *trans*-diesters (3.101-3.103) with PLE enzyme proceeds with high selectivity, although hydrolysis rate and selectivity are associated with certain well defined structural requirements. Thus PLE enzyme fails to hydrolyze models with two methylene groups in the bridge, whilst in addition, for its effectiveness a carbonyl function in a *trans* position to the *exo*-ester group is indispensable [299].

Of great practical importance is the hydrolysis of the racemic esters (3.106) in order to obtain (*S*)-ketorolac, an analgesic and anti-inflammatory agent (cf. also Section 2.3.2.6). Lipases were found to be highly *R* selective (e.g. MmL giving $E > 100$), whilst proteases were *S* selective (SgP, $E > 100$, BsP, $E=97$) [306].

CTR enzyme catalyzed hydrolysis of the ester (3.107) and its homolog (3.108) was remarkable in the fact that a reversal of stereochemical preference was reported [84,307]. In contrast to two other series of cyclic esters (3.109-3.111 and 3.113-3.116), there was no change in the sense of stereoselectivity in CTR catalyzed hydrolysis.

Despite the influence of steric hindrance, the tricyclic ester (3.117) was readily hydrolyzed by PLE enzyme, in a highly enantiomer selective manner, to a product which was then converted to optically active cyclopentenones [316].

Although it can be conducted in an enantiomer selective way, owing to practical difficulties (low rates and conversions, enzyme inhibition due to the acidity of the free acids) esterification of racemic acids is rarely used. CcL enzyme catalyzed esterification of α-halogeno carboxylic acids in an organic solvent is one example of such a slow, but sufficiently selective transformation (Fig.3.15).

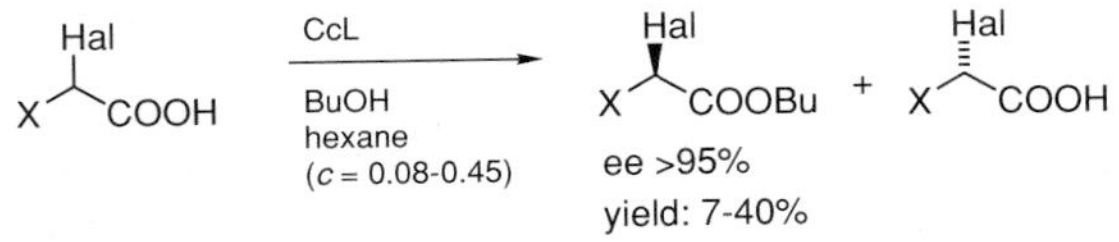

Fig. 3.15.   Enzymic esterification of racemic α-halogenated carboxylic acids [136]

3.3.5.3 Derivatives of Racemic Alcohols: Ester Hydrolysis, Transesterification, Acylation

In recent years it has perhaps been enantiomer selective transformation of racemic alcohol derivatives that has developed at the fastest pace. This has been due primarily to an ever widening selection of available lipases (cf. Section 3.1.1 and Table 3.1) which often proved to be both highly selective and tolerant to experimental conditions not the least to organic solvents.

This chapter will discuss hydrolase catalyzed transformations carried out on a preparative scale. Reactions have been classified according to substrate structure (primary, secondary, and tertiary, as well as cyclic alcohols) and to the type of reaction (hydrolysis, alcoholysis or direct esterification, and transesterification, (cf. Figs. 3.7.a-c in Section 3.3.5). This latter distinction is needed because in enantiomer selective hydrolysis and alcoholysis, the alcohol and ester fraction respectively, and furthermore the corresponding fractions in transesterification and acylation by the same enzyme, are enantiomers. This is a consequence of the fact that a given enzyme shows preference for the same enantiomer in hydrolysis and alcoholysis, as well as in acylation and transesterification. In the first case the product is an alcohol and the unchanged substrate an ester, whilst in the second case it is the other way round.

i) Derivatives of racemic primary alcohols
Hydrolysis and alcoholysis of esters of racemic primary alcohols (Fig. 3.16 and Table 3.10)

From the data presented it transpires that enantiomer selective hydrolysis of esters of racemic primary alcohols is an efficient method applicable to substrates with widely diverging structures. It is effective even with a compound (3.151) in which the ester function and the chiral center are separated by two methylene groups.

Good results can be obtained with partially purified pancreatin and steapsin (the latter is also known as commercial PPL). These inexpensive enzymes, readily accessible from pig pancreas, were used for the resolution of the chloroalcohols (3.118,3.119), the aminoalcohols (3.120, 3.121), and of the epoxyalcohols (3.122-3.134). This latter procedure is really a viable enzymic alternative to the asymmetric epoxidation of allylic alcohols via the Sharpless method [331]. The enzyme catalyzed transformation proceeds with both *cis-* and *trans*-disubstituted oxiranes (*cis*-3.123-3.126 and *trans*-3.127-3.129) with medium or high selectivity. Whilst selectivity is satisfactory even with oxirane-ethanols, with oxirane-methanols selectivity drops if R^1 is anything other than hydrogen (e.g. 3.131, 3.132).

Enzymic preparation of glycidol enantiomers (3.122) has now been developed into an industrial process and the products are converted, via tosylates, to optically active β-blockers [333]. The 2-epoxypentanol derivatives (3.124) and (3.134) were used for

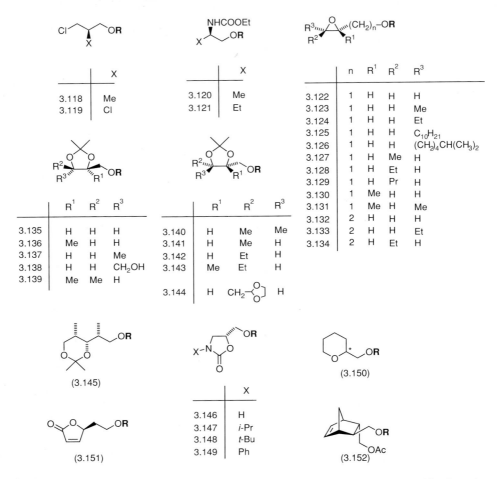

Fig. 3.16.   Enantiomer selective enzymic hydrolysis of racemic primary alcohol esters [Configurations of the alcohol products (3.X a, **R**=H) are shown. Remaining esters (3.X b, **R**=acyl) have enantiomeric structure. For details and references cf. Table 3.10]

the preparation of *anti-* and *syn-*(−)-pentanetriols [321], whilst the oxiran-2-ylmethanols (3.125) and (3.126) were converted to optically active pheromones [322].

The enantiomer selectivity of penicillin acylase (PA) has been studied with 1,3-dioxolanyl-4-methanols (3.135-3.144), i.e. with compounds thoroughly different from their natural substrates [323,324]. Selectivity was acceptable, and occasionally (with 3.136, 3.142, 3.143) even quite good. Sense and magnitude of the selectivity was much influenced by the position of substituents (cf. the series of compounds 3.135-3.139 and 3.140-3.144). The enzyme was sufficiently selective even in the hydrolysis of the phenylacetate of glycerol acetonide (3.135) which is in contrast to the poor selectivity experienced in the lipase catalyzed hydrolysis [320,332] or transesterification [139] of such esters.

Table 3.10. Enantiomer selective enzymic hydrolysis of racemic primary alcohol esters.

Substrate[a] (racemic)	R	Enzyme	E value	Conv. (%)	Alcohol[a] (3.X a:**R**=H) Y(%)	ee(%)	Rem. ester[b] (3.X b:**R**≠H) Y(%)	ee(%)	Ref.
3.118	COPr	PPL[c]		38		80[d]			[317]
3.119	Ac	P[e]		75				90	[318]
3.120	Ac	P[e]		52	30	90		>95	[319]
3.121	Ac	St[f]		53	35	87		>95	[319]
3.122	COPr	PPL		60				92	[320]
3.123	COPr	PPL		60				70	[320]
3.124	COPr	PPL[g]		70				83	[321]
3.125	Ac	PPL		38	32	>95			[322]
3.126	Ac	PPL		25	20	>95			[322]
3.127	COPr	PPL		60				95	[320]
3.128	COPr	PPL[g]		58				>96	[321]
3.129	COPr	PPL		60				56	[320]
3.130	COPr	PPL		60				50	[320]
3.131	COPr	PPL		60				73	[320]
3.132	COPr	PPL		60				77	[320]
3.133	COPr	PPL		60				70	[320]
3.134	COPr	PPL		60				82	[320]
3.135	COBn	PA		50				60	[323,324]
3.136	COBn	PA		50				90	[323,324]
3.137	COBn	PA		50				50	[323,324]
3.138	COBn	PA		50				62	[323,324]
3.139	COBn	PA		50				65	[323,324]
3.140	COBn	PA		50				33	[323,324]
3.141	COBn	PA		50				46	[323,324]
3.142	COBn	PA		50				80	[323,324]
3.143	COBn	PA		50				90	[323,324]
3.144	COBn	PA		50				70	[323,324]
3.145	Ac	PPL	18				30	>95	[325,326]
3.146	COC$_5$H$_{11}$	PaL		50		97		99	[327]
3.147	COC$_7$H$_{15}$	PaL		50		>97		>97	[328,329]
3.148	COC$_5$H$_{11}$	PaL		50		>97		>97	[328,329]
3.149	COPr	PaL		50		96		100	[327]
3.150	COPr	PPL		60				95[d]	[320]
3.151	Ac	PfL			56	51	43	64	[330]
3.152	Ac	PsL	>100	48	45	91	47	>95	[303]

a: Structures of the faster reacting enantiomers are shown in Fig. 3.16; b: **R** is the same as in the substrate. Configuration is enantiomeric to that shown in Fig. 3.16; c: immobilized; d: configuration is not given; e:pancreatin; f: steapsin; g: 10% t-BuOH.

The ester (3.145.b), remaining unchanged in high enantiomeric purity after the PPL catalyzed hydrolysis of the 1,3-dioxane, was used for the synthesis of the triene precursor of monensin A, a macrocyclic antibiotic [314,326].

In the hydrolysis of 2-oxazolidon-5-methanols (3.146-3.149) used in the preparation of β-blockers, a lipase from *Pseudomonas aeruginosa* (PaL) proved to be highly selective [327-329].

It is worth noting the contrast that exists between the high selectivity in PPL catalyzed hydrolysis of the tetrahydropyrane-2-methanol ester (3.150) and the complete lack of selectivity in the same transformation applied to its tetrahydrofuran analog [320].

Acylation and transesterification of racemic primary alcohols (Fig. 3.17 and Table 3.11; for additional very recent results, see [591-593]).

It was established that working within the same group of compounds and using the same enzyme meant that enantiomer preference for hydrolysis and acylation (transesterification) was the same. Note that selectivity in favor of (3.125*) and (3.126*)[5] is retained in transesterification whilst it is significantly increased with the aminoalcohols (3.120) and (3.121). In the case of glycerol acetonide (3.157), lipase catalyzed hydrolysis and acylation were equally low in selectivity, but changing to the crowded 2,2-diphenyl-1,3-dioxolanes much improved selectivity.

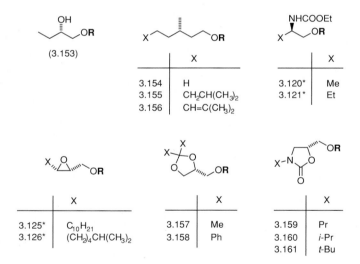

Fig. 3.17.   Enantiomer selective enzyme catalyzed acylation of racemic primary alcohols [Configurations of the ester products (3.X a, **R**=acyl) are shown. Remaining alcohols (3.X b, **R**=H) have enantiomeric structure. For details and references cf. Table 3.11]

---

[5] Here, and in several of the cases following, the numbers marked with an asterisk refer to compounds having a skeleton already mentioned earlier. The original numbers are preserved for the sake of comparison.

Table 3.11 Enantiomer selective enzyme catalyzed acylation of racemic primary alcohols.

Substrate[a] (racemic, R=H)	Enzyme	Acylating agent	Conversion (%)	Ester product[a] (3.X a:R≠H)			Rem. alcohol[b] (3.X b:R=H)		Ref.
				R	Y(%)	ee(%)	Y(%)	ee(%)	
3.120	St[d]	EtOAc	51	Ac	31	90[c]	30	>95[c]	[319]
3.121	St[d]	EtOAc	50	Ac	31	>95[c]	32	92[c]	[319]
3.125	PPL	EtOAc	65	Ac			31	>95[c]	[322]
3.126	PPL	EtOAc	60	Ac			36	>95[c]	[322]
3.153	CcL[e]	tributyrin	50	COPr	40	90	36	89	[334]
3.154	PLE[e]	MeOOCEt	50	COEt	22		38	94	[334]
3.155	PLE[e]	MeOOCEt	50	COEt	20		31	97	[334]
3.156	PLE[e]	MeOOCEt	50	COEt	29		35	96	[334]
3.157	PPL	vinyl acetate	80	Ac	40			65	[139]
	PfL	f		CO~~~COOH	40	60	40	61	[141]
3.158	PfL	f		CO~~~COOH	41	92	51	70	[141]
3.159	PfL	f		CO~~~COOH	46	75	38	98	[141]
3.160	PfL[g]	Ac$_2$O[h]	50	Ac	40	>95	42	>95	[335]
3.161	PfL[g]	Ac$_2$O[h]	50	Ac	43	>95	44	>95	[335]

a: Structures of the faster reacting enantiomers are shown in. Fig.3.17; b: Structures of remaining alcohols are enantiomeric to that appearing in Fig. 3.17; c: The same enantiomer preferences were observed as for hydrolysis of racemic esters, so the resulting esters (and alcohols) are enantiomers of esters (and alcohols) from hydrolysis; d: Steapsin on Celite; e: In Chromosorb or Sepharose beads; f: Succinic anhydride g: On Celite; h: Propionic or butyric anhydride gave similar results.

Enzymic acylation of the important 2-oxazolidon-5-methanols (3.159-3.161) with acid anhydrides also proved to be highly selective.

Additionally, enzyme catalyzed transesterification was used with a good degree of success in the kinetic resolution of 1,2-butanediol (3.153). (In our hands, however, transesterification of 1,2-diols with vinyl acetate using PPL and CcL enzymes proved to be at least three magnitudes slower than the hydrolysis of the corresponding diacetates (unpublished results)).

It should not go unnoted that the transesterification of some long chain 3-methyl substituted alcohols (3.154, 3.155) with citronellol (3.156) using PLE enzyme and methylpropionate is highly selective despite the fact that the alcohol function is separated from the chiral center by two methylene groups.

ii) Derivatives of racemic secondary alcohols
Hydrolysis and alcoholysis of esters of racemic secondary alcohols (Fig. 3.18 and Table 3.12)

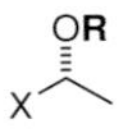

	X
3.162	Et
3.163	Ph
3.164	4-Me-C$_6$H$_4$
3.165	3-Me-C$_6$H$_4$
3.166	4-MeO-C$_6$H$_4$
3.167	Bn
3.168	4-pyridyl
3.169	2-naphthyl
3.170	Ph$\sim\sim$
3.171	2-furyl

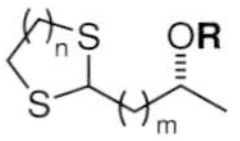

	m	n
3.172	0	1
3.173	0	2
3.174	1	1
3.175	1	2

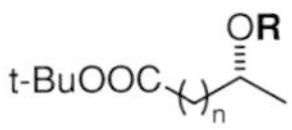

	m
3.176	0
3.177	1
3.178	5
3.179	10

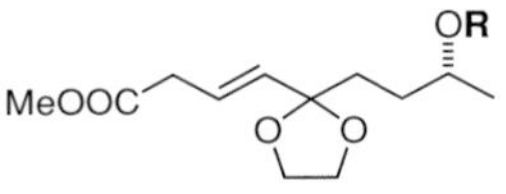

(3.180)

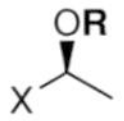

	X
3.181	Ph
3.182	CH$_2$Ph
3.183	CH$_2$CH$_2$Ph
3.184	$\sim\sim$Ph
3.185	CH$_2$COOEt
3.186	CH$_2$CO(CH$_2$)$_5$CH$_3$

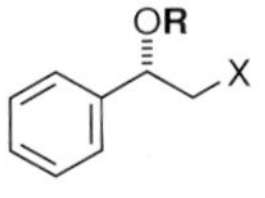

	X
3.187	Cl
3.188	Br
3.189	C$_5$H$_{11}$
3.190	C$_6$H$_{13}$
3.191	C$_7$H$_{14}$

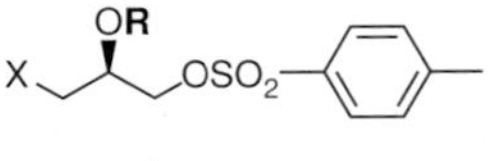

	X
3.192	Me
3.193	Et
3.194*	CH$_2$Cl
3.195*	C$_{10}$H$_{21}$
3.196*	Ph

Fig. 3.18. Enantiomer selective enzymic hydrolysis (or alcoholysis*) of racemic secondary alcohol esters [Configurations of the alcohol products (3.X a, **R**=H) are shown. Remaining esters (3.X b, **R**=acyl) have enantiomeric structure. For details and references cf. Table 3.12]
* molecules undergoing alcoholysis are marked with an asterix; see also note j in Table 3.12
** Σ Si=Thexyl(Me)$_2$Si

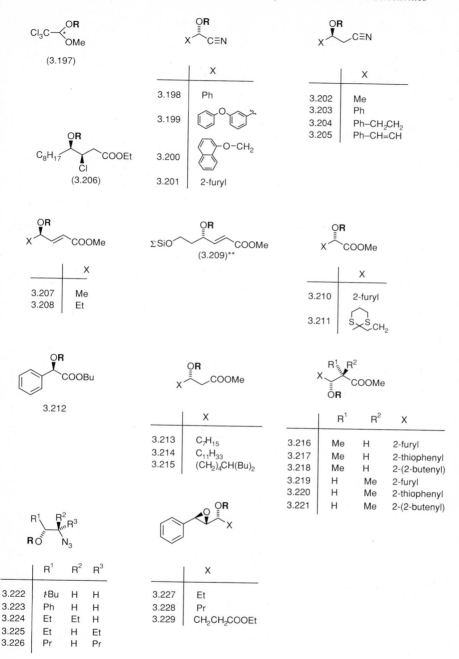

	X
3.198	Ph
3.199	(diphenyl ether)
3.200	(naphthyl-O-CH₂)
3.201	2-furyl

	X
3.202	Me
3.203	Ph
3.204	Ph–CH₂CH₂
3.205	Ph–CH=CH

	X
3.207	Me
3.208	Et

	X
3.210	2-furyl
3.211	(dithiane–CH₂)

	X
3.213	C₇H₁₅
3.214	C₁₁H₃₃
3.215	(CH₂)₄CH(Bu)₂

	R¹	R²	X
3.216	Me	H	2-furyl
3.217	Me	H	2-thiophenyl
3.218	Me	H	2-(2-butenyl)
3.219	H	Me	2-furyl
3.220	H	Me	2-thiophenyl
3.221	H	Me	2-(2-butenyl)

	R¹	R²	R³
3.222	tBu	H	H
3.223	Ph	H	H
3.224	Et	Et	H
3.225	Et	H	Et
3.226	Pr	H	Pr

	X
3.227	Et
3.228	Pr
3.229	CH₂CH₂COOEt

Fig. 3.18. (contd.)

Table 3.12. Enantiomer selective enzymic hydrolysis (or alcoholysis) of racemic secondary alcohol esters.

Substrate[a] (racemic) R	Enzyme	E value	Conv. (%)	Alcohol[a] (3.X a:R=H) Y(%)	ee(%)	Rem. ester[b] (3.X b:R≠H) Y(%)	ee(%)	Ref.
3.162 COPr	CcL			39	90	40	88	[289]
3.163 Ac	PsL	>1000	50	48	>99	48	>99	[336]
3.164 Ac	PsL	392	50	47	97	45	99	[336]
3.165 Ac	PsL	>1000	50	46	>99	46	97	[336]
3.166 Ac	PsL	22	50	46	80	47	80	[336]
3.167 Ac	PsL	270	50	47	97	48	97	[336]
3.168 Ac	PsL	117	50	46	95	47	89	[336]
COEt	PLE[c]			44	88	43	98	[337]
3.169 Ac	PsL	>1000	50	45	>99	46	>99	[336]
3.170 COCH$_2$Ph	PA		50		85			[324]
3.171 COCH$_2$Ph	PA[c,d]		60				80	
3.172 Ac	PfL		46	37	99	40	95	[338]
3.173 Ac	PfL		46	35	99	40	95	[338]
3.174 Ac	PfL		48	36	95	40	95	[338]
3.175 Ac	PfL		48	35	99	40	95	[338]
3.176 Ac	PsL[e]	>100			>99		96	[339]
3.177 Ac	PsL[e]	>100			>99		68	[339]
3.178 Ac	PsL[e,f]	>100			>99			[339]
3.179 Ac	PsL[e,f]	>100			>99			[339]
3.180 COCH$_2$Cl	PsL	>100	51		93		>99	[324]
3.181 Ac	CcL		40		57			[340]
COCH$_2$Cl	PsL	129[g]	50	50	94[g]	44	96[g]	[336]
3.182 Ac	CcL		46		94		98[h]	[340]
3.183 Ac	CcL		44		98		99[h]	[340]
3.184 Ac	CcL		37		93			[340]
3.185 Ac	CcL		41		96			[340]
3.186 Ac	CcL		28		94			[340]
3.187 Ac	PfL[i]		50	24	100	29	100	[341]
3.188 Ac	PfL[i]		50	24	94	11	100	[341]
3.189 Ac	PsL		42		92		>99	[342]
3.190 Ac	PsL		50		98		>99	[342]
3.191 Ac	PsL		53		97		98	[342]
3.192 Ac	PaL		50	40	>99	35	>99	[343]
3.193 Ac	PaL		50	48	>99	44	>99	[343]
3.194 COPr	PaL	>100	50	46	>99	45	>99	[344]
3.194* COPr	PfL[j]	>100	50		96		96	[345]
3.195* COPr	PfL[j]	92	26		>98		34	[345]
3.196* COPr	AlcL[j,k]	>100	42		>98		72	[345]
3.197 Ac	CcL			e		39	>97[m]	[346]
Ac	STC			e		30	>97[n]	[346]

Table 3.12. (contd.)

Substrate[a] (racemic) R	Enzyme	E value	Conv. (%)	Alcohol[a] (3.X a:**R**=H) Y(%)	ee(%)	Rem. ester[b] (3.X b:**R**≠H) Y(%)	ee(%)	Ref.
3.198 Ac	PsL[o]	71	53	1		42	>98	[347]
3.199 Ac	PsL	>100	50	1		40	>98	[347]
Ac	ArtL[p]		49	p	98		87	[348]
3.200 Ac	PsL					39	87	[349]
3.201 COCH$_2$Ph	PA[c,d]						72	[350]
3.202 COCH$_2$SMe	PfL	29	64				>98	[351]
3.203 COCH$_2$SPh	PfL	74	63				>98	[351]
3.204 COCH$_2$SPh	PfL	36	68				>98	[351]
3.205 COCH$_2$SPh	PfL	55	60				>98	[351]
3.206 Ac	PfL			31	94			[352]
3.207 Ac	PfL			37	>95	39	91	[353]
3.208 Ac	PfL			44	>95	45	>95	[353]
3.209 Ac	PfL			57	72	35	>95	[353]
3.210 COCH$_2$Ph	PA[c,d]						82	[354]
3.211 Ac	PfL			41	97	26	96	[355]
3.212* Ac	CcL[j]		45		92		82	[356]
3.213 COPr	CcL				74[q]		42[r]	[357]
3.214 COPr	CcL				84[q]		75[r]	[357]
3.215 COPr	CcL				92[q]		50[r]	[357]
3.216 Ac	AnL			56	66	33	>99	[358]
3.217 Ac	AnL			64	67	35	91	[358]
3.218 Ac	AnL			51	64	38	>99	[358]
3.219 Ac	AnL			51	75	32	98	[358]
3.220 Ac	AnL			53	85	47	>99	[358]
3.221 Ac	AnL			51	79	44	>99	[358]
3.222 COPr	CcL	>100	52				>98	[359]
3.223 COPr	PfL[s]	>100	53				>98	[359]
3.224 COPr	PfL[s]	>100	60				>98	[359]
3.225 COPr	PfL[s]	>100	40		>98			[359]
3.226 COPr	PfL[s]	>100	30		>98			[359]
3.227 COPr	PPL		49	50	100			[360]
3.228 COPr	PPL		30	50	60			[360]
3.229 COPr	PPL		48	22	56			[360]

a: Structures of the faster reacting enantiomers are shown in Fig. 3.18; b: In remaining ester fractions, **R** groups are the same as in the substrates. Configurations are enantiomeric to that shown in Fig. 3.18; c: on Eupergit C; d: 10% acetone; e: Amano AK lipase; f: Amano K-10 lipase; g: Enantiomer preference opposite to that shown in Fig. 3.18; h: Remaining esters were hydrolyzed to 55-60% conversion by *Trichoderma viridae* cellulase showing *S* preference. Enantiomeric purities refer to hydrolyzed alcohols; i: CvL gave similar results; j: Alcoholysis by butanol as nucleophile in organic solvent; k: Lipase from *Alcaligenes* sp.; l: Hydrolyzed product is not stable; m: (-) rotation; n: (+) rotation; o: This reference lists 32 compounds having *E* values higher than 30 with PsL (Amano SAM-2); p: Lipase from *Arthrobacter* sp. can operate at pH=4-5 where the cyanohydrin is stable; q: 40% conversion r: 60% conversion; s: Enantiomer selectivity with CcL is low.

Highly enantiomer selective hydrolysis and alcoholysis of esters of a wide structural variety of secondary alcohols is possible using several enzymes, as shown in Fig. 3.18 and Table 3.12. (Additional recent reports have appeared on hydrolysis of cyanohydrin esters [594-596], esters of hydroxystannanes [597] and fluorine containing alcohols [598-600], acetates of homoallylic alcohols [601], azido esters [602] and an arylpiperidine carbinol ester [603]). It is worth noting that with secondary methyl carbinols (3.162-3.180) it is *R* selectivity that prevails. Very high selectivities can be achieved in the hydrolysis of arylmethyl carbinol esters using *Pseudomonas* sp. lipases and similary, selectivity was also excellent with the esters of 2-(1,3-dithiolane- or dithiane)-alkylmethyl carbinols (3.172-3.175) and *tert*-butoxycarbonyl-alkyl-methyl carbinols (3.176-3.179). The methyl carbinol (3.180) which by hydrolysis could be obtained optically pure, is an intermediate in the synthesis of the antifungal macrolide (−)-pyrenophorin [324]. When the side chain is an alkyl [336,339] or a benzyloxyalkyl group [339], the selectivity of *Pseudomonas* lipases drops dramatically. It would seem that in such cases, CcL and PPL enzymes should be preferred (cf. Fig. 3.19 and Table 3.13) and indeed with certain alkyl- and aralkylmethyl carbinol esters PLE (3.168) and penicillin acylase (3.170, 3.171) catalyzed hydrolysis was also found to be quite selective.

In the face of phenyl-trifluoromethyl carbinol esters (3.163-3.169), it was found that the enantiomer preference of PsL enzymes is the same as it is with other aryl-methyl carbinols (3.163-3.169), whilst that of CcL enzyme is the opposite, the latter requiring the presence of an aromatic ring β-carbonyl groups for good selectivity. PfL enzyme catalyzed hydrolysis of chloromethylphenyl and bromomethylphenyl carbinol esters (3.187, 3.188) is also highly selective, and selective in the same sense as that shown by *Pseudomonas* lipases in the case of arylmethyl carbinol esters[6].

The hydrolysis and alcoholysis of the acyloxytosylates (3.192-3.196) deserves some comment. PaL enzyme catalyzed hydrolysis of the α-acyloxytosylates (3.192-3.194), as well as PfL or AcL catalyzed butanolysis of (3.194-3.196) are both highly selective, whereas acylation of the corresponding alcohols with vinyl acetate (Table 3.13) shows a selectivity that is much lower [345]. Thus the tosylates obtained by the former methods are of high enantiomeric purity and can be readily converted to epoxides [345].

Esters of some labile alcohols are also amenable to enzymic hydrolysis. Racemic chloral acetyl methyl acetal (3.197) can be hydrolyzed by CcL, PPL, and ChE enzymes leaving behind the optically pure (−)-enantiomer of the acetal, in contrast to use of STC enzyme where the pure (+)-enantiomer was obtained [346]. The hydrolysis product itself is unstable. This situation resembles to the enzymic hydrolysis of racemic cyanohydrin acetates, where the product also decomposes under the usual experimental conditions [347]. One exception to this rule is the (*S*)-cyanohydrin of 3-phenoxybenzaldehyde

---

[6] This preference was also conserved with phenylethyl carbinol acetate [336] and with the phenylalkyl carbinol acetates (3.189-3.191)

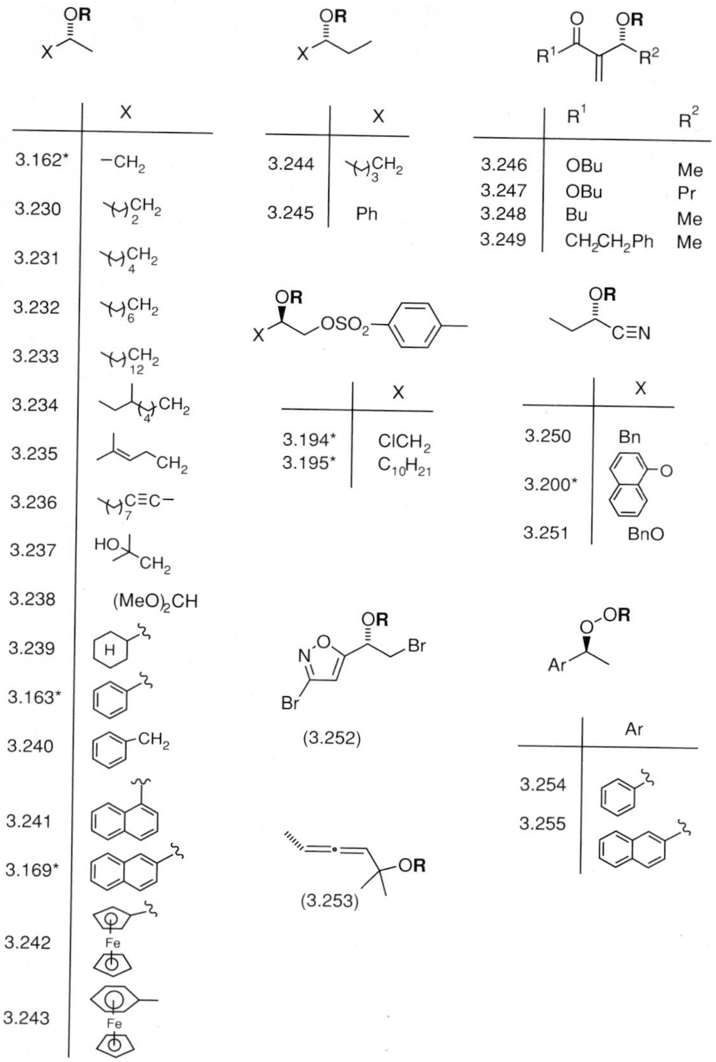

Fig. 3.19. Enantiomer selective enzyme catalyzed acylation of racemic secondary (tertiary) alcohols [Configurations of the ester products (3.X a, **R**=acyl) are shown. Remaining alcohols (3.X b, **R**=H) have enantiomeric structure. For details and references cf. Table 3.13]
* see note c in Table 3.13

(3.199.a), the alcohol component of a synthetic pyrethroid, which can be isolated in high optical purity after hydrolysis with *Arthrobacter* sp. lipase, an enzyme that can tolerate the acidity (pH 4-5) that secures the survival of the product [348].

Table 3.13. Enantiomer selective enzyme catalyzed acylation of racemic secondary alcohols.

Substrate[a] (racemic, R=H)	Enzyme	Acylating agent	E value	Conv. (%)	Ester product[a] (3.X a:R≠H)			Rem. alcohol[b] (3.X b:R=H)[b]		Ref.
					R	Y%	ee%	Y%	ee%	
3.162[c]	CcL	tributyrin		50	COPr	38	93[c]	35	89[c]	[334]
	CcL	PhOCOOPh		50	CO-OPh[d]		>80[d]		>80[c]	[361]
3.163[c]	CcL	tributyrin		50	COPr	41	85[c]	40	88[c]	[334]
	PPL	Cl_3C⌒OCOPr	>500	45	COPr	38	95[c]	28	90[c]	[136]
	PsL	CH_2=CH-OAc		48	Ac	45	>99[c]	41	93[c]	[362]
	PfL	succinic anhydride			CO⌒COOH	45	99[c]	41	97[c]	[141]
	PfL	Ac_2O		49	Ac	39	>95[c]	43	>95[c]	[335]
3.169[c]	PsL	CH_2=CH-OAc	>400	43	Ac	41	>99[c]	48	95[c]	[362]
3.194[c]	PfL	⟍OAc	24	47	Ac		83[c]		75[c]	[345]
3.195[c]	PfL	⟍OAc	21	30	Ac		87[c]		37[c]	[345]
3.200[c]	PsL	CH_2=CH-OAc	24	59	Ac				98[c]	[363]
3.230	MmL[e]	C_7H_{15}COOH			COC_7H_{15}	48	83	35	87	[364]
3.231	PPL	Cl_3C⌒OCOPr		47	COPr	35	95	30	90	[136]
	PPL	CH_2=CH-OAc		58	Ac	41	71	38	>98	[139]
	CcL	tributyrin		50	COPr		92		95	[334]
3.232	PPL	Cl_3C⌒OCOPr		46	COPr	44	>99	44	95	[136]
	MmL[e]	C_7H_{15}COOH			COC_7H_{15}	40	87	53	83	[364]
3.233	PPL	Cl_3C⌒OCOPr		46	COPr	42	98	43	>99	[136]

Table 3.13. (contd.)

Substrate[a] (racemic, R=H)	Enzyme	Acylating agent	E value	Conv. (%)	Ester product[a] (3.X a:R≠H)			Rem. alcohol[b] (3.X b:R=H)		Ref.
					R	Y%	ee%	Y%	ee%	
3.234	MmL	$C_7H_{15}COOH$			COC_7H_{15}	43	93	41	95	[365]
3.235	CcL	tributyrin			COPr	42	91	39	85	[334]
	PPL	tributyrin	9		COPr					[366]
	PPL	$CH_2=CH\text{-}OAc$		62	Ac				>98	[139]
	PPL	Cl_3C–OCOPr		65	Ac				>99	[367]
	PPL[f]	CF_3–$OCOC_{11}H_{23}$	100		$COC_{11}H_{23}$	38	90	43	>97	[366]
3.236	MmL[e]	$C_7H_{15}COOH$	>50		COC_7H_{15}		g		g	[364]
3.237	PPL	Cl_3C–OCOPr		48	COPr	31	87	26	92	[136]
3.238	PsL	$CH_2=CH\text{-}OAc$		30	Ac		70			[139]
3.239	MmL[e]	$C_7H_{15}COOH$	>50	30	COC_7H_{15}	37	86	42	67	[364]
3.240	PsL	$CH_2=CH\text{-}OAc$	>300		Ac	30	>99	66	43	[362]
	PfL	$O(COEt)_2$		48	COEt	39	>95	43	92	[335]
3.241	PPL	Cl_3C–OCOPr		50	COPr	47	95	46	>95	[368]
3.242	PPL	$CH_2=CHOCOEt$		60	COEt				84	[139]
3.243	PfL	$CH_2=CHOAc$		50	Ac		96		92	[369]
3.244	PPL	Cl_3C–OCOPr		45	COPr	38	70	28	57	[136]
3.245	PsL	Ac_2O	180	26	Ac	24	>98			[362]
	PfL	Ac_2O		49	Ac	39	>95	43	92	[335]
3.246	PfL	$CH_2=CH\text{-}OAc$	>56	50	Ac	41	>95			[370]
3.247	PfL	$CH_2=CH\text{-}OAc$	>56	50	Ac	23	>92			[370]
3.248	PfL	$CH_2=CH\text{-}OAc$	>56	52	Ac	23	>95			[370]
3.249	PfL	$CH_2=CH\text{-}OAc$	>24	56	Ac	41	>95			[370]

Table 3.13. (contd.)

Substrate[a] (racemic, **R=H**)	Enzyme	Acylating agent	E value	Conv. (%)	Ester product[a] (3.X a:**R≠H**)			Rem. alcohol[b] (3.X b:**R=H**)[b]		Ref.
					R	Y%	ee%	Y%	ee%	
3.250	PsL	CH$_2$=CH-OAc	24	59	Ac				98	[363]
3.251	PsL	CH$_2$=CH-OAc	12	63	Ac				95	[363]
3.252	PsL	CF$_3$‿OCOC$_7$H$_{15}$		30	COC$_7$H$_{15}$		97			[371]
3.253	CcL	C$_{11}$H$_{23}$COOH			COC$_{11}$H$_{23}$	35	72			[372]
3.254	PfL	⤳OAc		62	Ac				>99	[373]
3.255	PfL	⤳OAc		64	Ac				62	[373]

a: Structure of the faster reacting enantiomers are shown in Fig. 3.19; b: Structure of remaining alcohols are enantiomeric to that shown in Fig. 3.19; c: The same enantiomer preferences were observed as for hydrolysis of racemic esters, so resulting esters (and alcohols) are enantiomers of esters (and alcohols) from hydrolysis; d: And other products like di(2-butyl) carbonate; e: Lipozyme and NOVO-225; f: Enzyme dried to less than 1% water content; g: data not given.

With the aid of PsL enzyme, several synthetically useful cyanohydrin acetates can be prepared. In a series of more than 40 compounds selectivity was high ($E > 50$) for the following models: X= Ph; 2- or 3-substituted phenyl with F, Cl, Br, Me, OMe, OPh, and PhCH$_2$ as substituents; 3,4-disubstituted phenyl with OMe, OCH$_2$Ph, OCH$_2$O as substituents; 2,3- or 3,5-dimethoxyphenyl; 2-naphthyl; benzyl; cyclohexyl; and cyclooctyl. In the situation where X= 2-substituted phenyl, 1-naphthyl, or an aliphatic chain, selectivity was seen to decrease sharply [347]. Among the products, the acetate (3.200) is used in the preparation of (S)-propranolol, an important β-blocker [349].

The PfL catalyzed hydrolysis of the β-acyloxy nitriles (3.202-3.205) demonstrated that not only lipases but also PA enzyme could be used successfully with this type of compound. Hydrolysis gave good selectivity for the phenyl- (or methyl)-mercaptoacetyl esters (Table 3.12), whilst the valeroyl esters of the same compounds were hydrolyzed with poor selectivity ($E < 6$) [351].

Combined enantiomer and chemoselectivity has been recorded with a variety of α-, β-, and γ-acyloxy carboxylic acids (3.206-3.221) in the presence of several enzymes. Lipases (in the case of 3.210 PA) produced the α- and β-hydroxy compounds (3.210-3.221.a), while the carboxylic ester group remained intact. In the hydrolysis of (3.207-209) allyl acetates (3.207-209) enantiomeric preference depends on the size of the X substituent. Of the products obtained, the chlorohydrin (3.206) was utilized for the synthesis of (−)-avenacolide, an antifungal agent [352], and the 1,3-dithianes (3.211.a and b) were converted to enantiomeric intermediates that lead to platelet activating factor (PAF) [355]. The azido compounds (3.222-3.226) are precursors for aminoalcohols [359].

Enzymic hydrolysis of the racemic epoxides (3.227-3.229) clearly demonstrates the utility of this method for the preparation of optically active oxiranes.

Acylation and transesterification involving racemic secondary alcohols (Fig. 3.19 and Table 3.13; recently published works are available on enzymatic acylation of various other secondary alcohols [604-612]).

A characteristic of acylations and transesterifications carried out in organic solvents is that the enantiomer preference of the hydrolytic process is conserved (3.162*, 3.163*, 3.169*, 3.194*, 3.195*, 3.200*). Additionally, with lipases and methyl carbinols (3.162*, 3.163*, 3.169*, and 3.230-3.243) R preference is the rule. It was previously mentioned in discussing the hydrolysis of methyl carbinol esters that when *Pseudomonas* lipases were used, the parameter that provided high selectivity was the aromatic character of substituent X (e.g. 3.163*, 3.169* or 3.240). In the present case, it was observed that in the use of alkyl methyl carbinols (e.g. 3.230-3.237) PPL, CcL, and HmL enzymes show high selectivity and thus the substrate specificity of the two groups of enzymes is complementary. Of the reactions in this sphere it is worth highlighting the preparation of (S)-sulcatol (3.235), a compound with pheromone activity, by PPL catalyzed acylation of the racemic alcohol precursor. Here, both the nature of the acylating agent and the water content of the system were imperative. Changing the former from tributyrin to trifluoroethyl laurate and reducing the water content yielded a sharply increased enantiomer selectivity [366].

The complementary role of *Pseudomonas* lipases and PPL can also be demonstrated with the ethyl carbinols (3.244, 3.245). High selectivity was secured for the aliphatic substrate with PPL, and for the aromatic one with PsL and PfL enzymes

Enzymic methods are, at present, still the most practical approach to optically active α-methylene-β-hydroxy esters and ketones (3.246-3.249), which are versatile synthons with many useful applications [370].

The advantage of enzymic acylation of racemic cyanohydrins is not only an occasional improvement in selectivity (e.g. for 3.200*) as compared to hydrolysis [363], but also that the stability of cyanohydrins under such conditions makes their isolation without racemization simple.

The enantiomer preference of enzymes does not always work in the sense desired, however. This is exemplified by the attempts that were made to prepare (*S*)-broxaterol, a selective β-receptor agonist, via enzymic acylation of the bromoisoxazol (3.252). Unfortunately, only the (*S*)- ester ((*S*)-3.252b) was yielded in sufficient optical purity which was of little practical use since on further reaction it leads only to the unwanted *R* enantiomer (cf. also Section 2.3.2.6).

An example of enantiomer selective enzymic acylation of a racemic alcohol of axial chirality is that of the allene (3.253). Transesterification of the racemic hydroperoxides (3.254) and (3.255) provides an almost unique access to the optically active form of these unstable compounds. In comparison to the analogous alcohols (3.163* and 3.169*), there is an interesting reversal of enantiomer preference to *S*, and as with the secondary methyl carbinols, an aromatic substituent is also required to facilitate high selectivity [373].

iii) Derivatives of racemic cyclic alcohols[7]
Hydrolysis and alcoholysis of derivatives of racemic cyclic alcohols (Fig. 3.20 and Table 3.14; for recent results on such hydrolyses, see [613-615]).

Looking at the available data, one interesting feature that arises is the dominant *R* enantiomer preference that occurs with lipases (for 3.266, 3.267 also with PLE), a characteristic which is almost independent of ring size and the nature of the R^1 substituent. This preference also applies to unsaturated substrates (3.282-3.284, 3.286) or silicon containing (3.287) six-membered ring substrates, as well as to certain polycyclic systems (3.287, 3.289, 3.293-3.298, 3.301-3.314). This is an observation that should not however be taken as a general rule since it may well involve a degree of coincidence. Other cases, in turn, involved a more detailed study of the transformations which went so far as to propose models of predictive power for the substrate – active site complex. Models such as that for the CcL catalyzed hydrolysis of norbornanols [397] and for the action of PfL enzyme [369, 376, 613] are good examples of this.

---

[7]   The hydroxy group is directly attached to the ring.

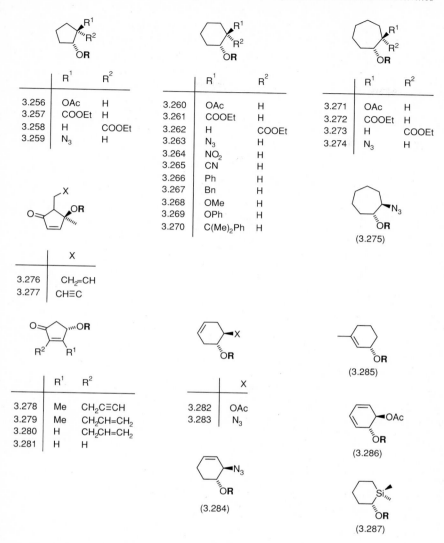

	R¹	R²
3.256	OAc	H
3.257	COOEt	H
3.258	H	COOEt
3.259	N₃	H

	R¹	R²
3.260	OAc	H
3.261	COOEt	H
3.262	H	COOEt
3.263	N₃	H
3.264	NO₂	H
3.265	CN	H
3.266	Ph	H
3.267	Bn	H
3.268	OMe	H
3.269	OPh	H
3.270	C(Me)₂Ph	H

	R¹	R²
3.271	OAc	H
3.272	COOEt	H
3.273	H	COOEt
3.274	N₃	H

(3.275)

	X
3.276	CH₂=CH
3.277	CH≡C

	R¹	R²
3.278	Me	CH₂C≡CH
3.279	Me	CH₂CH=CH₂
3.280	H	CH₂CH=CH₂
3.281	H	H

	X
3.282	OAc
3.283	N₃

(3.284)

(3.285)

(3.286)

(3.287)

Fig. 3.20.a   Enantiomer selective enzymic hydrolysis (or alcoholysis*) of racemic secondary alcohol esters [Configurations of the alcohol products (3.X a, **R**=H) are shown. Remaining esters (3.X b, **R**=acyl) have enantiomeric structure. For details and references cf. Table 3.14]
* Molecules undergoing alcoholysis are marked with an asterisk; see also note p in Table 3.14

Enantiomer preference with a single substrate can be inverted with a change of enzyme. This occurs when the cyclohexanol acetate (3.285) is hydrolyzed with PLE or PsL (Fig. 3.21, Table 3.15) [381]. The products can be used for the preparation of the enantiomers of the pheromones seudenol and 1-methyl-2-cyclohexen-1-ol.

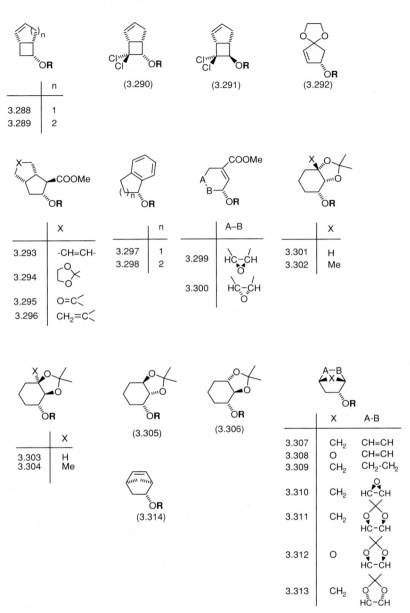

	n
3.288	1
3.289	2

(3.290)   (3.291)   (3.292)

	X
3.293	-CH=CH-
3.294	(dioxolane)
3.295	O=C<
3.296	CH₂=C<

	n
3.297	1
3.298	2

	A–B
3.299	HC–CH (epoxide)
3.300	HC–CH (carbonyl)

	X
3.301	H
3.302	Me

	X
3.303	H
3.304	Me

(3.305)   (3.306)

(3.314)

	X	A-B
3.307	CH₂	CH=CH
3.308	O	CH=CH
3.309	CH₂	CH₂-CH₂
3.310	CH₂	HC–CH (epoxide)
3.311	CH₂	HC–CH (dioxolane)
3.312	O	HC–CH (dioxolane)
3.313	CH₂	HC–CH (dioxolane)

Fig. 3.20.b

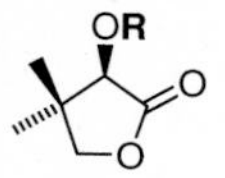

(3.315*) 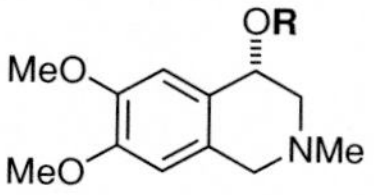 (3.316)  (3.317)

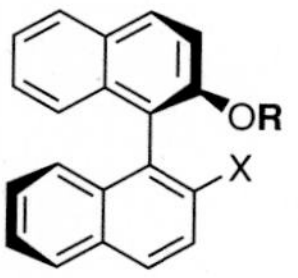

	X
3.318*	H
3.319*	OH

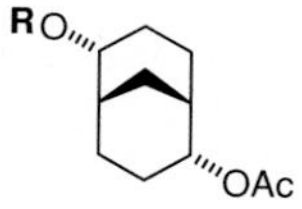

(3.320) 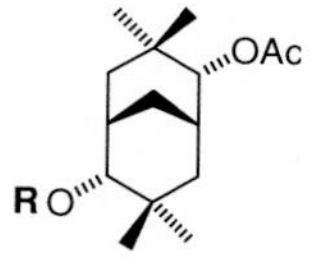 (3.321)

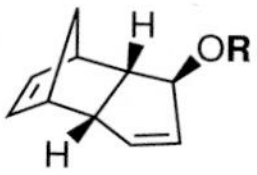

(3.322)

Fig. 3.20.c

Table 3.14. Enantiomer selective enzymic hydrolysis (or alcoholysis) of racemic cyclic alcohol esters.

Substrate[a] (racemic)	R	Enzyme	E value	Conv. (%)	Alcohol[a] (3.X a:**R**=H) Y(%)	ee(%)	Rem. ester[b] (3.X b:**R**≠H) Y(%)	ee(%)	Ref.
3.256	Ac	PfL			30	>99	65	30	[374]
3.257	Ac	PfL			42	>99	50	90	[374]
3.258	Ac	PfL			43	>99	42	95	[374]
3.259	COPr	CcL	1.4						[375]
	COPr	PsL	100		40	92		>98	[375]
3.260	Ac	PfL			33	>99	51	48	[376]
	Ac	PsL	120	56[c]	48	84	41	94	[377]
3.261	Ac	PfL			41	>99	59	55	[376]
3.262	Ac	PfL			32	>99	63	70	[376]
3.263	COPr	CcL	150		40	96		>98	[375]
3.264	COPr	CcL	80		40	>98		85	[375]
3.265	COPr	CcL	40		40	93		93	[375]
	COPr	PsL	100		38	>98		95	[375]
3.266	Ac	PLE				d		d	[378]
	Ac	PLE[e]			40	98	53	89	[379]
	COCH₂Cl	PsL	180	50	44	95	43	97	[377]
3.267	Ac	PsL	145	51	47	>95	45	>95	[377]
3.268	Ac	PsL	400	44	45	98	49	96	[377]
3.269	Ac	PsL	790	48	42	>99	45	96	[377]
3.270	Ac	PLE[e]		55	36	67	44	>95	[379]
3.271	Ac	PfL			45	>99	55	55	[376]
3.272	Ac	PfL			38	>99	58	68	[376]

Table 3.14. (contd.)

Substrate[a] (racemic) R	Enzyme	E value	Conv. (%)	Alcohol[a] (3.X a:**R**=H) Y(%)	ee(%)	Rem. ester[b] (3.X b:**R**≠H) Y(%)	ee(%)	Ref.
3.273 Ac	PfL			36	>99	64	45	[376]
3.274 COPr	CcL	32		40	89		91	[375]
3.275 COPr	CcL	40		40	93		92	[375]
3.276 Ac	PLE				98			[348]
3.277 Ac	PLE				98			[348]
3.278 Ac	ArtL[f]		50		99			[380]
3.279 Ac	ArtL[f]		48		98			[380]
3.280 Ac	ArtL[f]		43		79			[380]
3.281 Ac	ArtL[f]		20		30			[380]
3.282 Ac	PfL							[330]
3.283 COPr	CcL	80		40	88		>98	[375]
3.284 COPr	CcL	100		40	>98		94	[375]
3.285 Ac	PLE[g]			21	65			
	PLE[g]			12	>95[h]	28	>95[i]	[381]
3.286 Ac	PPL		40		63			
	PPL				97[h]		96[i]	[382]
3.287 Ac	CcL		38	27	95	50	57	[383]
3.288 Ac	PfL		50	46	>98		>98[i]	[384,385]
3.289 Ac	PPL		30[(i)]		>98		>98[i]	[386]
3.290 COPr	PfL	25	50		81		82	[385]
3.291 COPr	PfL	95	48		94			[385]
3.292 Ac	PfL			42	>95		>95	[387]
3.293 Ac	PfL[j]			50	96	48	87	[374]
3.294 Ac	PfL[j]			13[j]	45		k	[374]
3.295 Ac	PfL[j]			12[j]	77		k	[374]
3.296 Ac	PfL[j]			13[j]	>99		k	[376]
3.297 Ac	PsL	>1000	50	46	>99	47	>99	[336]
3.298 Ac	PsL	>1000	50	47	>99	47	>99	[336]
3.299 COPr	ChE[l]	14						[325]
COPr	PfL[m]	26	37		88			[325]
3.300 COC_5H_{11}	ChE[l]	54	54				97	[325]
3.301 COPr	CcL		50	43	>95	40	>95	[388]
3.302 COPr	CcL		42	40	>95	69	44	[388]
3.303 COPr	CcL		54	48	31	46	36	[388]
3.304 COPr	CcL		47	37	84	50	77	[388]
3.305 COPr	CcL		50	35	>95	50	>95	[388]
3.306 COPr	CcL		50	35	>95	48	>95	[388]

Table 3.14. (contd.)

Substrate[a] (racemic) R	Enzyme	E value	Conv. (%)	Alcohol[a] (3.X a:**R**=H) Y(%)	ee(%)	Rem. ester[b] (3.X b:**R**≠H) Y(%)	ee(%)	Ref.
3.307 Ac	CcL		n,o		90[n]		>96[o]	[389]
COPr	CcL		n,o		88[n]		89[o]	[390]
COPr	PsL		n,o		>97[n]		87[o]	[390]
3.308 COPr	CcL		n,o		93[n]		>97[o]	[350]
3.309 Ac	CcL		n,o		75[n]		52[o]	[390]
3.310 COPr	CcL		n,o		94[n]		>97[o]	[390]
3.311 COPr	CcL		n,o		85[n]		83[o]	[390]
3.312 COPr	CcL		n,o		>97[n]		85[o]	[350]
3.313 COPr	CcL		n,o		22[n]		14[o]	[390]
3.314 COPr	CcL		n,o		0[n]		0[o]	[390]
3.315[p] Ac	CcL[p,q]	34			70		36	[356]
3.316 Ac	CcL[r]			44	93	40	94	[391]
3.317 Ac	CcL	52			67		73	[392]
Ac	CcL				96[h]		95[i]	[392]
3.318[p] COBn	PsL[p,s]	50		40	99	49	>99	[393]
3.319[p] COPr	PsL[p,t]	49		47	98	50	96	[394]
3.320 Ac	CcL	50		36	81	46	95	[395]
Ac	PLE	50		47	30	43	31	[395]
3.321 Ac	CcL	50		5[u]	66	40	53	[395]
Ac	PLE	50		43	96	46	86	[395]
3.322 Ac	CcL[v]	33			x			[396]

a: Structures of the faster reacting enantiomers are shown in Fig.3.20; b: In the remaining ester fractions, **R** groups are the same as in the substrate, structures are enantiomeric to that shown in Fig.3.20; c: Over-hydrolysis. The (*R,R*)-diol (yield 10%) has ee=97%; d: Nearly optically pure. e: Pig liver acetone powder, 13% acetone; f: Lipase from *Arthrobacter sp.*; g: -10°C, 20% MeOH; h: After 50-70% hydrolysis of the reacylated alcohol; i: After further 30-50% hydrolysis of the remaining ester fraction; j: Slow hydrolysis; k: Low enantiomeric purity; l: Cholesterol esterase from bovine pancreas; m: 0-5°C; n: 40% conversion; o: 60% conversion; p: Alcoholysis with NuOH in an organic solvent; q: Nu=Bu in note p; r: Hydrolysis in isooctane saturated with water. In water fast chemical hydrolysis also takes place so enantiomeric purities are low; s: Nu=Et in note p; t: Nu=Me in note p; u: Overhydrolysis. 40% (*1S,2R,5S,6R*)-diol (ee=55%) can be isolated as main product; v: 10% acetone; x: After three recrystallizations from hexane a product of ee=99% was obtained in 30% yield.

Enantiomer preference may also be reversed by modification of the substrate, as reported in the PLE and CcL enzyme catalyzed hydrolysis of the bicyclic diacetate (3.320) with C_2 symmetry and that of its tetramethyl analog (3.321).

Several of the products of the transformations presented in Fig.3.20 and Table 3.14 are very useful synthetic intermediates. The optically active alcohols obtained from (3.276-3.280) are components of synthetic pyrethroids [348]. The azido compounds (3.259, 3.263, 3.274, 3.283, 3.284) can be used to prepare cyclic aminoalcohols [375,398], the alcohols (3.292-3.296.a,b) can be converted to natural cyclopentanoids, e.g. two prostaglandins [387,374], and the cyclohexanols (3.266-3.270) are efficient chiral auxiliaries which may substitute (−)-phenylmenthol [377-379]. The chiral binaphthyls (3.318, 3.319) accessible by enzymic alcoholysis can also be exploited as chiral auxiliaries, whilst the enantiomers of the norbornanol (3.307) are intermediates in the synthesis of carbocyclic nucleosides [399]. The tricyclic alcohol (3.322) was used in the synthesis of (+)-α-cuparenone, a bicyclic sesquiterpene [396].

Acylation of cyclic alcohols (Fig. 3.21 and Table 3.15; for further recent acylations, see [616-619]).

The *R* enantiomer selectivity experienced in the lipase catalyzed hydrolysis of 2-substituted cyclohexanols (3.323-3.330) is also conserved in the CcL catalyzed esterification of this class of substrates. Both the hydrolysis and alcoholysis of menthyl esters, as well as the acylation of menthol were the subject of numerous studies [113,401] which helped to establish that restriction of the water content to less than 1%, so inducing a more rigid conformation of the enzyme, resulted in better selectivity. The same was found to be true in the esterification of the silicon containing alcohol (3.287*) [283].

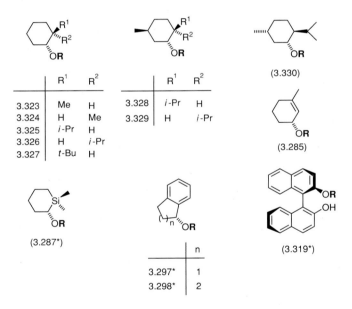

	R^1	R^2		R^1	R^2
3.323	Me	H	3.328	*i*-Pr	H
3.324	H	Me	3.329	H	*i*-Pr
3.325	*i*-Pr	H			
3.326	H	*i*-Pr			
3.327	*t*-Bu	H			

	n
3.297*	1
3.298*	2

Fig. 3.21. Enantiomer selective enzyme catalyzed acylation of racemic cyclic alcohols [Configurations of the ester products (3.X a, **R**=acyl) are shown. Remaining alcohols (3.X b, **R**=H) have enantiomeric structure. For details and references cf. Table 3.15]
* see note c in Table 3.15

iv) Transformation of derivatives of racemic diols (Fig. 3.22, see also [620, 621])

This type of reaction is somewhat complicated by the fact that hydrolysis may affect both functional groups. Enantiomer selective processes which stop after the hydrolysis of the first ester group have already been discussed. Fig. 3.22 illustrates the lipase catalyzed

Table 3.15. Enantiomer selective enzyme catalyzed acylation of racemic cyclic alcohols.

Substrate[a] (racemic, R=H)	Enzyme	Acylating agent	E value	Conv. (%)	Ester product[a] (3.X a:R≠H)			Rem. alcohol[b] (3.X b:R=H)		Ref.
					R	Y%	ee%	Y%	ee%	
3.285	PsL	CH₂=C(Me)OAc		32	Ac	48	67			[139]
3.287[c]	CcL[c]	triacetin		52	Ac		95[c]	36	96[c]	[383]
3.297[c]	PsL[c]	CH₂=CH-OAc	>500	49	Ac	46	>99[c]	48	95[c]	[362]
3.298[c]	PsL[c]	CH₂=CH-OAc	>300	50.5	Ac	49	97[c]	44	>99[c]	[362]
3.319[c]	PsL[c]	CH₂=CH-OAc			Ac	52	95[c]	48	89[c]	[394]
3.323	CcL	C₁₁H₂₃COOH		44	COC₁₁H₂₃		98		80	[400]
3.324	CcL	C₁₁H₂₃COOH		52	COC₁₁H₂₃		80		86	[400]
3.325	CcL	C₁₁H₂₃COOH		37	COC₁₁H₂₃		>99		68	[400]
3.325	CcL	C₁₁H₂₃COOH		30	COC₁₁H₂₃		88			[400]
3.327	CcL	C₁₁H₂₃COOH		30	COC₁₁H₂₃		>99			[400]
3.328	CcL	C₁₁H₂₃COOH		42	COC₁₁H₂₃		87		70	[400]
3.329	CcL	C₁₁H₂₃COOH		34	COC₁₁H₂₃		96		55	[400]
3.330	CcL	C₁₁H₂₃COOH		45	COC₁₁H₂₃		95			[113,400]
	CcL	triacetin		48	Ac	34	96	41	82	[402]

a: Structures of the faster reacting enantiomers are shown in Fig. 3.21; b: Structures of the remaining alcohols are enantiomeric to that shown in Fig. 3.21; c: The same enantiomer preferences were observed as for hydrolysis of racemic esters, so the resulting esters (and alcohols) are enantiomers of esters (and alcohols) from hydrolysis.

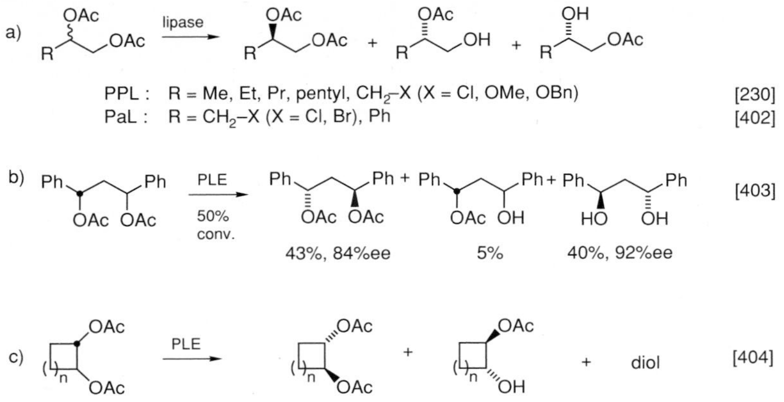

a)

PPL :   R = Me, Et, Pr, pentyl, CH$_2$–X (X = Cl, OMe, OBn)   [230]
PaL :   R = CH$_2$–X (X = Cl, Br), Ph   [402]

b)

PLE
50%
conv.

43%, 84%ee   5%   40%, 92%ee   [403]

c)

PLE   + diol   [404]

n	Conv.(%)	Y(%)	ee(%)	Y(%)	ee(%)	Y(%)	ee(%)	Config.
1	30	41	>95	49	>95	10	>95	S,S
2	25	54	54	43	50	<3		
2	74	<1		46	63	53	47	S,S
3	25	33	>95	26	0	41	>95	R,R

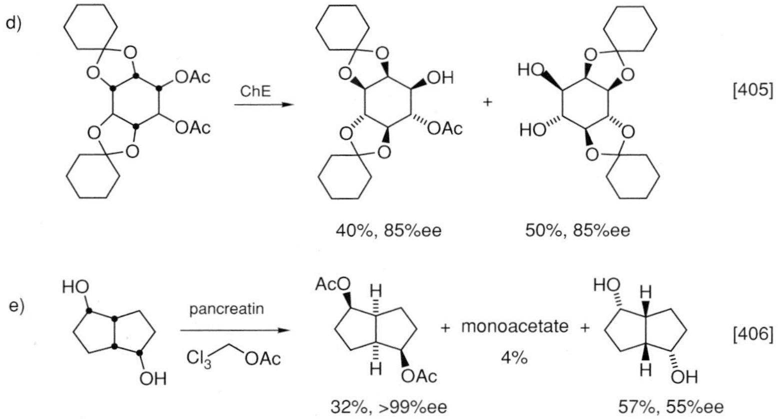

d)

ChE

40%, 85%ee   50%, 85%ee   [405]

e)

pancreatin

+ monoacetate +   [406]

4%

32%, >99%ee   57%, 55%ee

Fig. 3.22.   Racemic diol derivatives: ester hydrolysis, acylation

hydrolysis of racemic open-chain 1,2-diol diacetates which may involve both enantiomer
and regioselectivity. The ratio of monoacetate regioisomers obtained using PPL enzyme
much depends on the constitution of the diol. Highest selectivity was achieved with
the diacetates of 1,2-butane- and 1,2-pentanediols [230][8]. With PaL enzyme highest
selectivity was observed in the hydrolysis of 1-phenyl-1,2-ethanediol diacetate (ee 73%
for the recovered diacetate after 50% conversion) [402].

---

[8]   Here, the ee's of recovered (*R*)-1,2-diacetoxybutane and -pentane were 91% and 95% respectively
after 50% conversion and repeated hydrolysis of the separated diacetates up to 25% conversion

In Fig. 3.22.b, PLE catalyzed hydrolysis of racemic *anti*-1,3-diacetoxy-1,3-diphenylpropane is shown. Enantiomer selectivity is high and therefore it is not surprising that both ester groups of the *R,R* enantiomer hydrolyze quickly and thus only a little of the monoacetates can be isolated. Hydrolysis with intact *Trichoderma viridae* cells gave similar selectivities, but opposite enantiomer preference [403].

PLE hydrolysis of racemic cycloalkane-1,2-diol diacetates (Fig. 3.22.c) gives somewhat unexpected results. It seems that the diols are partly formed from the diacetate in a single step. This is supported by the observation that in the hydrolysis of *trans*-1,2-diacetoxycyclobutane, an *R,R* configured monoacetate was formed in high enantiomeric purity, whilst the diol although of similar purity was of *S,S* configuration. Moreover, the hydrolysis of racemic *trans*-cyclohexane-1,2-diol diacetate was 40 times slower than that of the diacetate and gave a diol of poor enantiomeric purity [404].

In a similar process, hydrolysis of the diacetate intermediate leading to D-myoinositol (shown in Fig. 3.22.d) does not stop after the hydrolysis of one of the ester groups and yields a monoacetate and a diol of opposite configuration in high enantiomeric purity.

In the enzyme catalyzed acylation of a racemic bicyclic diol with homotopic hydroxy groups (Fig. 3.22.e), owing to high enantiomer selectivity (and presumably a higher acylation rate of the monoacetate, compared to that of the diol) the diacetate was isolated in high enantiomeric purity whilst the monoacetate could only be obtained both in poor yield and low purity. The products were used for the synthesis of prostanoids [406].

3.3.5.4 Enantiomer Selectivity with Lactones, Epoxides, Amines and other Compounds

i) Lactones

Since lipases and esterases are useful for the selective transformation of acids and alcohols, it is not surprising that hydrolysis and the formation of lactones has also been studied.

Hydrolysis of racemic lactones can be conducted in an enantiomer selective way (Fig. 3.23). In the hydrolysis of monocyclic lactones (Fig. 3.23.a) [23,407], HLE enzyme showed the highest selectivity except with β-propiolactone for which PPL enzyme was more suitable. An interesting detail observed with 5-substituted δ-valerolactones was that both PLE and HLE enzymes hydrolyzed the *S*-5-methyl compound selectively, whilst with the 7-heptyl analog HLE was *R* selective and PLE *S* selective. With lactones of larger ring size, HLE enzyme exhibits high enantiomer selectivity.

For the hydrolysis of γ-substituted α-aminobutyrolactones (Fig. 3.23b) [408] PPL enzyme proved to be the optimal choice of catalyst. As substrates, the *N*-carbamoyl derivatives provided better selectivity than the corresponding *N*-benzoyl compounds.

HLE is also the enzyme of choice for the hydrolysis of racemic bicylic lactones (Fig. 3.23.c and d) [409]. In general this is a highly or fairly selective process except when it involves lactones condensed to a five-membered ring, when selectivity is generally quite poor.

a)

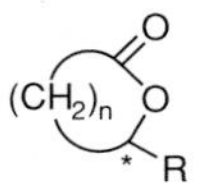

n	R	Enzyme	Y(%)	ee(%)	Config.
2	Et	PPL		75	R
	C_7H_{15}	PPL		76	R
3	Me	HLE	36	95	S
	C_7H_{15}	HLE		92	S
4	Me	HLE	30	76	R
		PLE	30	83	R
	C_7H_{15}	HLE		72	S
		PLE		72	R
5	Me	PLE	33	>95	S
6	Me	HLE	40	>95	S
7	Me	HLE	40	>99	S

c)

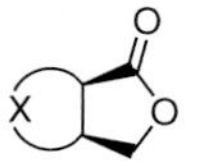

X	Enzyme	Y(%)	ee(%)
CH_2	HLE	41	98
$(CH_2)_2$	HLE	42	80
$(CH_2)_3$	HLE		racemic
$(CH_2)_4$	HLE	40	47
$CH_2-CH=CH-CH_2$	HLE	34	95

b)

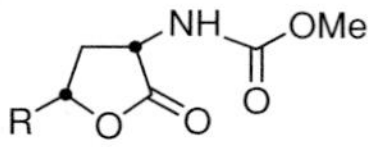

R	Enzyme	Conv.(%)	ee(%)	Config.
H	PPL	50	62	S
CH_2=CH	PPL	50	95	**
Ph	PPL	27*	32	**

d)

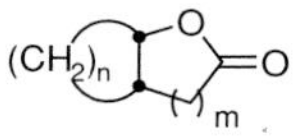

m	n	Enzyme	ee(%)
1	3	HLE	racemic
2	4	HLE	60

*   accompanied by hydroxy acid with ee = 86%;
** not determined

Fig. 3.23. Data for remaining lactones from enantiomer selective enzymic hydrolysis of racemic lactones after 50 - 65% conversion

Lactone formation from esters of racemic hydroxy acids in organic solvents is affected by lipase (Fig. 3.24). (For enzymic lactonization of achiral hydroxyacids see refs. [125,414 and 411], for very recent enantiomer selective lactonizations see refs. [622-624] and for enantiotope selective lactonization of 4-hydroxypimelate esters see ref. [415]). Lactone formation may be accompanied by the formation of oligomers. (For other enantiomer and enantiotope selective enzymic oligomerisations see refs. [361,416 and 417]. Oligomerisation and/or oligolactonisation occur between achiral diols and diacids [418].) Macrocyclic lactones can thus be obtained with high selectivity but poor yields by lactonisation using *Pseudomonas* lipases, whilst for the preparation of 4-substituted-γ-butyrolactones PPL enzyme was found to be most efficient. With suitable substrates

a)

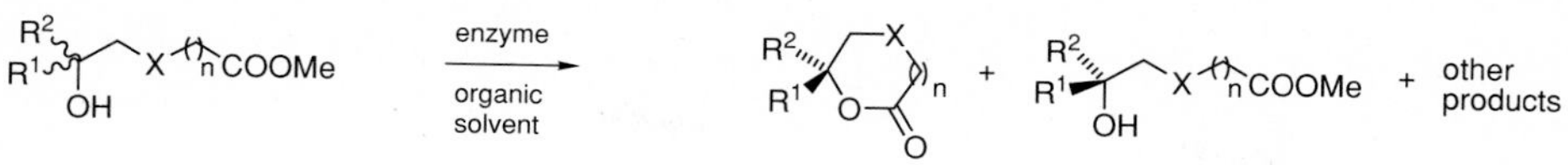

R^1	R^2	X	n	Enz.	Conv.* (%)	Y.* (%)	ee (%)	Y. (%)	ee (%)	Ref.
H	CH_3	CH_2–CH_2	7	PsL	57	6	–			[410]
H	C_6H_{13}	CH_2–CH_2	7	PsL	91	14	>99			[410]
H	C_6H_{13}	CH=CH (E) CH=CH (Z)	7	PsL	87	18	>99			[410]
H	C_6H_{13}	CH_2 CH_2	7	PsL	70	20	98			[410]
H	CH_3		7	PfL			>99			[411]
CH_3	H	CH_2	0	PPL	36		>95			[412]
					60				>95	[412]
Ph	H		0	PPL			high		high	[412]

* differences refer to oligomers and oligolactones

b)

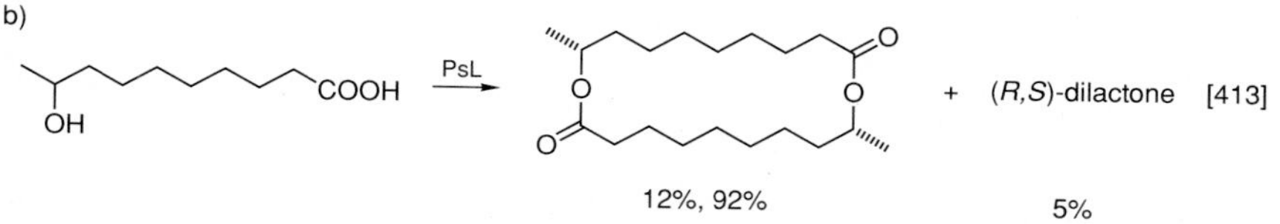

12%, 92%   5%

Fig. 3.24. Enantiomer selective enzymic lactonization of racemic hydroxy acids

Pseudomonas lipase is also able to catalyze, with high enantiomer selectivity, the formation of dilactones (Fig. 3.24b).

ii) Amines and aminoalcohols

Hydrolysis of the phenylacetyl amides of racemic amines and aminoalcohols with penicillin acylase enzyme proceeds in an enantiomer selective manner [418,420]. This is exemplified by optically active 2-hydroxy-3-(1-naphthyl)-propanamine which can be prepared in this way:

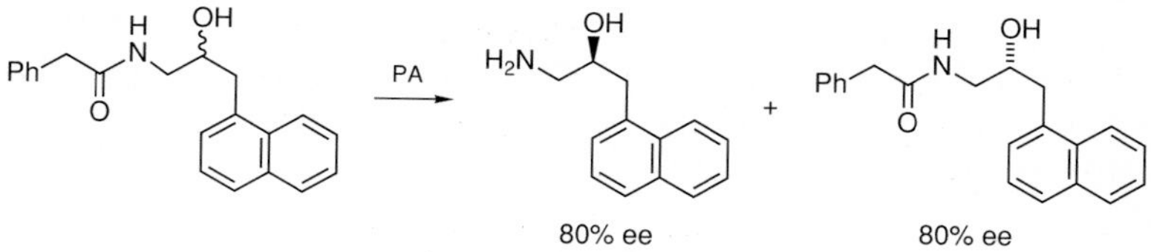

80% ee   80% ee

Acylation of racemic 2-amino-butan-1-ol and of 2-hydroxypropanamine with PPL enzyme in ethyl acetate gave the *N*-acetyl derivatives in >95% ee [421]. It is not however clear whether amide formation was due to lipase itself, to protease contamination of PPL or to acyl migration [422]. Recently, in connection with enantiomer selective acylation of amines [625] it has become clear that lipases are able to catalyze amide formation.

Subtilisin may be selected for the enantiomer selective acylation of racemic amines [423]. Selectivity is much dependent on the solvent used and the best results were attained with trifluoroethyl butyrate in 3-methyl-3-pentanol. In this way, the following amines were obtained in >95% ee.

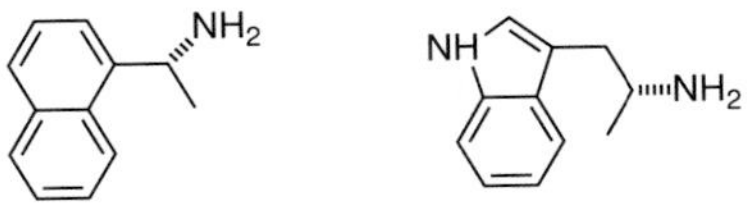

iii) Enzymic ring opening of racemic epoxides

Interesting ring opening reactions can be realized with epoxide hydrolase (EH), a microsomal hydrase from rabbit liver (EC 3.3.2.3). Using alkyl substituted oxiranes as substrate [424,425], concurrent enantiomer and regioselectivity was observed, as shown below:

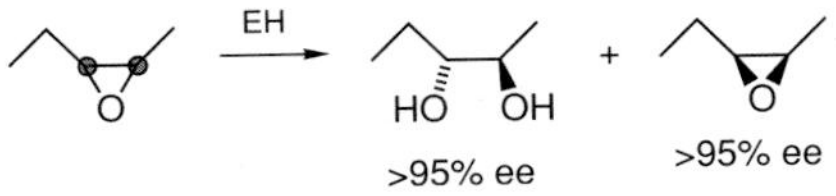

The hydrolysis of the achiral cyclohexene oxide yielded an optically active diol in an enantiotope selective process [426]:

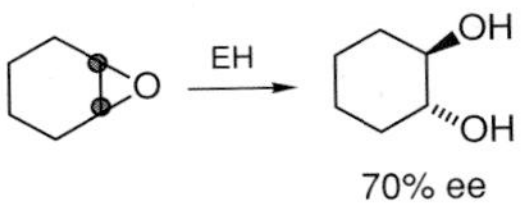

The above result would suggest that ring opening of the oxa-analog of cyclohexene oxide should be enantiomer selective (Fig.3.25.a). Quite surprisingly, however, both enantio-mers underwent hydrolysis yielding a single *R,R*-diol of high enantiomeric purity [427]. This result can be interpreted by assuming that the enzyme is specific for the conformers of *M* helicity of both enantiomers. Thus, both in the 2*S*,3*R* and the 2*R*,3*S* enantiomers the *S* centre is attacked giving rise, in processes of opposite regioselectivity, to the same 2*R*,3*R*-diol, whereas the conformers with *P* helicity are not substrates for the enzyme.

A similar phenomenon was recorded with an epoxy hydrase isolated from the antennae of gypsy moth that plays a vital role in the pheromone recognition of this insect. The observation that this enzyme cleaves both enantiomers of the epoxide type pheromone specifically to the same *R,R*-diol (Fig. 3.25.b) can be explained in a similar way [428].

Regioselectivity associated with enantiomer or enantiotope selectivity of EH enzyme was observed in the hydrolysis of aromatic and polyaromatic epoxides [429-431]. Hydro-lysis of substituted cyclohexene oxides proceeds with regio- and enantiomer selectivity [432-436], a selectivity that is especially outstanding with dihydropyrane-3,4-epoxides yielding both the diols and the unchanged epoxides in over 95% ee [424,437-439]:

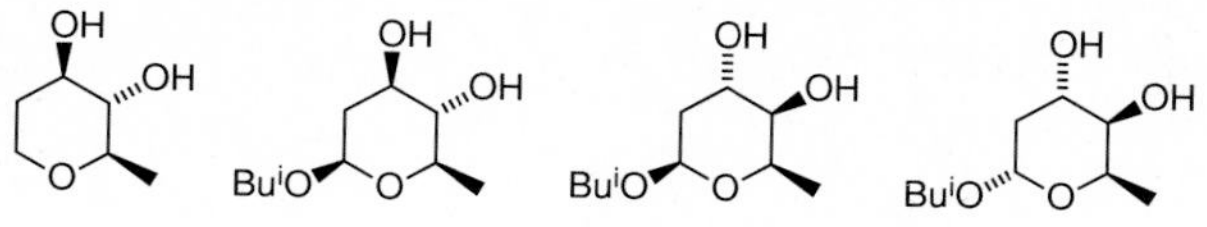

OH ...OH   OH ...OH   OH ...OH   OH ...OH

Bu^{i}O   Bu^{i}O   Bu^{i}O

iv) Enantiomer selectivity in other enzyme catalyzed transformations

The enantiomer selectivity of monophosphate-5'-nucleosidase [440] and of adenosine aminohydrolase [441] was made use of in the preparation of optically active forms of the antiviral carbocyclic analogs of purine nucleosides.

Enantiomer selectivity was also reported in the β-glucosidase catalyzed glycosidation of glycidol and glycerol acetonide [442], in hydrolytic release of chlorine from 2,3-dichloropropionic acid by 2-haloacid dehalogenase [626] or in thiotransesterification of mercapto esters by lipases [627].

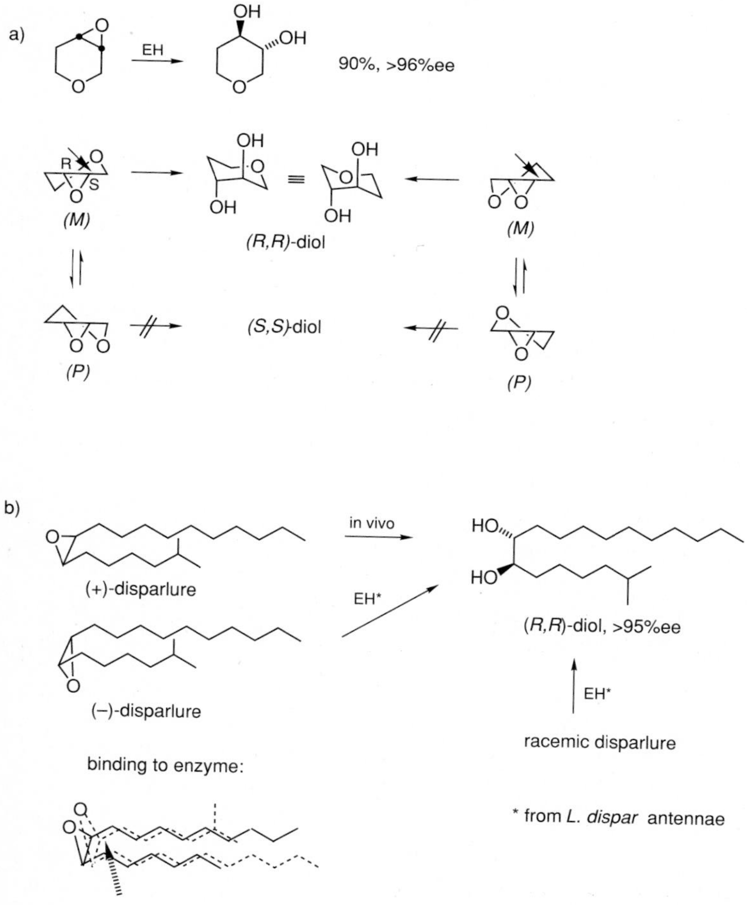

a)

   EH →   OH ...OH   90%, >96%ee

R / S
(M)

OH OH
(R,R)-diol   (M)

(P)   ─//→   (S,S)-diol   ←//─   (P)

b)

(+)-disparlure

(−)-disparlure

binding to enzyme:

in vivo →

EH*

HO...
HO
(R,R)-diol, >95%ee

↑ EH*

racemic disparlure

* from *L. dispar* antennae

Fig. 3.25. Formation of single enantiomers from racemic substrates by epoxy-hydrases

3.3.6   Hydrolases: Enantiotope Selectivity

Selective catalysis by hydrolytic enzymes can be exploited most efficiently for the transformation of enantiotopic groups, whilst examples involving enantiotopic faces are rare (cf. Section 3.3.6.3). In contrast to enantiomer selectivity, in a completely enantiotope selective transformation the total amount of the achiral substrate, at least in principle, can be transformed to an enantiomerically pure chiral product.[9] In this process, enantiotopic groups are converted to constitutionally different groups. This sometimes facilitates the transformation of the product by a special sequence of reactions to both enantiomers of a desired product (cf. Fig. 2.13.c).

Enantiotope selectivity of hydrolases can be utilized in hydrolysis, esterification, and transesterification.[10] Enantiotopic groups can be found either in achiral compounds containing a single prochiral centre, or in both open chain and cyclic *meso* compounds. In the following sections we discuss enantiotope selective hydrolytic transformations according to the structural features of their substrates.

3.3.6.1   Transformations of Compounds Containing a Single Prochiral Center

In this group, it was predominantly derivatives of dicarboxylic acids and diols that proved to be fruitful as substrates, though there is an important difference between the two. Hydrolysis of dicarboxylic esters can be regarded as irreversible, since in neutral solution the half ester is present as its anion which is a very poor substrate for the enzyme. Hydrolysis therefore stops after the first ester group has been hydrolyzed thus enantiomeric purity is practically independent of conversion (see Section 2.3.2.7.)

Hydrolysis of diol esters is also generally irreversible, but it does not stop at the half ester stage and diol formation decreases the yield. In other words, enantiomeric purity can be improved at the expense of yield by increasing conversion.

Alcoholysis of diol esters, as well as acylation and transesterification of diols is reversible (see Section 3.2.2.) except, of course, when "irreversible" transesterification agents are used (see Section 3.2.2.).

i) Hydrolysis of various prochiral dicarboxylic esters with certain enzymes has been characterized in full detail, whilst alcoholysis of diesters in organic solvents has not been studied, despite the fact that employment of the latter may be an efficient way to

---

[9] For a general definition of this type of selectivity and for the consequences of incomplete selectivity cf. Section 2.3.2.7.

[10] For a general discussion of such reactions cf. Sections 3.2.1 and 3.2.2.

increase enantiotope selectivity. On the other hand however, enzymic esterification of dicarboxylic acids does not seem promising.

Table 3.16 summarizes data on enzymic hydrolysis of *prochiral malonic esters*. With monosubstituted malonic esters, hydrolysis can be conducted in an enantiotope selective manner although the optically active product racemizes rapidly, as shown by the CTR catalyzed hydrolysis of diethyl acetamido malonate [452]. One exception to this is hydrogen methyl 2-fluoromalonate (3.345.b), which is not liable to racemization [448].

Table 3.16. Enantiotope selective enzymic hydrolysis of prochiral malonates.

$$\underset{\text{(3.Xa)}}{\text{ROOC}\overset{R^1\quad R^2}{\diagup\diagdown}\text{COOR}} \quad \xrightarrow[\substack{H_2O \\ 70\text{-}100\%}]{\text{enzyme}} \quad \underset{\text{(3.Xb)}}{\text{HOOC}\overset{R^1\quad R^2}{\diagup\diagdown}\text{COOR}}$$

No.	Substrate (3.X a) R^1	R^2	R	Enzyme	Cosolvent	Product[a] (3.X b) ee(%)	Ref.
3.331	C_2H_5	CH_3	Me	PLE	25% DMSO	73	[443]
3.332	C_3H_7	CH_3	Me	PLE	25% DMSO	52	[443]
3.333	C_4H_9	CH_3	Me	PLE	25% DMSO	58	[443]
3.334	CH_3	C_5H_{11}	Me	PLE	25% DMSO	46	[443]
3.335	CH_3	C_6H_{13}	Me	PLE	25% DMSO	87	[443]
3.336	CH_3	C_7H_{15}	Me	PLE	25% DMSO	88	[443]
					50% DMSO	97	[89]
3.337	Ph	CH_3	Et	PLE		84	[444]
			Et	PLE	25% DMSO	23	[443]
			Me	PLE		53	[89]
			Me	PLE	25% DMSO	16	[89]
	CH_3	Ph	Me	PLE	50% DMSO	44	[89]
			Me	CTR	25% DMSO	>98	[89]
3.338	Ph	C_2H_5	Me	PLE		86	[444]
3.339	CH_3	$CH_2C_6H_3(3,4)\text{-}(MeO)_2$	Me	PLE	50% DMSO	93	[89,445]
			Me	CTR	50% DMSO	>98	[445]
3.340	CH_3	CH_2OCH_2Ph	Me	PLE		67	[446]
3.341	CH_3	CH_2OBu^t	Me	PLE		96	[446]
3.342	CH_3	$CH_2OSiMe_2Bu^t$	Me	PLE		95	[446]
3.343	CH_3	CH_2SiMe_3	Me	PLE	50%DMSO	98[b]	[447]
3.344	CH_2Br	CH_3	Me	PLE		46[b]	[446]
3.345	F	H	Me	CcL		62	[448]
3.346	F	CH_3	Me	CcL		95	[448]
			Et	CTR		70	[449]
3.347	F	C_2H_5	Me	CcL		99	[448]
3.348	F	C_3H_7	Me	CcL		33	[448]
3.349	NHAc	CH_3	Et	PLE		81[b]	[450]
3.350	$-CH_2NHCH_2CH_2-$		Et	PLE		20	[451]

a: R^1 and R^2 are the same as given for the substrate; b: Configuration is not given.

For synthetic purposes it was predominantly the hydrolysis of disubstituted malonates yielding non-racemizing products that was explored, most often using PLE enzyme. The following observations deserve mention:

– With disubstituted malonates in which one of the substituents is a methyl group, enantiotope preference depends on the bulk of the other substituent [443] (see Table 3.16). Note that compounds (3.331-3.333) and (3.337) contain small groups, whilst compounds (3.334-3.336) and (3.339-3.342) possess bulky groups as the second substituent.

– Selectivities with methyl esters are generally higher than with ethyl esters. (The diesters (3.337) are exceptions to this rule).

– Enantiotope selectivity can be much influenced by the addition of DMSO which promotes *pro-S* selectivity [89]. (Note that here the hydrolysis of the *pro-S* ester group gives rise to a product of *R* configuration). This means that optical purity decreases (or may even be inverted, e.g. with (3.337)), when the second substituent is small, and increases when the latter is bulky.

ArCH$_2$ substituted malonates (3.337, 3.339) can be hydrolyzed with CTR enzyme with very high *pro-S* selectivity [445].

CcL enzyme showed good selectivity with 2-fluoromalonates (3.345-3.349) and it is interesting that whilst CTR fails to hydrolyze dialkyl malonates at all [445], 2-fluoro-2-methylmalonate (3.346) is a substrate for this enzyme.

Chiral dialkyl or alkyl-phenyl malonates are intermediates in the preparation of barbiturates [444]. The half ester of (3,4-dimethoxyphenyl)-methyl malonic acid (3.339), which can be produced in high enantiomeric purity by CTR, or PLE catalyzed hydrolysis can be used to manufacture l-α-methyl-DOPA [445,450].

In Table 3.17, data for the enzymic hydrolysis of *3-substituted, or 3,3-disubstituted prochiral glutarates* were compiled. PLE and CTR enzymes were the enzymes of choice for their transformation, although PLE has a broader substrate tolerance and the sense of selectivity of the two enzymes is often opposite (e.g. 3.363: R=Me, 3.364-3.368).

In PLE catalyzed hydrolysis, again it was often found that better selectivity could be achieved with the methyl esters [315,455,470], although in the hydrolysis of 3-hydroxyglutarates (3.363) hydrolysis of the propyl ester was much more selective than that of the methyl ester.

With glutarates monosubstituted with apolar groups, the *pro-S* selectivity characteristic of small 3-substituents (3.351-3.354) changed to *pro-R* with larger groups (3.355-3.362) [88][11].

PLE catalyzed hydrolysis of 3-hydroxy- and 3-aminoglutarates (3.363 and 3.371) is rather unselective (in the case of diethyl 3-hydroxyglutarate, hydrolysis by a bacterial esterase proved to be quite selective [628]), but by suitable protection of these groups selectivity can be much enhanced. When the protecting group is relatively small, e.g. acetyl (e.g. 3.364 or 3.372) the sense of selectivity is the same as with glutarates bearing

---

[11]   An earlier report claiming substituent independent high *pro-S* selectivity [471] has not been confirmed.

Table 3.17. Enantiotope selective enzymic hydrolysis of prochiral glutarates.

$$\text{ROOC}\underset{(3.Xa)}{\overset{R^1 \quad R^2}{\diagdown\diagup}}\text{COOR} \quad \xrightarrow[\substack{H_2O \\ 70\text{-}100\%}]{\text{enzyme}} \quad \text{HOOC}\underset{(3.Xb)}{\overset{R^1 \quad R^2}{\diagdown\diagup}}\text{COOR}$$

No.	Substrate (3.X a) R¹	R²	R	Enzyme	Cosolvent	Product[a] (3.X b) ee(%)	Ref.
3.351	CH₃	H	Me	PLE		79-95	[34] [453-455]
			Me	PLE[b]		84(>95[c])	[22]
			Me	PLE	20%MeOH[d]	97	[88]
3.352	C₂H₅	H	Me	PLE		50	[88]
3.353	C₃H₇	H	Me	PLE		25	[88]
3.354	cyclohexyl	H	Me	PLE		17	[88]
3.355	H	i-Pr	Me	PLE		38	[88]
3.356	H	Ph	Me	PLE		42	[88]
3.357	H	CH₂Ph	Me	PLE		54	[88]
			Me	PLE	10% acetone	73	[456]
3.358	H	(CH₂)₃Ph	Me	PLE	10% acetone	88	[456]
3.359	H	CH:CHPh(E)	Me	PLE	10% acetone	93	[456]
3.360	H	CH₂CH:CHPh(E)	Et	PLE	10% acetone	91	[456]
3.361	H	CH₂CH₂OCH₂Ph	Me	PLE	10% acetone	54	[456]
3.362	H	CH:CHCH₂OTHP	Me	PLE	10% acetone	74	[456]
3.363	H	OH	Me	PLE		10-20	[451,457]
	H	OH	Pr	PLE	e	76	[458]
	OH	H	Me,Et	CTR		69-100	[457] [459-461]
3.364	OAc	H	Me	PLE[f]		90	[230,462]
			Et	CTR		95	[462]
3.365	OCH₂Ph	H	Me	CTR		84	[457]
	H	OCH₂Ph	Me	PLE		12	[457]
3.366	OCH₂Ph	H	Me	CTR		92	[457]
	H	OCH₂Ph	Me	PLE		59	[457]
3.367	OSiMe₂Bu^t	H	Me	CTR		93	[457]
	H	OSiMe₂Bu^t	Me	PLE		14-50	[230,457]
3.368	OCH₂OCH₃	H	Me	CTR		93	[457]
	H	OCH₂OCH₃	Me	PLE		0-14	[230,457]
3.369	CH₃	OH	Me	PLE		>95	[464]
3.370	CF₃	OH	Me	PLE		88	[465]
3.371	NH₂	H	Me	PLE[g]		40[g]	[466]
3.372	NHAc	H	Et	CTR		h	[467,468]
			Me	PLE[b]		93	[21]
3.373	H	NHCOBu^t	Me	PLE[b]		93	[21]
3.374	H	NHCOCH:CHCH₃	Me	PLE[b]		>95	[21]
3.375	H	NHCOOCH₂Ph	Me	PLE		>96	[466]
			Me	PLE[b]		93	[21]
3.376	H	Cl	Me	PfL		83	[469]

a: R¹ and R² are the same as for the substrate; b: As a crude extract of liver acetone powder; c: After one recrystallization with (-)-cinchonidine; d: At -10 °C; e: At 0 °C; f: Chemical yield is low (40-50%) due to concurrent hydrolysis of the acetyl ester; g: Concurrent chemical hydrolysis. In the blank test 30% conversion was observed [466]; h: Optically pure.

a small apolar substituent. Large protecting groups (*O*-protected 3.365-3.368, *N*-protected 3.373-3.375), in turn, give results analogous to those observed with bulky nonpolar substituents. That high selectivity occurs in the hydrolysis of 3-hydroxy-3-methyl (3.369) and 3-hydroxy-3-trifluoromethyl glutarate (3.370) is of note.

Enantiotope selectivity in glutarates can also be influenced by the addition of cosolvents [88,456] or by decreasing the temperature [88,458].

The optically active glutarates prepared by enzymic methods have found diverse synthetic applications. Thus, the 3-ethyl glutarate (3.351.b) prepared by PLE catalysis has been used as a bifunctional synthon in the synthesis of a precursor of monensin A [314,326,453] and other natural macrocyclic compounds [454,455,470], as well as in the synthesis of (+)-faranal, an insect pheromone [470,472]. The hydroxy compounds (3.363.b) and their protected derivatives (3.364-3.368.b) are intermediates in the synthesis of compactin and its analogs [458,461], and similarly in the synthesis of pimaricin, a macrocyclic antibiotic [473]. The protected amine (3.375.b) was used in the synthesis of β-lactames [466,474-476] and the antibiotic negamycin [460]. The 3-hydroxy-3-methyl and trifluoromethyl half esters (3.369.b and 3.370.b respectively) are starting materials for the preparation of *R*-mevalolactone [464] and its trifluoromethyl analog [465].

In Fig. 3.26, hydrolase catalyzed enantiotope selective transformations of some miscellaneous compounds containing a single prochiral center are shown. A special feature of the hydrolysis of the prochiral sulfoxide diester shown in Fig. 3.26.a is that here, it is the sulphur atom of the sulphoxide which becomes the chiral center in the half-ester product.

In Fig.3.26.b, enzyme catalyzed alcoholysis of 3-alkylglutaric anhydrides is shown. Enantiomeric purity of the half esters obtained in this way is higher than that prepared by PLE hydrolysis of the corresponding dimethyl esters and the configuration of the products is consistently *R*.

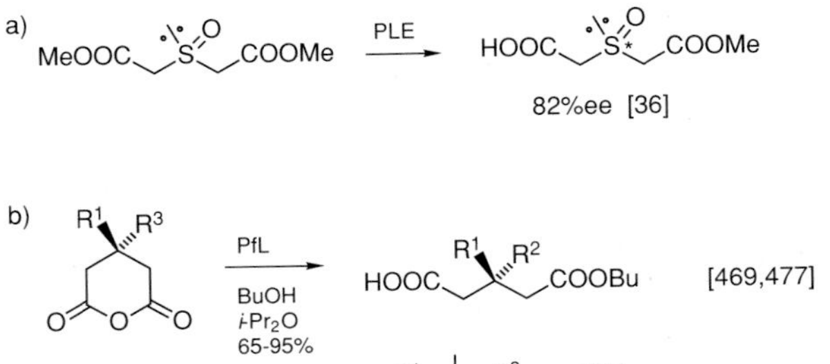

a)

MeOOC—S(=O)—COOMe →(PLE)→ HOOC—S(=O)—COOMe

82%ee [36]

b)

(3-substituted glutaric anhydride) →(PfL, BuOH, *i*-Pr$_2$O, 65-95%)→ HOOC—C(R^1)(R^2)—COOBu   [469,477]

R^1	R^2	ee(%)
Me	H	93
Et	H	87
Pr	H	60
i-Pr	H	76
H	Cl	62

Fig. 3.26.   Enantiotope selective enzymic transformation of a prochiral sulphone diester (a) or prochiral glutaric anhydrides (b)

ii) Transformation of prochiral diol derivatives (For recent additional results on the hydrolysis of 2-substituted 1,3-diacyloxypropanes, see refs. [629-631] and on the acylation of 2-substituted 1,3-propanediols see refs. [632-634]. An interesting example for enantiotope selective hydrolysis of a prochiral bicyclic spirodiester [639] has also been reported.)

Table 3.18 contains information about the hydrolase catalyzed enantiotope selective transformation of 2-substituted 1,3-propanediols (results with 2,2-disubstituted propanediol derivatives are poor [488]). It can be seen that using the same enzyme groups of the same prochiral configuration are preferred both in the hydrolysis of diesters and the acylation of the corresponding diols. As a consequence, the products are enantiomers. Enantiotope selective alcoholysis of the diesters in organic solvents has not been reported.

Table 3.18. Enantiotope selective hydrolase catalyzed transformations of prochiral 1,3-propanediol derivatives: hydrolysis of diesters or acylation of diols.

| No | Substrate | | Enzyme | Process: H (cosolvent) A (acylating agent) | Product[a] | | | Ref. |
	R^1	R^2				R	Y% ee%	
3.377a	CH_3	H	PfL	H	b^b	Ac	33 >99	[478]
c			PfL	A, CH_2=CHOAc	b^b	Ac	70 60	[479]
3.378a	H	$CH_2CH=CH_2$	PPL	H	b	Ac	34 95	[480]
a			PPL	H	b	Ac	76 55	[481]
a			CcL	H, 30% acetone	b	Ac	48 89	[481]
c			PfL	A, CH_2=CHOAc	d^d	Ac	89 81	[479]
c			PPL	A, CH_2=CHOAc	d	Ac	37 30	[481]
3.379a	H	i-Pr	PfL	A, CH_2=CHOAc	d^c	Ac	85 61	[479]
3.380a	H	$CH_2CH_2CH=CH_2$	PPL^d	H	b^c	Ac	80 >95	[24]
c			PPL^d	A, MeOAc	d^c	Ac	70 90	[24]
3.381a	H	$CH=CHC_5H_{11}$	PPL^d	H	b^c	Ac	49 84	[24]
c			PPL^d	A, MeOAc	d^c	Ac	63 96	[24]
3.382a	H	cyclohexyl	PPL^d	H	b^c	Ac	96 60	[24]
c			PPL^d	A, MeOAc	d^c	Ac	90 98	[24]
3.383a	H	Ph	PPL^d	H	b^c	Ac	91 >95	[24]
c			PPL^d	A, MeOAc	d^c	Ac	98 92	[24]

Table 3.18. (contd.)

No	Substrate R^1	R^2	Enzyme	Process: H (cosolvent) A (acylating agent)	Product[a] R	Y%	ee%	Ref.
3.384a	H	CH$_2$Ph	PPL[e]	H, 30% acetone	b Ac	57	83	[481]
a			PPL[e]	H	b[c] Ac	65	61	[24]
c			PfL	A, CH$_2$=CHOAc	d Ac	96	97	[479]
c			PPL[d]	A, MeOAc	d[c] Ac	90	13	[24]
3.385c	H	CH$_2$(1-naphthyl)	PfL	A, CH$_2$=CHOAc	d[c] Ac	95	90	[479]
3.386c	OCH	H	CvL	A, CH$_2$=CHOAc	d Ac	88	92	[483]
3.387c	OC$_2$H$_5$	H	CvL	A, CH$_2$=CHOAc	d Ac	87	89	[483]
c			PfL	A, CH$_2$=CHOAc	d Ac	90	90	[483]
c			PfL	A, PhOAc	d Ac	90	90	[483]
3.388a	OCH$_2$Ph	H	PPL	H	b Ac	40	60	[484]
a			PPL	H, 15% THF	b Ac	40	80	[484]
a			PPL	H, i-Pr$_2$O	b Ac	45	88	[485]
a			PsL	H	b Ac	75	91	[484]
c			PsL	A, CH$_2$=CHOAc	d Ac	53	96	[139,486]
c			PfL	A, CH$_2$=CHOAc	d Ac	92	94	[483,487]
3.389a	NHCOOCH$_2$Ph	H	PPL	H	COBu	55	>97	[139,486]
c			PPL	A, CH$_2$=CHOCOBu	d COBu	77	97	[139,486]
3.390a	COOEt[f]	H	PPL	H	b Ac	58	~85	[488]
3.391a	-COCH$_2$CH$_2$CH$_2$-		AChE	H	b Ac	30	90	[489]

a: R^1 and R^2 are the same as for the substrate; b: There is a contradiction here; for both products *R*-configuration was assigned; c: Configuration is not given; d: A purified fraction of the crude PPL enzyme. With the crude enzyme enantiomeric purities are significantly lower; e: At -5 °C; f: Absolute configuration is not confirmed.

Although the above propanediols are substrates for several enzymes (among them PLE), it is usually lipases that have provided the best selectivities.

A survey of the data in the table reveals that the enantiomeric purity of the monoacetate fraction (e.g. in the hydrolysis of 3.388.a) fluctuates even when using the same enzyme preparation. One reason for this may be that hydrolysis does not stop at the monoacetate stage and therefore enantiomeric purity depends greatly on conversion.

It was reported that in the case of lipases and hydrophobic substituents (3.377-3.385), transformations are generally *pro-R* selective. When the substituent is linked by a heteroatom (O, N) to the main chain (3.386-3.389), enantiotope preference is inverted in a geometrical sense, although as a result of the sequence rules the affected group is still labelled *pro-R*. In the case of PPL enzyme, an active center model constructed in conjunction with the hydrolysis of diacetate (3.390.a) [488] can help to predict the results:

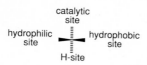

For a given diol derivative, several methods are available for improving selectivity. This is illustrated by the transformation of the benzyloxy substituted compounds (3.388). Both hydrolysis [484] and acylation [483] of these substrates was studied to observe the influence conferred by the type of enzyme used, further to examine the correlation between conversion and enantiomeric purity [139,486]. Using a single enzyme (PPL) it was also found that selectivity could be influenced by cosolvents [484,485].

The optically active half esters found their way into several interesting syntheses. From the benzyloxy substituted half esters (3.388.b or d), after suitable functionalisation and crystallisation it was possible to prepare enantiomerically pure chiral synthons [484,485] which were then used, amongst other things, in the synthesis of platelet activating factor (PAF) [490] and (S)-propranolol [487]. The methyl substituted monoester (3.377.b) can be used for the synthesis of the macrocyclic (R)-muscone [478], the alkyl monoester (3.378.b) for that of (R)-factor A acting on microbial growth [480], and the cyclic half-ester (3.391.b) for the synthesis of (-)-malyngolide, a substance which possesses antibiotic activity [489].

3.3.6.2   Transformation of Meso Compounds[12]

Using hydrolases, one can carry out enantiotope selective transformations of both cyclic and open chain *meso* compounds. The product, of course contains more than one element of chirality (e.g. centers of asymmetry). Whilst primarily this reaction is most important for the *meso*-diols and *meso*-dicarboxylic acids, this type of selectivity is not solely restricted to this circle of compounds. Hydrolases are also, for instance, able to transform both *meso*-anhydrides [477] and *meso*-oxiranes [491-494].

i) Enantiotope selective hydrolysis of meso-dicarboxylic esters (Table 3.19)

The enzyme most frequently applied in this transformation is PLE, thus, unless otherwise mentioned, any results refer to hydrolyses with this enzyme although in principle, other hydrolases may also be as efficient and could well be employed in the future.

It is interesting to note that the hydrolysis of both dimethyl 2,4-dimethylglutarate (3.392.a) and its 3-hydroxy analog (3.393.a) (cf. Section 3.3.6.1) proceeded with somewhat modest selectivity, whereas hydrolysis of a compound that combines the two structures (3.393) was highly selective [34].

---

[12] As in Section 3.3.6.1, only transformations showing high (ee >70%) selectivity have been discussed, except in cases where examples of lower or no selectivity were quoted for the sake of comparison or to illustrate specific trends.

Table 3.19. Enantiotope selective enzymic hydrolysis of *meso*-dicarboxylic acid esters.

Substrate (3.X a) No		R¹,R²	Enzyme	Product(s) (3.X b:R¹=H,R² ≠H) (3.X c:R¹ ≠H,R²=H) R¹	R²	Y(%)	ee(%)	Ref.
3.392	R¹OOC‑CH₂‑CH₂‑COOR²	Me	PLE	H	Me	85	64	[495]
		Me	CTR	H	Me	48	100	[34]
		Me	G.r.	Me	H	75	>98	[495]
3.393	R¹OOC‑CH(OH)‑COOR²	Me	PLE	H	Me	95	98	[34]
3.394	X:CH₂ (X COOR¹/COOR²)	Me	PLE	Me	H	90	>95	[34,274][496,497]
3.395	X:C(CH₃)₂	Me	PLE	Me	H	42	80	[274]
3.396	X:O	Me	PLE	H	Me	69	31	[496]
3.397	cyclobutane COOR¹/COOR²	Me	PLE	Me	H	>87	86-100	[34,274][496,497]
3.398	X:CH₂ (cyclopentane)	Me	PLE	H	Me	80-92	9-17	[34,274][496,497]
3.399	X:O=C	Me	PLE[a]	H	Me		82	[27]
3.400	X: HO/H	Me	PLE	H	Me		80	[27]
3.401	X: t-BuO/H	Me	PLE	Me	H		84	[27]
3.402	cyclohexane COOR¹/COOR²	Me	PLE	H	Me	75-98	80-100	[34,274][496,497]
		Me	PLE[b]	H	Me		>97	[301]
		Et	PLE	racemic				[301]
3.403	cyclohexene COOR¹/COOR²	Me	PLE[c]	H	Me	95	85	[34]
		Me	PLE[d]	H	Me	98	96	[498]
		Me	PLE[e]	H	Me	98	98	[21]
		Et	PLE[e]	H	Et	67	27	[21]

Table 3.19. (contd.)

Substrate (3.X a)		Enzyme	Product(s) (3.X b:R¹=H,R² ≠H) (3.X c:R¹ ≠H,R²=H)				Ref.
No	R^1,R^2		R^1	R^2	Y(%)	ee(%)	

No	Substrate	R^1,R^2	Enzyme	R^1	R^2	Y(%)	ee(%)	Ref.
3.404	ferrocene COOR¹/COOR²	Me	PLE	Me[f]	H[f]	88	74[f]	[499]
3.405	Ph–N–C(=O)–N–Ph imidazolidinone, R¹OOC, COOR²	Pr	PLE	H	Pr	85	75	[450]
		Me	PLE	H	Me	71	38	[450]
3.406	O epoxide COOR¹/COOR²	Me	PLE	H	Me	90	99	[500]
3.407	X: O=C (cyclopentane COOR¹/COOR²)	Me	PLE[d]	Me	H	86	88	[27,501]
		Me	CTR	H	Me		83	[27]
3.408	X: (dioxolane)	Me	PLE	Me	H	99	90	[501]
3.409	cyclohexene COOR¹/COOR²	Me	PLE	Me	H	91	76	[501]
		Me	CTR	H	Me		86	[27]
		Me	PPL[d]	H	Me	97	97	[501]
	R¹OOC–X–COOR²							
3.410	X:CH₂	Me	PLE	H	Me	82	34	[502]
3.411	X:O	Me	PLE	Me	H	98	42	[502]
		Me	PPL	H	Me		91	[503]
3.412	X:S	Me	PLE	Me	H	83	46	[502]
3.413	X:NHCH₂Ph	Me	PLE[d]	Me	H	85	80	[504]
		Me	PLE[g]	Me	H	39	100	[505]

Table 3.19. (contd.)

Substrate (3.X a)		Enzyme	Product(s) (3.X b:R^1=H,$R^2 \neq$H) (3.X c:$R^1 \neq$H,R^2=H)				Ref.
No	R^1,R^2		R^1	R^2	Y(%)	ee(%)	

| 3.414 | X:CH_2 | Bu PLE | Bu | H | | 47 | [503] |
| 3.415 | X:O | Me PLE | Bu | H | | 71 | [503] |

3.416	X:H	Me PLE	H	Me		60	[506]
3.417	X:OH	Me PLE	Me	H	90	>95	[35]
3.418	X:OBu^t	Me PLE	no hydrolysis				[35]

| 3.419 | 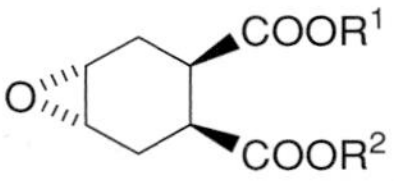 | PLE | 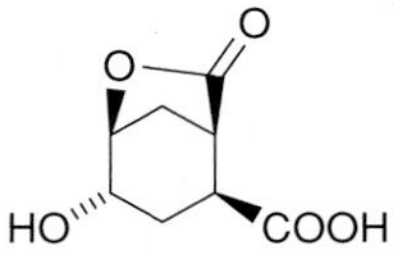 | | 72 | 96 | [507] |

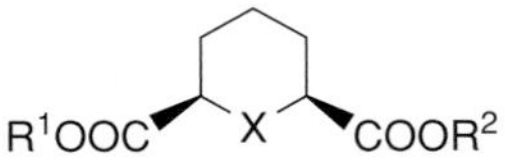

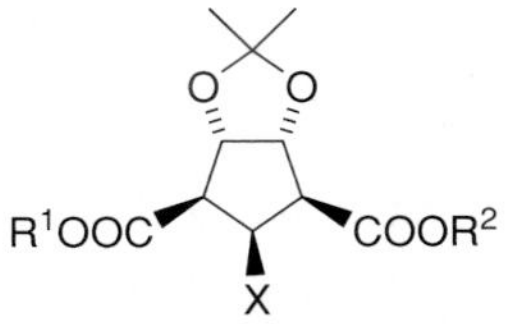

3.420

| 3.421 | 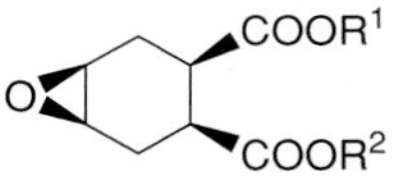 | PLE | no hydrolysis | | | | [507] |

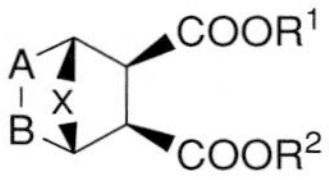

	X	A-B						
3.422	CH_2	CH=CH	Me	PLE	H	Me	85	<10	[508]
3.423	O	CH=CH	Me	PLE	H	Me	86	75	[508]
			Me	PLE^h	H	Me		82	[90]
3.424	O	CH_2-CH_2	Me	PLE	H	Me	82	>98	[508]

| 3.425 | O | 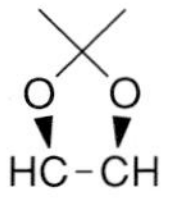 HC–CH | Me | PLE | very slow hydrolysis | | | [509] |

Table 3.19. (contd.)

Substrate (3.X a)		Enzyme	Product(s) (3.X b:R¹=H,R² ≠H) (3.X c:R¹ ≠H,R²=H)			Ref.
No	R^1, R^2		R^1	R^2	Y(%) ee(%)	

	X	A-B					
3.426	CH₂	CH=CH	Me	PLE	no hydrolysis		[508]
3.427	O	CH₂-CH₂	Me	PLE	H	Me	87 64	[508]
3.428	O	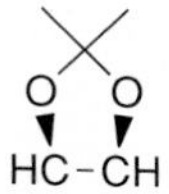 (acetonide)	Me	PLE	no hydrolysis		[510]

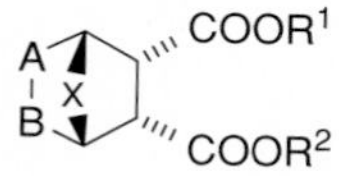

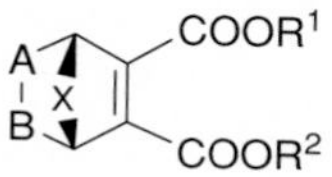

	X	A-B						
3.429	CH₂	CH=CH	Me	PLE	Me	H	77 36	[511]
3.430	CH₂	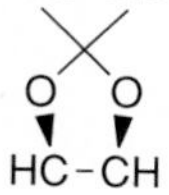 (acetonide)	Me	PLE^e	H	Me	100 80	[512]
			Et	PLE^e	H	Et	30 100	[21]
3.431	O	 (epoxide)	Me	PLE	H	Me	100 77	[509,513]
3.432	O	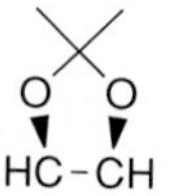 (acetonide)	Me	PLE	H	Me	96 77	[509,513]

| 3.433 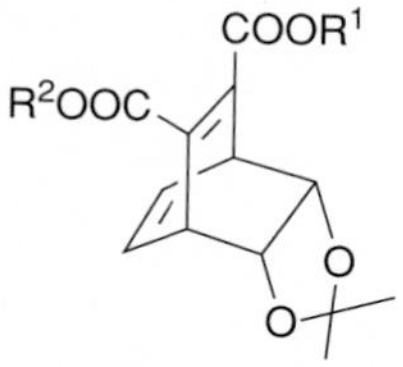 | | Me | PLE | Me^i | H^i | 88 i,j | [514] |

---

a: 10% MeOH; b: 25% DMSO; c: At pH 8; d: 10% acetone; e: Pig liver acetone powder; f: Configuration is not given; g: 25% DMSO; h: 10% t-BuOH; i: Configuration is tentative; j: Enantiomeric purity is not given, $[\alpha]_D^{25} = -34°(CH_2Cl_2)$.

A study on 1,2-*cis*-cycloalkane dicarboxylic esters revealed that ring size exerted a strong influence on selectivity. With the cyclopropane and cyclobutane derivatives (3.394) and (3.397) respectively, high *S* selectivity was found to occur, i.e. preferential hydrolysis of the *S* configured ester group. Hydrolysis of the corresponding cyclopentane (3.398) however, proceeds with opposite but very poor selectivity and finally the cyclohexane ester (3.402) showed high *R* selectivity (3.398 and 3.402 respectively) [34,496]. The sense of selectivity yielded in these and other examples can be explained using models of the active center of PLE (Fig. 3.27).[13]

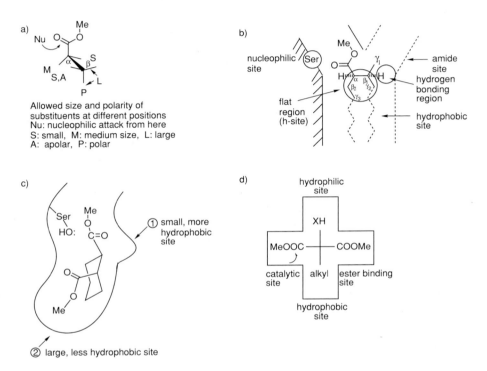

a)

Allowed size and polarity of substituents at different positions
Nu: nucleophilic attack from here
S: small, M: medium size, L: large
A: apolar, P: polar

b)

nucleophilic site (Ser)
flat region (h-site)
Me
amide site
hydrogen bonding region
hydrophobic site

c)

Ser
HO:
Me
C=O
O
Me

① small, more hydrophobic site

② large, less hydrophobic site

d)

hydrophilic site
XH
MeOOC — COOMe
catalytic site | alkyl | ester binding site
hydrophobic site

Fig. 3.27.   Active center models for PLE: a) *Tamm* and coworkers [34]; b) *Ohno* and coworkers [13,33,35]; c) *Johnes* and coworkers [496,503]; d) *Zemlicka* and coworkers [35]

It is important to note that selectivity is much dependent on the alcohol component of the ester. Thus with cyclohexane and 4-cyclohexene-1,2-dicarboxylic acid derivatives

---

[13] Very recently a new and detailed active center model for PLE was published [565]. The validity of these models are limited by the fact that PLE is a mixture of isoenzymes and the manner in which it functions is much influenced by experimental conditions [27,507]. Primarily, these models serves to provide some orientational guidance about the fitting of substrates to active site, and to provide a qualitative prediction about the sense of selectivity likely for substrates that do not differ too greatly from those used to set up the model.

(3.401 and 3.402), taking the ethyl esters instead of the methyl esters dramatically reduces selectivity [21,301], whereas with the tricyclic esters (3.430) and the 2-imidazolidon esters (3.405) hydrolysis of the propyl esters was more selective than that of the methyl esters. On the other hand, with respect to reaction rates the trend is clear: increasing the bulk of the alcohol component diminishes rates of raction [21].

Selectivity is also pH dependent, being generally better at pH 7 than at pH 8 [274].

For a given ring size, substitution at the ring significantly modifies the outcome of the hydrolysis of 1,2-disubstituted dicarboxylic esters. Thus, the 3,3-dimethylcyclopropane ester (3.395) hydrolyzes several orders of magnitude more slowly than the parent compound (3.394), and in addition selectivity is much lower. Introduction of an oxygen into the ring, as in the case of oxirane (3.396.a), owing to a change of polarity, inverts the sense of selectivity. Selectivity in the hydrolysis of cyclopentane-1,2-dicarboxylic ester itself (3.398.a) is rather poor, and the situation is not much changed by the introduction of a small and moderately polar substituent (e.g. methylene, acetoxy, methoxy) at C-4, especially if it is disposed *trans* to the ester groups [27]. A sharp increase in *R* selectivity is effected, however, if this substituent is small and polar (e.g. 4-oxo, 3.399) and *cis* oriented (e.g. *cis*-4-hydroxy, 3.400). With the *trans*-4-hydroxy and 4-acetoxy compounds, selectivity was low and of the opposite sense. When polarity of the 4-substituent is reduced or its size larger, *R* selectivity decreases, or may even invert to *S*. For example, the *cis*-4-*tert*-butoxy compound (3.401.a) hydrolyzes with high *S* selectivity.

With the six-membered cycles, *R* selectivity increases in comparison to the parent compound (3.402.a) when a double bond is introduced at position 4 (3.403.a).

High enantiotope selectivity was recorded in the hydrolysis of the ferrocene ester (3.404.a) providing a product (3.404.b) that possesses chirality of the rare planar type.

It is also possible to attain good selectivity with cyclic *meso*-bis(methoxycarbonyl-methyl) compounds (3.406-3.409). Interestingly, in the case of the cyclopentanone (3.407) and the cyclohexene (3.409), CTR[14] and PLE enzymes acted with opposite selectivity, whilst PPL was more selective than either of them.

Cyclic 1,3-dicarboxylic esters are substrates for both PLE and PPL enzymes and the two often show opposite, but generally low selectivities in hydrolysis [503] such as that reported with the tetrahydrofuran ester (3.411.a). In PLE catalyzed transformations, substituent polarity is again an important factor. Thus hydrolysis of the cyclopentane diester (3.410.a) and its heterocyclic analogs (3.411-3.413) proceeds with opposite selectivity. The situation is similar with analogous bicyclic systems; with the apolar model (3.416a, X=H) selectivity corresponds to that of the cyclopentane esters, whilst the analog possessing a free hydroxy group (3.410.a) hydrolyzes with opposite selectivity.[15] It is also interesting to note, that of the two diastereomeric 4,5-epoxides of 1,2-*cis*-cyclohexanedicarboxylic esters, the *cis*-epoxide (3.380.a) is not a substrate for PLE, but

---

[14] In general cyclic 1,2-dicarboxylic esters are not substrates of CTR enzyme.

[15] Owing to a change of priority in substituents, the stereodescriptor remains unchanged.

the methoxycarbonyl group of the *trans*-epoxide (3.419.a) hydrolyzes with excellent *R* selectivity. After lactone formation from the monoester by nucleophilic attack the second ester group takes up an equatorial position (in the case of cyclohexanecarboxylic esters, PLE hydrolysis of equatorial esters is preferred [515]) and is now hydrolyzed by the enzyme giving rise finally to a bicyclic lactone (3.420) [507].

Stereostructure and polarity of the substituents also play an important role in the hydrolysis of the bi- and tricyclic esters (3.422-3.433). Thus selectivity is poor with the methylene bridged *endo*-ester (3.422.a), but high with the oxygen bridged analog (3.423.a). Steric factors may help to explain the fact that, while the bicyclic diesters (3.423) and (3.424.a) are good substrates, the tricyclic compound (3.425.a) fails to hydrolyze at all. Similar effects are operative in the hydrolysis of the *exo*-dicarboxylic esters (3.426-3.428.a). Compounds containing an apolar bridge, like (3.426.a), or a bulky tricyclic structure are not substrates for PLE.

It is notable that analogs where the ester groups are attached to the sp^2 carbons of a double bond (e.g. 3.429-3.433.a) are less susceptible to steric and polar effects. One possible explanation may be the thoroughly different conformations of these models.

In connection with carbacyclin synthesis an interesting enantiotope selective demethoxycarbonylation of a bicyclic keto diester was observed [635].

Half-esters obtained in high enantiomeric purity by the procedures above are valuable and versatile synthetic intermediates. The half esters (3.392.c) and (3.393.b) are synthons in the pathway to a precursor of monensin A [314,326] and various antibiotics, [517,518,636] respectively.

The half ester (3.403.b) is a striking example of the versatility of enzymatically produced chiral building blocks. It was used in the synthesis of β-lactam antibiotics [504,519,520], (−)-fortamine [521,522], vitamin D intermediates [523] and metabolites [637], prostaglandin precursors [512,524,525], carbacyclin [526-528], isocarbacyclin [529], brefeldin A [530,531], pentalenolactone [530,532], and intermediates of compactin [638]. Hydrolysis products of the diesters (3.399-3.401.a) and (3.407-3.409.a) are also valuable intermediates in the synthesis of cyclopentanoids.

The imidazolidone (3.405.b) is a precursor of (+)-biotin [450] and the oxirane (3.406.c) that of nonactinic acid [500].

ii) Enantiotope selective transformations of meso-diol derivatives (Table 3.20)

Hydrolase catalyzed enantiotope selective transformation of *meso*-diol derivatives can be carried out either by the hydrolysis of the diesters or by (trans)esterification and acylation of the diols. As with diol derivatives containing a prochiral centre, enantiomeric

Table 3.20. Enantiotope selective enzyme catalyzed transformations of *meso*-diol derivatives.

Substrate (3.X a:R¹, R² ≠H) (3.X b:R¹, R²=H) No	R¹,R²	Enzyme	Reaction partner[a] Cosolvent[b]	Product(s) (3.X b:R¹=H, R²≠H) (3.X c:R¹ ≠H,R²=H) R¹	R²	Y%	ee%	Ref.
3.434 R¹O〜〜OR²	Ac	PPL		H	Ac		90	[533]
	Ac	PLE		Ac	H	15	95	[533]
3.435 R¹O (NO₂) OR²	Ac	PLE[c]		H[d]	Ac[d]	50	>90[d]	[534]
3.436 n= 1	Ac	PfL		racemic				[374]
	H	EcβG[e]	GlucOPh[f]	Gluc[f]	H	30	89[g]	[535]
3.437 n= 2	Ac	PfL		racemic				[376]
	Ac	PLE		racemic				[404]
	H	PPL	MeOAc	H[d]	Ac[d]		84[d]	[536]
	H	EcβG[e]	GlucOPh[f]	Gluc[f]	H	10	90[g]	[535]
3.438 A-B: CH=CH	H	EcβG[e]	GlucOPh[f]	Gluc[f]	H	24	75[g]	[535]
3.439 A-B: CH₂CH₂	H	EcβG[e]	GlucOPh[f]	Gluc[f]	H	28	63[g]	[535]
3.440	Ac	PPL		Ac	H	94	72	[537]
	Ac	PPL	Ca²⁺,h	Ac	H	74	93	[503,538]
	Ac	PLE		Ac	H	54	44	[537]
	Ac	PSL		Ac	H	83	90	[539]
	H	PsL	CH₂=CHOAc	H	Ac	80	>95	[539]
3.441	Ac	PPL		Ac	H	75	40	[537]
	Ac	PLE		H	Ac	69	20	[537]
3.442	Ac	PPL		Ac	H	97	88	[537]
	Ac	PPL	Ca²⁺,h	Ac	H	78	96	[503,538]
	Ac	PLE		Ac	H	62	0	[540]
	Ac	PsL		Ac	H	87	>95	[539]
	H	PsL	EtOAc[i]	H	Ac	82	>95	[539]

Table 3.20. (contd.)

Substrate (3.X a:R^1, R$^2 \neq$H) (3.X b:R^1, R^2=H) No		R^1,R^2	Enzyme	Reaction partner^a Cosolvent^b	Product(s) (3.X b:R^1=H, R$^2 \neq$H) (3.X c:R$^1 \neq$H,R^2=H) R^1	R^2	Y%	ee%	Ref.
3.443	OR^1 / OR^2 (cyclobutene)	Ac	PPL^j		Ac	H	57	86	[536]
3.444	OR^1 / OR^2 (cyclopentane)	Ac	PPL		Ac	H	94	88	[537]
		Ac	PPL	Ca$^{2+,h}$	Ac	H	90	89	[503,538]
		Ac	PLE		H	Ac	40	8	[537]
		Ac	PsL		Ac	H	86	>95	[537]
		H	PsL	CH$_2$=CHOAc	H	Ac	85	>95	[539]
3.445	O= OR^1 / OR^2 (cyclopentanone)	Ac	PPL		Ac	H	70	50	[536]
3.446	O,O-dioxane spiro OR^1 / OR^2	Ac	PPL^k		Ac	H	78	94	[536]
		H	PPL^j	MeOAc	H	Ac	74	68	[536]
3.447	Cl'''' OR^1 / OR^2	Ac	PPL		Ac	H	81	88	[536]
		H	PPL^j	MeOAc	H	Ac	79	84	[536]
3.448	AcO''' OR^1 / OR^2	Ac	PPL		Ac	H	60	90	[536]
3.449	PhS''' OR^1 / OR^2	Ac	PPL		Ac	H	65	96	[536]
3.450	OR^1 / OR^2 (cyclohexane)	Ac	PPL		Ac	H	81	78	[537]
		Ac	PPL	Ca$^{2+,h}$	Ac	H	67	87	[503,538]
		Ac	PLE		Ac	H	31	4	[537]
		Ac	PsL		Ac	H	54	50	[539]
3.451	OR^1 / OR^2 (cyclohexene)	Ac	PPL		Ac	H	96	>99	[320,537]
		Ac	PsL		Ac	H	90	92	[539]
		H	PsL	CH$_2$=CHOAc	H	Ac	67	88	[539]
		Ac	PLE		H	Ac	40-55		[90,537]
		Ac	PLE	20% DMSO	H	Ac	62	59	[90]
		Ac	PLE	20% DMF	H	Ac	74	84	[90]
		Ac	PLE	5% t-BuOH	H	Ac	76	94	[90]
		Ac	PLE	10% t-BuOH	H	Ac	78	96	[90,540]

Table 3.20. (contd.)

Substrate (3.X a:R¹, R² ≠H) (3.X b:R¹, R²=H) No	R¹,R²		Enzyme	Reaction partner[a] Cosolvent[b]	Product(s) (3.X b:R¹=H, R²≠H) (3.X c:R¹ ≠H,R²=H) R¹	R²	Y%	ee%	Ref.
3.452	Ac	Ac	PLE		H	Ac	57		[90]
	Ac	Ac	PLE	10% t-BuOH	H	Ac	68		[90]
3.453	Ac		PLE	10% MeOH	H	Ac	70	92	[480]
3.454	Ac		PLE		H	Ac	85	81	[533]
	Ac		PLE[k]		H	Ac	87	75	[38]
	Ac		S.c[l]		H	Ac	87	74	[541]
	Ac		MmL		Ac	H	45	97	[542]
	Ac		PPL		Ac	H	44	92	[542]
	COPr		PPL		COPr	H	44	95	[542]
	Ac[m]		PPL		Ac	H	75[n]	100°	[543]
	Ac[m]		PPL	EtOH/hexane	Ac	H	53[n]	100°	[543]
	Ac		AChE		Ac	H	94	96	[544]
	H		P[p]	Cl₃CCH₂OAc	H	Ac	48	95	[545]
3.455	Ac		AChE		Ac	H	80	98	[546,547]
3.456 X: C=O	COPr		PLE		racemic				[548]
	COPr		CcL		racemic				[548]
3.457 X: C(OBn)(H)	Ac		PLE		Ac	H	81	95	[548]
3.458 X: C(OBn)(H)	Ac		PLE		H	Ac	70	80	[548]
	COPr		PLE		racemic				[548]
3.459	COPr		CcL		H	COPr	60	98	[549]

Table 3.20. (contd.)

Substrate (3.X a:R¹, R² ≠H) (3.X b:R¹, R²=H) No	R¹,R²	Enzyme	Reaction partner[a] Cosolvent[b]	Product(s) (3.X b:R¹=H, R²≠H) (3.X c:R¹ ≠H,R²=H) R¹	R²	Y%	ee%	Ref.
3.460 (cyclohexane: OBn, BnO, OBn, R¹O, OR²)	COPr	CcL		H	COPr		95	[549]
3.461 (NO₂-cyclohexane: R¹O, OR²)	Ac	PLE[c]		H	Ac	50	>90	[534]
(cycloheptene: R¹O, OR², X, Y) X Y								
3.462 H H	Ac	AChE		H	Ac	39	100	[550]
3.463 CH₃ H	Ac	CcL		Ac	H	61	100	[550]
3.464 H OH	Ac	AChE		H	Ac	79	>95	[551]
(5-membered ring with X: R¹O, OR²)								
3.465 X: CH₂	COPr	PPL		H	COPr		35	[503]
3.465 X: C(butyl)H	Ac	RdL		H	Ac	62	96	[552]
3.467 X: C(alkenyl)H	Ac	RdL		H	Ac	69	>99	[552]
3.468 X: O	Ac	PLE		H	Ac	86	96	[553]
	Ac	CcL		Ac	H	20	80	[553]
	COPr	MjL		H	COPr	75	>99	[554]
3.469 (tetrahydrofuran ring with O: R¹O, OR²)	COPr	MjL		H	COPr	60	>99	[554]
	COPr	PfL		H	COPr	84	99	[554]
	COPr	RnL		H	COPr	78	96	[554]
(6-membered ring with X: R¹O, OR²)								
3.470 X: CH₂	COPr	PLE		COPr	H		30	[503]
3.471 X: O	COPr	PPL		COPr	H		55	[503]
	COPr	PLE		COPr	H		41	[503]

Table 3.20. (contd.)

Substrate (3.X a:R^1, R^2 ≠H) (3.X b:R^1, R^2=H) No	R^1,R^2	Enzyme	Reaction partner^a Cosolvent^b	Product(s) (3.X b:R^1=H, R^2≠H) (3.X c:R^1 ≠H,R^2=H) R^1	R^2	Y%	ee%	Ref.
3.472	Ac	PfL		H	Ac	79	96	[556]

a: For acylations; b: For hydrolyses; c: Pig liver acetone powder; d: Configuration is not given; e: Glucosidation by β-glucosidase from *Escherichia coli*; f: Diastereotope selective process; g: Diastereomeric excess; h: Additive; i: Addition of vinyl acetate gave lower enantiomeric purity; j: Adsorbed on SiO$_2$; k: Purified and immobilized enzyme; l: *Saccharomyces cerevisiae*; m: *cis/trans*=58:42 mixture; n: Based on the *cis*-content; o: After recrystallyzation; p: Pancreatin.

purity of the product depends on conversion. As exemplified by the cyclic *meso*-diols (3.436-3.439), enzymic glycosidation can also be conducted in an enantiotopic manner with respect to the diol component.

With the open-chain diester (3.434.a), selectivity of PLE and PPL enzymes is complementary leading to enantiomeric products (3.434.c or .d).

Enzymic hydrolysis of the diacetates (3.434.a) and (3.437.a) failed to produce anything other than racemic material with any of the enzymes tried, but this was probably due to racemization by fast acyl migration in the monoacetate products rather than to an absence of selectivity in the hydrolysis step [404]. Transesterification [536] or glycosidation [535] of the appropriate diols (e.g. of 3.437.b) can, however, give rise to optically active monoalcohols.

Diacetates of cyclic *cis*-1,2-bis(hydroxymethyl) compounds may be substrates of PLE [537], PPL [537,538] and PsL enzymes [539]. In the case of the cycloalkanes of this type (3.440, 3.442.a, 3.444.a, 3.450.a) both chemical and optical yields of the monoalcohols were poor when using PLE enzyme, in contrast to the lipases (PPL, PsL) which showed high *S* selectivity (cf. Section *i*) in this chapter) with such substrates[16]. With PsL enzyme, both ester hydrolysis and acylation of diols has been studied [539] showing that the two processes led in all cases to enantiomeric monoacetates. Substituent effects have been reported in these reactions too. With PPL enzyme, for instance, dimethyl substitution of a three-membered ring (3.441) decreases both rate and selectivity, whilst in a four-membered ring unsaturation of the ring (3.443) only slightly impaired selectivity. In the hydrolysis of five-membered ring substrates, introduction at C-4 of a carbonyl group (3.445) or other polar groups (HO, AcO, MeO, EtO) in *cis* disposition to the ester group decreases selectivity, and whilst in *trans* configuration small substituents (Cl: 3.447.a,

---

[16] It has been reported that under optimized conditions hydrolysis of *cis*-1,2-bis(butyryloxymethyl)cyclopropane by PPL enzyme was highly enantiotope selective [640]. Enzymatic alcoholysis of *meso*-bis(acetoxymethyl) aziridines [641], and acylation of *cis*-endo-5-norbornene dimethanol [644] have also been reported recently.

AcO: 3.448.a, 3.449.a) have no influence, bulky ones improve selectivity [536]. In ketals and thioketals, which can be regarded as 4,4-disubstituted compounds, selectivity is variable, e.g. with (3.446.a) it is seen to increase [536]. In six-membered cylic compounds, PPL catalyzed hydrolysis of the saturated model (3.450.a) is less selective than that of its unsaturated counterpart (3.451.a), the latter giving an enantiomerically almost pure product probably due to a more rigid conformation of the cyclohexene ring. Note that in the four-membered ring series, where unsaturation essentially does not influence conformation, it has no effect on selectivity in PPL catalyzed hydrolysis either.

The effect of cosolvents on selectivity may be illustrated by PLE hydrolysis of the diacetate (3.451.a).

Enantiotope selectivity of hydrolases can also be put to use in the transformation of cyclic *meso*-1,3-diol derivatives. Data are mainly available for the hydrolysis of diesters, here the monoesters are generally obtained in high enantiomeric purity, since in this case intramolecular acyl migration, occurring with the 1,2-diols, is precluded. An intensively studied substrate was *cis*-cyclopent-1-en-3,5-diol (3.454). After testing several enzymes it was concluded that from the point of view of yields and purity AChE enzyme was the agent of choice, whilst from an economic standpoint PPL enzyme was the most advantageous. With PPL enzyme all the possible types of transformations, i.e. hydrolysis, alcoholysis and transesterification were tested.

Hydrolysis of 5-substituted *cis*-1,3-cyclohexanediol esters revealed some interesting features. While the 5-oxo derivative (3.456.a) gave a racemic product with both PPL and CcL, the benzyloxy compound (3.456.a) could be hydrolyzed with high selectivity using PLE enzyme [642]. Note that the nature of the acyl component is significant, e.g. hydrolysis of the corresponding dibutyrate was non-selective. Unexpectedly PLE catalyzed hydrolysis of the 5-benzyloxymethyl substituted diester (3.458.a) was also highly selective but in the opposite sense.

When the 1,3-diol derivatives were highly substituted (e.g. 3.459, 3.460), CcL was deemed the enzyme of choice for enantiotope selective hydrolysis. It was surprising that enantiotope selectivity remained high even when the relative configurations of neighboring bulky benzyloxy groups were inverted.

In the case of 2-nitro-1,3-diacetoxy-cyclohexanes PLE proved to be highly selective [643].

cis-Cycloheptenediol derivatives (3.462 and 3.464) featuring, as *cis*-3,5-diacetoxy-1-cyclopentene (3.454.a), a bis(allylester) functionality, were also well accepted by AChE enzyme, in contrast to the dimethyl analog (3.463.a) which was not. For the latter, CcL enzyme proved to be a more useful catalyst.

For the transformation of cyclic *cis*-1,3-bis(acyloxymethyl) compounds (3.465-3.472), microbial lipases are promising, although in the hydrolysis of the tetrahydrofuran diester (3.468a) PLE proved to be highly selective too. Additionally, opposite selectivities can also be realized in this class of substrates by hydrolysis with either PLE or CcL. Choice of the proper ester is sometimes imperative. Hydrolysis of the diacetate (3.468a) with CcL enzyme for example proceeded with good *S* selectivity, whereas with the corresponding dibutyrate, poor *R* selectivity was recorded [554].

The optically active half esters obtainable from cyclic *cis*-1,2-bis(hydroxymethyl) derivatives can be used for the same purposes as those resulting from the corresponding dicarboxylic esters (Fig. 3.28). The half esters (3.444-3.449), (3.454), (3.455) and (3.465-3.467.c or d) are intermediates in the syntheses of biologically active cyclopentanoids. Thus enzymic transformation of the diacetate (3.454.a) provides intermediates for prostanoids [542-544], so for example, using the half ester (3.455.d) a triply convergent synthesis of (−)-prostaglandin E$_{2\alpha}$ can be realized [546,547].

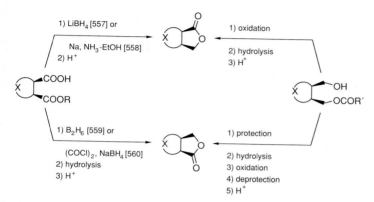

Fig. 3.28.   Transformation of optically active *cis*-1,2-dicarboxylic acid monoesters and *cis*-1,2-bis(hydroxymethyl) monoacyl compounds to the same enantiomeric lactones via chemical steps

The cyclohexene-dimethanol-monoacetate (3.455.d) is a chiral building block in the synthesis of a yohimbine alkaloid [540] and the imidazolidon monoacetate (3.453.c) serves in another approach to (+)-biotin [480]. The monoacetate (3.464.c) is a synthon for the lactone moiety of anticholesteremic HMG-CoA inhibitors [551], whilst the monobutyrates (3.468) and (3.469.c) are synthons for the preparation of enantiomerically pure PAF antagonists [554]. Another monobutyrate (3.471.d) was processed to the perfume of civet cat [555], and the monoacetate (3.472.c) converts to one of the stereoisomers of 2-hydroxy-4-hydroxymethyl-4-butanolide, a hunger modulating agent [556].

3.3.6.3   Transformation of Enantiotopic and Diastereotopic Faces

In this section some hydrolase catalyzed additions to activated carbon-carbon double bonds are discussed. Such reactions can be enantiotope selective with achiral substrates possessing enantiotopic faces and enantiomer or diastereotope selective when the substrate is chiral (Fig. 3.29).

In Fig. 3.29.a, a number of enantiotope selective Michael-additions are presented [85]. Hydrolase catalyzed addition to 2-trifluoromethylacrylic acid seems to have rather a wide scope of reagent combinations showing high enantiotope selectivity with a variety

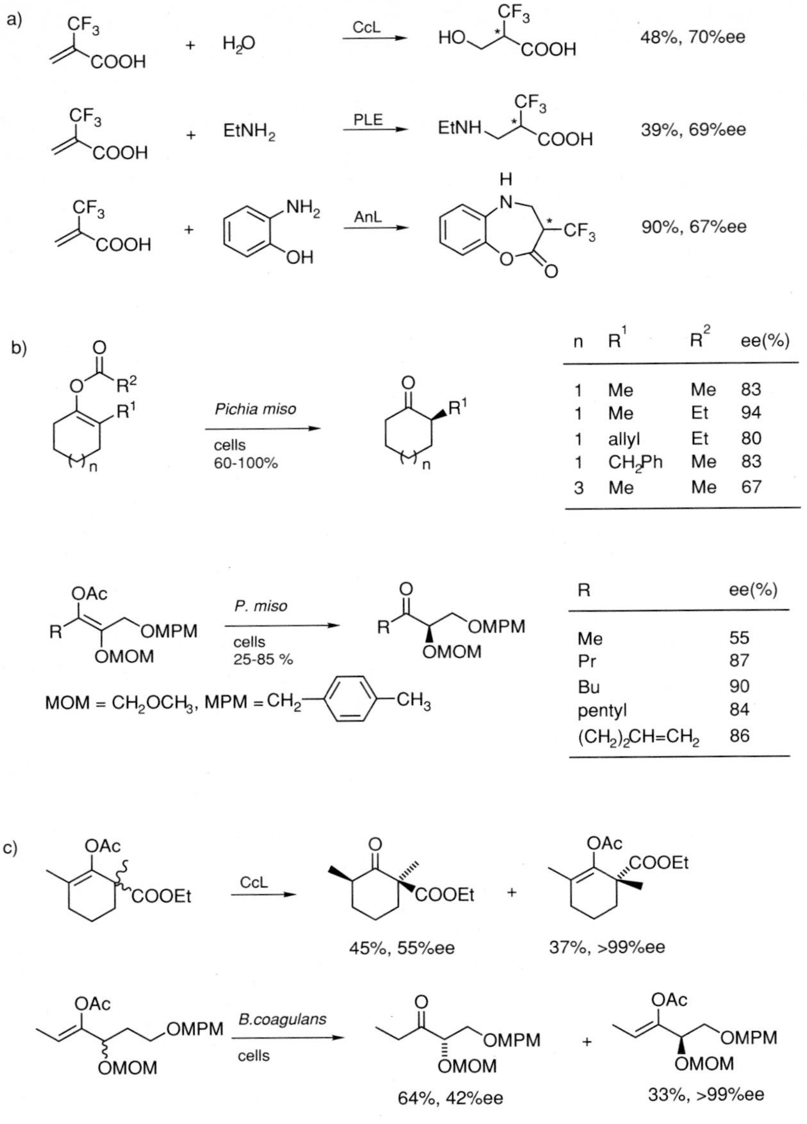

n	R^1	R^2	ee(%)
1	Me	Me	83
1	Me	Et	94
1	allyl	Et	80
1	CH_2Ph	Me	83
3	Me	Me	67

R	ee(%)
Me	55
Pr	87
Bu	90
pentyl	84
$(CH_2)_2CH{=}CH_2$	86

Fig. 3.29. Hydrolase catalyzed selective transformations of enantiotopic (a), b) or diastereotopic (c) faces of sp^2 carbons

of nucleophiles and hydrolases. The preference for one of the faces in this and the following examples may be explained by assuming that the enolate produced by nucleophilic attack against the acyl-enzyme intermediate is protonated when it is still bound to the enzyme.

In Fig. 3.29.b, enantiotope selective microbial hydrolyses of achiral cyclic [563, 645] and open chain [564] enolesters are listed. In these processes, protonation of the enolate again proceeds preferentially from one direction (the proton source may be one of the acidic groups of the enzyme catalyzing the transformation [563,564]).

Fig. 3.29.c shows the microbial and enzyme catalyzed hydrolysis of chiral but racemic cyclic [561] and open chain [562, 646, 647] enolesters. These reactions are both enantiomer and diastereotope selective. Here, CcL catalyzed hydrolysis of the cyclic model indicates that such transformations are indeed catalyzed by a single hydrolase.

4 Oxidoreductases: Preparative Biocatalysis

Although many synthetically useful reactions can be carried out with oxidoreductases, it seems that their use for preparative purposes has been much less popular than that of hydrolases or baker's yeast. The reasons behind this are probably related to both the scarcity of easy to handle microorganisms performing oxidoreductive transformations and the fact that the target compounds are often more readily accessible using other biocatalytic approaches.

The general class of oxidoreductases (EC 1) has been subdivided into subclasses according to the nature of the group serving as hydrogen donor, the subclasses being coded in the form EC 1.m, where m represents the type of donor group. E.g. m=1: CH-OH; m=2: C=O; m=3: CH-CH; m=4: CH-NH$_2$; m=5: CH-NH; and m=6: NAD(P)H. Further subdivision (i.e. EC 1.m.n) is determined by the nature of the acceptor group. Thus n=1: NAD(P)+; n=2: cytochrome; n=3: molecular oxygen; n=5: quinones; n=7: iron-sulfur; n=8 flavin [1].

In terms of practical applications oxidoreductases can be divided into two large groups namely enzymes which require added cofactors for their *in vitro* functioning and those which do not [2]. In the latter group cofactors are bound either covalently to the enzyme protein or are automatically regenerated in the catalytic process.

4.1 Oxidoreductases Not Requiring Added Cofactors

Oxidoreductases functioning without added cofactors can be found predominantly among oxidases, peroxidases, and oxygenases. Oxidases use molecular oxygen as the electron acceptor and produce hydrogen peroxide, peroxidases act on hydrogen peroxide as electron acceptor, and oxygenases reductively activate oxygen for insertion into the substrate. Dioxygenases and monooxygenases can be distinguished on the grounds of whether both or only one of the oxygen atoms of molecular oxygen is incorporated into the substrate [3].

The enzymes all mentioned above contain in a more or less firmly bound state, either a redox-active transition metal such as copper or iron (the latter in heme or non-heme environment), or a non-metal cofactor (pterin of flavin coenzyme, e.g. FAD or FMN). Sometimes other ions are also needed.

R = H: riboflavin (vitamin B$_2$)
R = PO$_3^{2-}$: FMN
R = ADP: FAD

The oxidoreductases discussed here, in order of ascending EC numbers, are those which are or may become in some way important in preparative chemistry.

Glucose oxidase (EC 1.1.3.4) is a flavoenzyme widely used in the food industry as an antioxidant [4] although it also has a number of analytical applications [5]. It is worth mentioning that under anaerobic conditions the enzyme catalyzes electron transfer from glucose to electron acceptors such as benzoquinone [6].

Galactose oxidase (EC 1.1.3.9) contains copper and, in addition to galactose, oxidizes several other alcohols too. 2-Methyl-1,3-propanediol and dihydroxyacetone are oxidized even more rapidly than galactose itself [7]. The enzyme can be utilized for the regio- and stereoselective oxidation of primary hydroxyls in glycerol and 3-halogeno-1,2-propanediols [8].

X = OH, Cl, Br

Several unusual L-sugars were prepared via selective polyol oxidation by galactose oxidase. However, this approach suffers from the serious drawback of product inhibition [9].

Xanthine oxidase (EC 1.2.3.2) is a flavoprotein catalyzing the transformation: xanthine + H$_2$O + O$_2$ → uric acid + H$_2$O$_2$. The enzyme is also capable of catalyzing the oxidation of other purines, pterines and aldehydes; as a special application of this being the purification of a mixture of substituted benzaldehydes by selective oxidation [11].

Both *L- and D-Amino acid oxidases* (EC 1.4.3.2 and EC 1.4.3.3) are flavoproteins and catalyze the oxidative conversion of the corresponding amino acid enantiomers to α-keto acids [1].

Snake venoms often exhibit L-amino acid oxidase activity, but rat kidney is also a good source of the enzyme. D-amino acid oxidase activity has been demonstrated in many vertebrates, hog kidney being the most commonly used source. These enzymes are useful for the enantiomer selective oxidation of racemic amino acids, as exemplified by the treatment of racemic *syn-β*-imidazolyl-serine with D-amino acid oxidase to yield as the useful product the L-amino acid, a precursor of bleiomycin [12]:

D-Amino acid oxidase coupled with leucine DH in a multienzyme system was used to convert racemic methionine, alanine or leucine to their L-enantiomers [169].

Catalase (EC 1.11.1.6) is a hemoprotein peroxidase catalyzing the reaction $2H_2O_2 \rightarrow 2H_2O + O_2$. Catalase activity is often present in mammalian liver but can also be seen in microorganisms [13]. As electron donors various alcohols, including ethanol, are utilized and this reaction controls H_2O_2 decomposition until the latter remains at low concentration [14], a property that leads to catalase frequently being used in coupled enzyme systems to decompose H_2O_2 that has formed and to regenerate NAD(P)H respectively. An interesting further application of the enzyme is to effect ring closure to yield heterocyclic compounds [15] (similar cyclizations having also been reported with the use of rat liver microsomes [16,17]).

X = NH, NMe, NPh, O

Horseradish peroxidase (EC 1.11.1.7, HPO) is also a hemoprotein acting on H_2O_2 as an electron acceptor. In nature, peroxidases catalyze the oxidative coupling of phenols to give lignin. In vitro, phenolic coupling in aqueous media by HPO leads to low molecular weight oligomers (e.g. 4,4'-dihydroxybiphenyl from phenol) due to poor solubility and precipitation of higher molecular weight products [18]. With the aid of HPO, biologically active compounds such as L-DOPA or L-epinephrine can be prepared by hydroxylation of the corresponding phenols (Fig. 4.1) [19]. In media containing a high percentage of organic solvents oxidative phenol coupling tends to yield higher molecular weight products, thus, through the manipulation of solvent composition, providing a means to control the degree of polymerization of the product [20]. HPO retains its activity even in media of very low water content [21]. Through chemical modification it can become soluble in organic solvents but still retain its activity [22].

Chloroperoxidase (EC 1.11.1.10) again is a heme (probably heme thiolate) protein and can also act on Br and I , but not on F. (It was however reported [23], that organic dihalide mixtures, including compounds containing C-F bonds, were generated in the reaction between olefins and inorganic halides catalyzed by iron containing peroxidases.) The enzyme catalyzes halonium cation transfer in a halide-H_2O_2 system and oxygen transfer from H_2O_2 to a wide selection of organic substrates [24]. Thus it can catalyze halogenation of phenyl-methylpyrazolone (Fig. 4.2.a) [25], the stereoselective formation of a bromohydrin from an unsaturated steroid (Fig. 4.2.b) [26] and transform various methyl sulphides into optically active sulphoxides (Fig. 4.2.c) [28]. (For a recent review on biocatalytic oxidation of organic sulphides see ref. [27]).

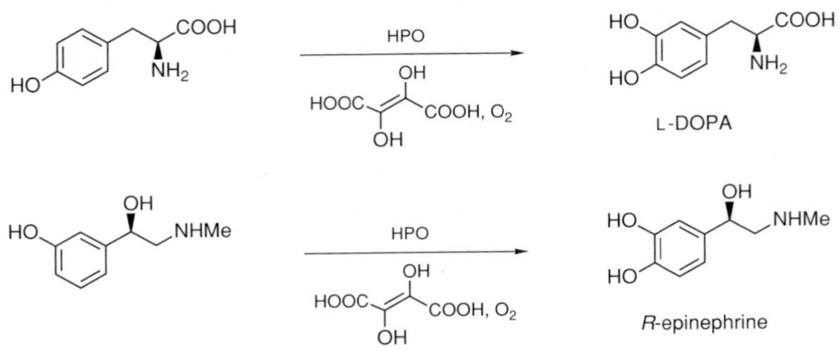

Fig. 4.1. Preparation of L-DOPA and L-ephinephrine by horseradish peroxidase catalyzed hydroxylations

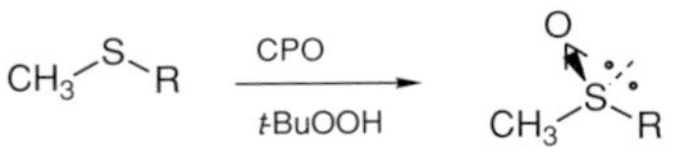

R	Yield (%)	ee(%)
Ph	~100	76
p-Me-Ph	60	86
p-MeO-Ph	70	92
Ph-CH₂	51	91
Bu	54	38

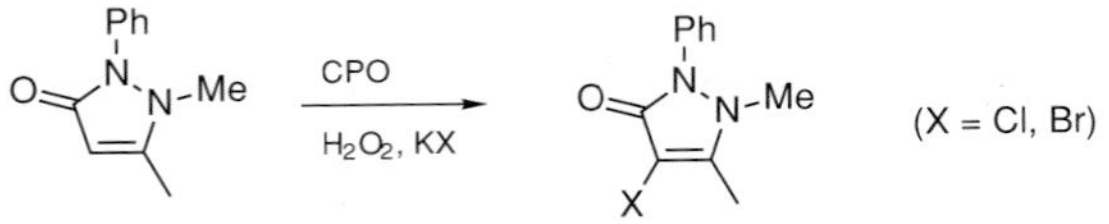

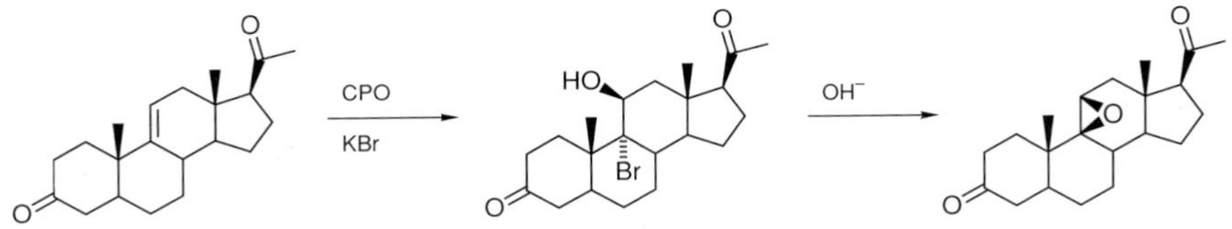

Fig. 4.2. Reactions catalyzed by chloroperoxidase

Lipoxygenase (EC 1.13.11.12) is a non-heme iron dioxygenase which catalyzes the addition of dioxygen into polyunsaturated fatty acids [29]. The synthetic applications of lipoxygenase are illustrated in Fig. 4.3.

Arachidonic acid is transformed in the presence of lipoxygenases to hydroperoxides in a regio- and enantiotope selective manner (Fig. 4.3.a). The site of oxygenation depends on the source of the enzyme. Thus with lipoxygenase extracted from potato, (5S)-hydroperoxy-eicosatetraenic acid, an intermediate in leucotriene synthesis could be obtained [30,31]. The enzyme from soybeans (SBLO) catalyzes attack at another position yielding the (15S)-hydroperoxide [32] and under the action of the same enzyme (SBLO) the (8S,15S)-bishydroperoxide can be obtained in good yield from arachidonic acid (Fig. 4.3.b) [33]. Regioselectivity of the second step can be influenced by modification of conditions. At pH 11, 35% of the(5S,15S)-bishydroperoxide appears in the mixture, which can be then readily converted to (6R)-lipoxin A [34].

Fig. 4.3.c depicts the SBLO catalyzed regio- and stereoselective reactions of linolenic acid [32,35]. Realization of the process in an organic solvent containing reverse micelles is advantageous in view of the hydrophobic nature of the substrate with the enzyme exerting its activity in the aqueous medium inside the micelles [112].

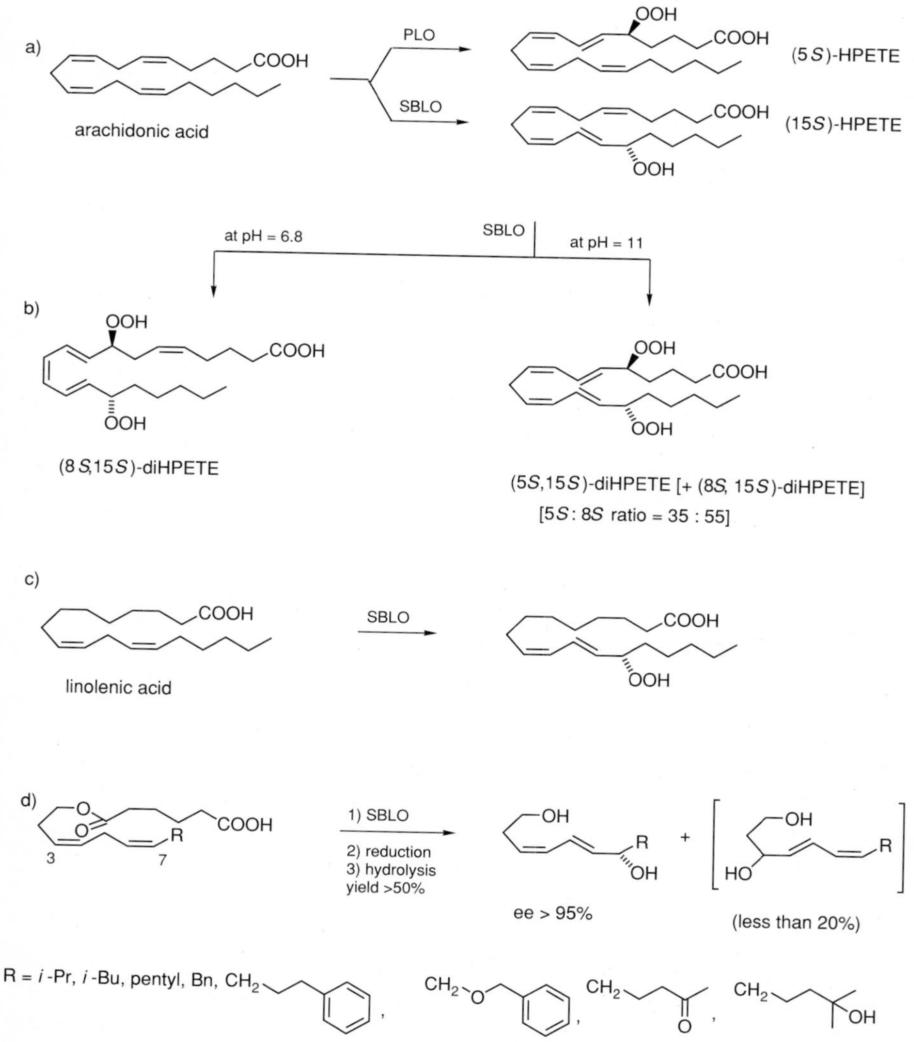

Fig. 4.3. Lipoxygenase (from soybean: SBLO, from potato: PLO) catalyzed transformations

As shown in Fig. 4.3.d, high regio- and stereoselectivity of SBLO enzyme can be exploited for the transformation of appropriately selected non-natural substrates as well. From the adipate monoesters containing a (Z,Z)-1,4-pentadienyl moiety, 1,7-dihydroxydienes were prepared in high isomeric and enantiomeric purity [37].

Dopamine-β-monooxygenase (EC 1.14.1.7) is a copper protein which, in mammalian tissues, catalyzes the hydroxylation of dopamine involving the *pro-R* hydrogen to give norepinephrine. It also readily catalyzes oxygenation adjacent to an aromatic ring of functionalities such as saturated or unsaturated carbon, carbinol, sulphur, or selenium in substrate analogs [38-40 and references therein]. The enzyme is capable of catalyzing hydroxylation in allylic positions too [40].

Tyrosinase (EC 1.14.18.1) is an iron containing enzyme, the physiological role of which is to convert tyrosine into L-DOPA. It can also be used for the coupling of hindered phenols [170]; as more importantly, for the synthesis of coumestans, benzofurans, and related compounds. In a mushroom tyrosinase catalyzed oxidation of catechol, *o*-quinone is generated which then reacts *in situ* with a variety of partners to give a number of coupling products [36]. One example is the preparation of 11,12-dihydroxycoumestan.

In addition to those discussed so far, several other enzymes or enzyme systems working without added cofactor may be of synthetic interest. One of them is isopenicillin-N cyclase, an iron containing enzyme assisting in the cyclization of the tripeptide precursor to isopenicillin-N [41].

The same enzyme catalyzes the transformation of several modified substrates too [41-43 and references cited therein], e.g. tripeptides modified in the valine [43] or cysteine unit [42] are also amenable to cyclization. In this way modified penicillins and cephames can be prepared.

Occasionally other oxidases are also used in isolated form, steroid hydroxylases [45] or ω-hydroxylase in epoxidation [44] for example, but it seems that it is more convenient to carry out such transformations with intact microorganisms.

The advantage that an oxidoreductase does not require an added cofactor is more or less compensated by the fact that substrate specificity of these enzymes is relatively narrow, and that they are often both unstable and of poor specific activity. Many of the transformations carried out with the above mentioned isolated enzymes can be better realized with living cells.

The synthetic applications of enzymes which require added cofactors, and primarily of alcohol dehydrogenases, are much more diverse and there are now several well established methods available for the *in situ* regeneration of cofactors. These techniques are discussed in the following section.

4.2 Oxidoreductases Requiring Added Cofactors

Oxidoreductases requiring an added cofactor, especially NAD(P) dependent alcohol dehydrogenases (EC 1.1.1), can be applied quite generally. The synthetic potential of alcohol dehydrogenases is discussed in the following sections, although their highly important, but rather special role in the stereoselective preparation of isotopically labelled compounds [46-48] will only be touched on occasionally.

Alcohol dehydrogenases play a central role in several metabolic transformations [48,49] and are therefore ubiquitous throughout both the animal and plant kingdoms. In the presence of nicotinamide-adenine dinucleotide cofactors [50,51] they catalyze the reaction shown in Fig. 4.4.a.

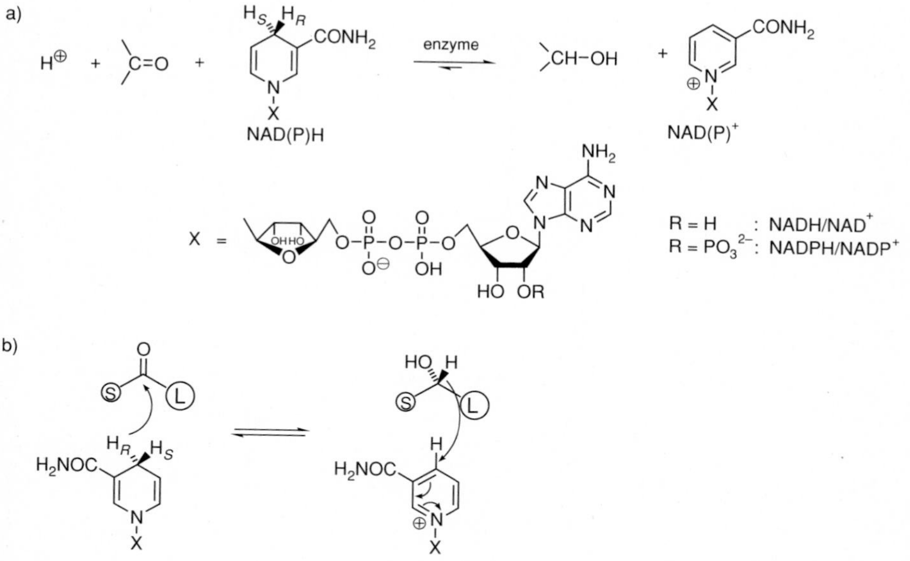

Fig. 4.4. Alcohol dehydrogenase catalyzed NAD(P)H dependent oxidoreductions
a) general scheme
b) interpretation of Prelog's rule for an A type enzyme

The *kinetics* of alcohol dehydrogenases and similar enzymes involving the binding of two substrates is very complex [3,52-55], as indicated, amongst others, by the following equation [52]:

$$E \underset{k_{2,1}}{\overset{k_{1,2}}{\rightleftharpoons}} E \cdot NAD^+ \underset{k_{3,2}}{\overset{k_{2,3}}{\rightleftharpoons}} E \cdot NAD^+ \cdot S_{OH} \underset{k_{4,3}}{\overset{k_{3,4}}{\rightleftharpoons}} E \cdot NADH \cdot S_{CO}$$

$$\underset{k_{5,4}}{\overset{k_{4,5}}{\rightleftharpoons}} E \cdot NADH \underset{k_{6,5}}{\overset{k_{5,6}}{\rightleftharpoons}} E$$

where S_{OH} and S_{CO} represent the alcohol and carbonyl substrates respectively.

In the nicotinamide moiety of the reduced cofactors *two diastereotopic hydrogen atoms are attached* to C-4. With type A enzymes it is the *pro-R* and with type B the *pro-S* hydrogen which is transferred [48,56,57]. In nature even the same substrate, e.g. dihydroxyacetone can be reduced by both type A and B glycerol dehydrogenase [172]. For simple carbonyl substrates, enantiotope selectivity of the reduction catalyzed by the most common alcohol dehydrogenases can be predicted by Prelog's rule [58] (Fig.4.4.b).[1] Attack from the *re* face is predominant, provided that the sequence of the ligands in the substrate in Fig. 4.4.b is O>L>S although in practice, distinction between large (L) and small (S) ligands is sometimes problematic. With the less frequently used B-type enzymes the sense of selectivity is, of course, inverted.

The alcohol dehydrogenase catalyzed transformations shown in Fig 4.4.a are reversible [p. 2 in 49]. For reduction it is the reduced form of the cofactors (NADH or NADPH) that is required, whilst for oxidation the oxidized form (NAD$^+$ or NADP$^+$) is needed.

When carrying out an enzymic oxidoreduction on a larger, or eventually industrial scale [59], a number of points discussed below should be considered.

In terms of economy and scaling up, *in situ* regeneration of the nicotinamide cofactors is of prime importance since the main cost component, apart from the price of the enzyme, is that of the cofactors. Cofactors are too expensive (several 100$ /mol) to permit their use in a stoichiometric quantity and as a result many solutions to this problem have been suggested [2,60-64 and references cited therein]. Thus enzymic regeneration may be coupled to the main process [2,60] or the task can be solved by electrochemical [61-63], photochemical [64] or purely chemical methods. In Tables 4.1 and 4.2 *methods for the regeneration of NAD(P)H and NAD(P)$^+$* respectively have been compiled though comparison and detailed discussion of these strategies you are referred to the original papers. It seems that the enzymic approach has the widest scope of uses [2,73,85], although the electrochemical or photochemical methods should not be disregarded. Purely chemical methods however appear to be the least efficient. When selecting a method of cofactor regeneration not only should its efficiency be considered but also its compatibility with the primary process.

---

[1] It is not surprising, however, that ADH enzymes with opposite facial selectivity, e.g. the synthetically promising ADH from *Pseudomonas* sp. [171] exist.

Table 4.1. Coupled systems for *in situ* NAD(P)H regeneration.

AH[a]	A	Regeneration[b] (E₂)	Rereduction	Reference
formate	CO_2	FDH[c]	no	[65-67]
glucose	gluconate	GDH	no	[68]
glucose-6-P	6-P-gluconate	6-P-GDH	no	[69-71]
$R^1CH(OH)R^{2d}$	R^1COR^{2d}	DH[e]	no	[72]
methanol	CO_2	ADH+AldDH+FDH[c]	no	[73]
ethanol	acetate	ADH+AldDH	no	[73]
$2MV^{+\cdot}$	$2MV^{2+}$	microorganisms	H_2/microorg.	[61]
$2MV^{+\cdot}$	$2MV^{2+}$	lipoamide DH[c] or ferredoxin reductase[f]	H/hydrogenase	[74-77]
$2MV^{+\cdot}$	$2MV^{2+}$	reductase[f]	electro-chemical	[61,78]
$2MV^{+\cdot}$	$2MV^{2+}$	reductase[f]	photochemical	[79]
$2Rh(bpySO_3H)^{3+}$	$2Rh(bpySO_3H)^{2+}$	chemical	electro-chemical	[80]
$2Rh(bpySO_3H)^{3+}$	$2Rh(bpySO_3H)^{2+}$	chemical	photochemical	[81]
$S_2O_4^{2-}$ +H_2O		chemical	no	[82,83]

a: Abbreviatons: MV= methyl viologen; $Rh(bypSO_3H)_3^{3+}$ Rh(III)-tris(2,2'-bipiridyl-5-sulphonic acid); b: Abbreviations: AldDH= aldehyde dehydrogenase; FDH= formate dehydrogenase; GDH= glucose dehydrogenase; 6-P-GDH= glucose-6-phosphate dehydrogenase; DH= different dehydrogenases; c: Acts on NAD^+/NADH; d: $R^1CH(OH)R^2$ is a good substrate for the ADH enzyme in excess compared to the substrate in the primary reduction. e: If DH is the same as E_1 the method is called "coupled substrate cofactor regeneration" [48]. Such systems, e.g.: HLADH, ethanol, NADH [48]; TbADH, isopropanol, NADPH [84]; f: Acts on $NADP^+$/NADPH.

Amongst the oxidoreductases working with added cofactors, *alcohol dehydrogenases* assume overwhelming importance [2,52,54,59], although the utilization of other enzymes, e.g. of monooxygenases for Baeyer-Villiger type oxidation also shows some promise [95,96]. (Enzymes catalyzing these latter reactions are mostly flavoprotein monooxygenases, but for the reoxidation of the flavin coenzyme it is usually coupled $NADP^+$ systems that are used, justifying their inclusion for discussion in the present chapter).

Although some oxidoreductases have a rather narrow substrate specificity they can nevertheless be used to advantage in certain syntheses if applied to the right substrates. Primarily these are various hydroxy acid dehydrogenase enzymes such as the inexpensive

Table 4.2. Coupled systems for *in situ* NAD(P)$^+$ regeneration.

A[a]	AH	Regeneration[b] (E$_2$)	Reoxidation	Ref.
2-oxoglutarate	L-glutamate	GluDH	no	[85]
pyruvate	L-lactate	L-LDH	no	[85,86]
acetaldehyde	ethanol	ADH[c]	no	[87,88]
MB$^+$	MBH$_2$	diaphorase	O$_2$	[85]
MB$^+$	MBH$_2$	hν	O$_2$	[89]
SnTMPyP^{4+}	SnTMPyPH^{3+}	hν	electrochemical	[90]
2Fe(tmphen)$_3^{3+}$	2Fe(tmphen)$_3^{2+}$	chemical	electrochemical	[91]
PQQ	PQQH$_2$	chemical	O$_2$	[92]
FMN	FMNH$_2$	chemical	O$_2$	[93,94]

a: Abbreviations: FMN= flavin mononucleotide; Fe(tmphen)$_3^{3+}$= Fe(III)-*tris*(3,4,7,8-tetramethyl-1,10-phenantroline); MB= methylene blue; PQQ= pyrroloquinoline quinone tricarboxylic acid; SnTMPyP^{4+}= Sn(IV)-*meso*-N-tetramethylpyridinium porphyrin; b: Abbreviations: GluDH=glutamate dehydrogenase; L-LDH= L-lactate dehydrogenase; ADH=alcohol dehydrogenase; c: May be the same enzyme as for the main transformation

[97] L-lactate dehydrogenase (L-LDH) [98,99] and its counterpart D-lactate dehydrogenase (D-LDH), or in addition amino acid dehydrogenases.

The NADH dependent dehydrogenase prepared from baker's yeast (YADH) unfortunately has a very narrow substrate tolerance which precludes its use for most applications save for the reduction of a few acyclic aldehydes and ketones [52].

For *general synthetic purposes*, it is the alcohol dehydrogenases prepared from horse liver (HLADH) and an enzyme isolated from the heat resistant microorganism *Thermoanaerobium brockii* (TbADH) that are most often used. HLADH is a mixture of isoenzymes [100] and can be used for the transformation of a very broad selection of acylic, mono- and bicyclic compounds [52,101,102]. The relative instability and heat sensitivity of HLADH are serious limitations which prompted the search for a more robust, heat resistant dehydrogenase that led to the discovery and introduction of TbADH [84,103]. A number of other heat resistant enzymes with relatively broad substrate specificities, like the NADPH dependent enzymes of *Thermoanaerobacter ethanolicus* (TeADH) [104] or the NADH dependent enzyme from *Sulfolobus solfataricus* (SsADH) [105] still however await synthetic exploitation. It is always advisable in these situations to compare the isolated enzyme and the parent, usually resting cell culture *in vivo*, since working with the latter is often found to be more economical [54,61,106].

An active center model, the so called diamond lattice model [58], has been proposed based on the results of experiments aimed at the characterization of the active centers in the alcohol dehydrogenase CfADH, MjADH, and PLADH [48,52]. CfADH and MjADH are of microbial origin whilst PLADH is obtained from pig liver although none of them is sufficiently stable for synthetic utilization. These active center models were constructed by measuring the conversion rates of various substrates and superimposing the reacting conformations of the substrates (Fig. 4.5).

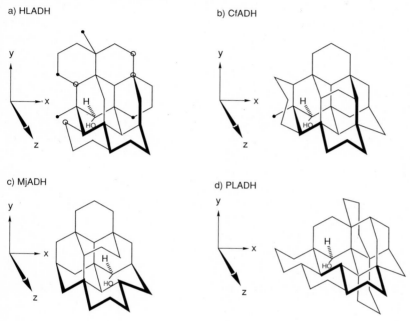

a) HLADH   b) CfADH

c) MjADH   d) PLADH

Fig. 4.5. Diamond lattice-section active site models for alcohol dehydrogenases [after refs. 48,52,58]: ● Forbidden positions; ○ Undesired positions; The others are the allowed positions of the "diamond lattice" model

With the aid of the diamond lattice and similar models (e.g. the spatial cube model for the active centre of HLADH [107]) it was subsequently feasible to predict, even for more complex substrates, the suitability of the enzyme in question for the transformation of a given substrate, indeed it was also possible to foresee the sense of stereoselectivity. It should be noted however, that enzyme reactions are highly sensitive to the reacting conditions and the reaction models are only valid under conditions that do not greatly differ from those used in the model experiments [108].

The influence of external parameters on the activity of oxidoreductases has been thoroughly studied, mainly on YADH and HLADH [49(p. 352),52].

Thus, regarding the *influence of pH*, it was found that with dehydrogenases no single pH optimum could be defined since the effect of pH is dependent on whether the substrate is an alcohol or a carbonyl compound and whether the enzyme is interacting with the reduced or the oxidized cofactor [109, 52(p. 264)]. Thus the pH dependencies of oxidation and reduction catalyzed by the same enzyme are significantly different, the two pH optima usually differing by about 2 pH units [109], that for oxidations being the higher. pH optimum may also depend on the actual substrate itself and in turn pH may influence the stereochemical outcome of the reaction [108].

Temperature can significantly influence selectivity [108]. Indeed with ketone reduction even the sense of selectivity may be temperature dependent [110].

Immobilization may affect catalytic properties. Thus with HLADH immobilized on two different carriers it was found that the conformation of the active center remained essentially unchanged in the vicinity of the catalytically active zinc atom, whilst in more remote regions it was significantly different in the two immobilized preparations [111]. This may explain the identical behavior of the two systems towards the small ethanol molecule, in contrast to the larger cyclohexanol, towards which the two preparations behaved differently.

When linked to the main transformation, *the cofactor regenerating system* can also be a modifying factor; (for references on coupled substrate-cofactor regeneration see Table 4.1). Thus, in the course of HLADH reduction, concentration of ethanol, i.e. of the coupled substrate alters the steric course of the reduction [108].

Product inhibition may arise as a serious problem in dehydrogenase catalyzed oxidations since the carbonyl compound formed is more tightly bound to the hydrophobic active site than the alcohol substrate [55]. This limits the maximum permissible initial concentration, unless removal of the product e.g. by extraction is catered for. Working in two-phase aqueous/organic solvent systems [113-115] (Fig. 4.6), in organic solvents of low water content [108,117], or in organic solvent/reverse micelle systems [112,116] may, however, significantly improve this situation.

4.2.1   Selective Reactions with Oxidoreductases Requiring Added Cofactors

In considering oxidoreductases, examples for practically all of the types of selectivities discussed in Section 2.3 can be found and with certain limitations, all that has been said about the outcome of such transformations and about the influence of external parameters in Sections 2.3.1 and 2.3.2 still remains valid.

Many different reactions can be catalyzed by oxidoreductases, and accordingly the selectivities that can be realized are also manifold.

With enoate reductase flavoenzymes which catalyze the reduction of the double bond in α, β-unsaturated carbonyl compounds, depending on the symmetry of the substrate, enantiotope or diastereotope selectivity at both ends of the double bond can be achieved (Fig. 4.7.a). Such enzymes can be isolated, from *Clostridium* or *Proteus* strains for example, and can be used as the crude cell-free extracts. Alternatively the resting intact cells may themselves be applied [61].

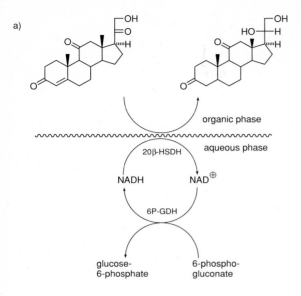

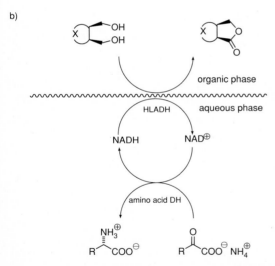

Fig. 4.6. Examples for oxidoreductions catalyzed by dehydrogenases in aqueous-organic biphasic system: a) reduction of a hydrophobic steroid [115]; b) oxidation of cyclic meso-diols [X=(CH₂)ₙ, n=1-4 or (CH₂-CH=)₂] with a coupled L-amino acid producing NADH regenerating system [Amino acid DH can be GluDH, AlaDH, LeuDH, PheDH; R can be (CH₂)₃COOH, Me, Et, t-Bu, (CH₂)₂Ph] [114]

Enantiotope selective Baeyer-Villiger oxidations can be carried out with a purified cyclohexanone monooxygenase enzyme obtained from an *Acinetobacter* strain (Fig. 4.7.b). This method, by which *meso*-ketones can be converted to chiral lactones with very high selectivity, clearly demonstrates the advantages of the enzymic over the chemical ap-

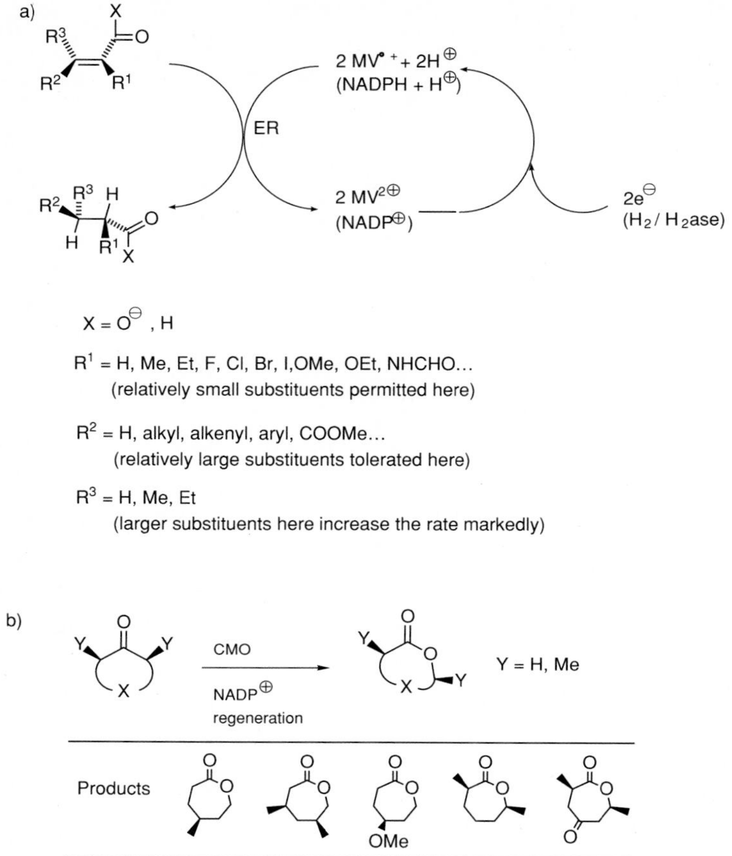
a)
ER
2 MV•+ + 2H⊕
(NADPH + H⊕)
2 MV2⊕
(NADP⊕)
2e⊖
(H2 / H2ase)
X = O⊖ , H
R1 = H, Me, Et, F, Cl, Br, I, OMe, OEt, NHCHO...
(relatively small substituents permitted here)
R2 = H, alkyl, alkenyl, aryl, COOMe...
(relatively large substituents tolerated here)
R3 = H, Me, Et
(larger substituents here increase the rate markedly)
b)
CMO
NADP⊕
regeneration
Y = H, Me
Products
OMe

Yield (%)	88	73	76	27	25
ee (%)	>98	>98	75	>98	>98

Fig. 4.7. Non-dehydrogenase catalyzed enantiotope selective transformations requiring coupled cofactor regenerating systems: a) enoate reductase (ER) catalyzed reductions of activated C=C bonds [62,118 and references cited therein]; b) cyclohexanone monooxygenase (CMO) catalyzed Baeyer-Villiger oxidations [119]

proach. Enzymic and chemical oxidation, the latter with peracids, however, show very similar regioselectivities [95].

Amino acid dehydrogenases and α-hydroxy acid dehydrogenases catalyze the enantiotope selective reduction of α-keto acids (Fig. 4.8) and also the reverse process [120], although the latter is of little synthetic relevance.

A variety of α-hydroxy acids can be obtained by enzyme catalysis (Fig. 4.8.a). It is primarily the L-LDH enzymes (one enzyme name may cover more than one protein, c.f. Section 1.1.1) that are useful in the preparation of L-α-hydroxy acids. [98,99 and references cited therein], while for the synthesis of the (*R*)-acids several enzymes are available. Whilst these enzymes provide enantiomerically almost pure products, unfor-

a)

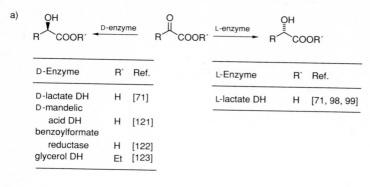

D-Enzyme		R`	Ref.
D-lactate DH		H	[71]
D-mandelic			
acid DH		H	[121]
benzoylformate			
reductase		H	[122]
glycerol DH		Et	[123]

L-Enzyme	R`	Ref.
L-lactate DH	H	[71, 98, 99]

b)

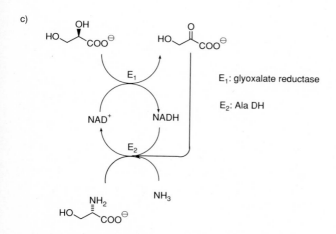

R	Enzyme	Ref.
CH₂F	Ala DH	[124]
CHFCH₂COOH	Glu DH	[125]

c)

E₁: glyoxalate reductase

E₂: Ala DH

Fig. 4.8. Enantiotope selective reduction of α-keto acids: a) preparatively used L- or D-α-hydroxy acid producing enzymes; b) examples for unnatural L-α-amino acid productions using amino acid dehydrogenases; c) enzymic interconversion of an α-hydroxy acid to an L-α-amino acid via α-keto acid

tunately their substrate specificity is relatively narrow and when the substituent R is bulky rates drop steeply. With the correct choice of enzyme, any desired enantiomer of a hydroxy acid can be prepared from a given keto acid [71]. Using an increased molecular weight NADH-type cofactor, the reaction can be realized continuously in a membrane reactor [121].

With the aid of amino acid dehydrogenase enzymes α-keto acids can be converted to α-amino acids [65,127-129,173], a transformation that is also quite feasible in a membrane reactor. Substrate specificity of individual amino acid dehydrogenases is rather narrow, although they can nevertheless assist in the transformation of alien substrates to non-natural amino acids (Fig. 4.8.b and Fig.4.6.b)

By combining the two types of enzymes mentioned above with a coupled enzyme - NADH recycling system, direct transformation of an α-hydroxy acid to the corresponding amino acid becomes feasible (Fig. 4.8c).

Alcohol dehydrogenases exhibit the widest substrate specificity, and depending on the structural features of the substrate a variety of selectivities can be realized with them.
i) Reduction of aldehydes does not generate a center of chirality except when a heavy isotope of hydrogen is involved as is the case in the preparation of selectively deuteriated mevalolactone (Fig. 4.9). With unsymmetrical dialdehydes one would expect regioselectivity to result, with racemic aldehydes enantiomer selectivity, and finally with prochiral or *meso*-aldehydes enantiotope selectivity.

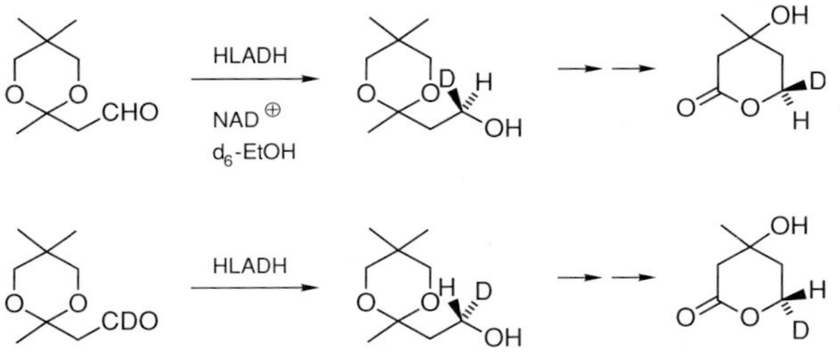

Fig. 4.9.   Synthesis of deuterium labelled mevalolactones [130]

ii) In the reduction of unsymmetrical ketones the enantiotope selective (or with chiral ones the diastereomer selective) character of enzymic reduction is expressed and the products are chiral. Reduction of achiral unsymmetrical ketones is enantiotope selective, whilst with more complex substrates other selectivities may also emerge. With racemic ketones simultaneous enantiomer and diastereotope selectivity can be expected (cf. Section 2.3.2.8 and Fig. 2.14). In molecules containing more than one constitutionally different oxo group, regioselectivity predominates, and with prochiral or *meso*-dioxo compounds enantiotope and diastereotope selectivity appear.
iii) In the oxidation of alcohols the center of chirality of unsymmetrically substituted carbinols is destroyed. This can be employed for useful syntetic purposes, for example using oxidations of polyhydroxy substrates where chemo- or regioselectivities predominate, or the oxidation of racemic alcohols in which enantiomer selectivity is favored and also with prochiral or *meso*-diols (or polyols) where enantiotope selectivity can be exploited.

4.2.1.1   Enantiomer Selectivity: Reduction of Racemic Aldehydes and Oxidation of Racemic Alcohols

Enantiomer selectivity of alcohol dehydrogenases can be put to use both in the reduction of racemic aldehydes and in the oxidation of racemic alcohols (Table 4.3) although none of these methods carries much importance in terms of suitability for syntheses.

HLADH reduction of racemic aldehydes (4.1-4.6) proceeds with poor selectivity, although reduction of the metal carbonyl complexes (4.7-4.8) seems to be an exception to this rule.

Enzymic oxidation of racemic alcohols has been relatively rarely exploited (e.g. 4.9-4.23). Hydrolase assisted esterification or hydrolysis may be more convenient for alternative kinetic resolutions (oxidation of racemic alcohols is also discussed in another context in Section 4.2.1.4.) An elegant method for the resolution of the metallocene alcohols (4.11-4.13) however, is enzymic enantiomer selective oxidation. Although at the cost of much slower rates, both the reduction of racemic aldehydes (e.g. of 4.1) and oxidation of racemic alcohols (4.9-4.10) can be carried out in organic solvents containing a small amount of water by adsorbing both the enzyme and the cofactor onto a carrier [131]. Of more direct synthetic impact however, is oxidation with an HLADH+AldDH enzyme system of the racemic diols (4.14-4.23) which provide 2-hydroxy carboxylic acids and 1,2-diols of high enantiomeric purity.

4.2.1.2   Enantiotope Selective Reduction of Achiral Carbonyl Compounds

Enantiotope selective enzymic carbonyl reductions of synthetic value were primarily realized with aliphatic substrates (Table 4.4) where Prelog's rule (cf. Section 4.2) is usually honored. An exception is TbADH, an enzyme convenient for preparative work, which with low molecular weight ketones (4.24-4.26) gives rise to products with configurations opposite to the rule, although with longer chain ketones (4.27-4.33, 4.35-4.40) the rule is again followed. Optical purity is often better when, instead of the living cells, the isolated enzymes are used. As shown by the transformation of ketones (4.34) and (4.35) for example; enantiotope selectivity can be inverted by taking another enzyme, i.e. it is possible to obtain both enantiomeric alcohols from the same ketone.

From a synthetic point of view the availability of enzymes capable of transforming carbonyl compounds bearing further functional groups (e.g. 4.30, 4.33-4.39) is important.

An example of the application of TbADH enzyme is the preparation of the pheromone (S)-(+)-sulcatol (4.31) in high enantiomeric purity [136].

Table 4.3. Enantiomer selectivity: reduction of racemic aldehydes or oxidation of racemic alcohols.

Substrate (racemic)[a] No	X	Enzyme (cofactor)	Conv. (%)	X	Product[a] (4.X a) Y(%)	ee(%)	Remaining substrate[b] (4.X b) X	Y(%)	ee(%)	Ref.
4.1	=O	HLADH[d] (NADH)	15		6	95[c]	=O			[131]
4.2	=O	HLADH (NADH)	43		37	82	=O	48	52	[132]
4.3	=O	HLADH (NADH)	52		48	87	=O	45	42	[132]
4.4 R=Me	=O	HLADH (NADH)	42		39	84	=O	52	62	[132]
4.5 R=Et	=O	HLADH (NADH)	42		37	79	=O	53	52	[132]
4.6	=O	HLADH (NADH)	60		23	73	=O			[132]

Table 4.3. (contd.)

No	Substrate (racemic)[a] X	Enzyme (cofactor)	Conv. (%)	Product[a] (4.X a) X	Y(%)	ee(%)	Remaining substrate[b] (4.X b) X	Y(%)	ee(%)	Ref.
4.7	(cyclopentadienyl–Mn(CO)₃ with CH=X, CH₃) X = =O	HLADH (NADH)	50	H⸍⸍OH	35	>95	=O	31	>95	[133,174]
4.8	(cyclopentadienyl–Cr(CO)₃ with CH=X, CH₃) X = =O	HLADH (NADH)	50	H⸍⸍OH	36	>95	=O	28	>95	[133]
4.9	(cyclopentanone deriv., H⸍⸍OH)	HLADH[d] (NAD⁺)	20	=O	11	96	OH⸍⸍H			[132]
4.10	(cyclohexanone deriv., H⸍⸍OH)	HLADH[d] (NAD⁺)	28	=O	16	>99	OH⸍⸍H			[132]

Table 4.3. (contd.)

No	Substrate (racemic)[a]	Enzyme (cofactor)	Conv. (%)	Product[a] (4.X a) X	Y(%)	ee(%)	Remaining substrate[b] (4.X b) X	Y(%)	ee(%)	Ref.
	(ferrocene-type structure with X)	HLADH (NAD$^+$)	40	=O			(structure OH,,,H)			
4.11	M=Fe							44	100	[133]
4.12	M=Rh							44	92	[133]
4.13	M=Os							33	92	[133]
4.14–4.23	(structure OH R^f—X)	HLADH$^+$ AlaDH[e] (NAD$^+$)		COOH	30–35	>99	CH$_2$OH			[134]

a: Structure of the faster reacting enantiomer is given; b: The majority of the remaining substrates are enantiomeric to the depicted substrate; c: Absolute configuration not reported; d: Reactions were carried out in isopropyl ether; e: HLADH/aldehyde DH/glutamate DH were coimmobilized in polyacrylamide gel; f: R= Me, Et, i-Pr, allyl, CH$_2$OH, CH$_2$NH$_2$, CH$_2$Hal (Hal= F, Cl, Br).

Table 4.4. Enantiotope selective alcohol dehydrogenase catalyzed reduction of achiral oxo compounds.

Product		ee(%)	Enzyme	Ref.
4.24		48	TbADH	[84]
4.25		86	TbADH	[84]
4.26		44	TbADH	[84]
4.27		79	TbADH	[84]
4.28		95	TbADH	[84]
4.29	n=3-7	>95	TbADH	[84]
4.30	n=3-5	>95	TbADH	[84] [103]
4.31		99	TbADH	[84] [136]
4.32	n=2,3	>95	TbADH	[84]
4.33		98	TbADH	[84] [103]
4.34		>99 >99	YADH HLADH	[68] [68]

Table 4.4. (contd.)

Product		ee(%)	Enzyme	Ref.
4.35	OH / OMe / OMe	51	TbADH	[68]
4.36	OH / CH(OMe)$_2$	>99 >89 91	YADH HLADH TbADH	[68] [68] [68]
4.37	OH / COOEt	84 88	HLADH TbADH	[68] [68]
4.38	OH / Cl COOEt	98 90 >99	HLADH TbADH GlycDH^a	[68] [68] [123]
4.39	OH / COOMe	98 98	TbADH HLADH	[115] [115]
4.40	OH / CF$_3$	94	TbADH	[68]
4.41	OH / S / n=1	33	HLADH	[135]
4.42	n=2	28	HLADH	[135]

a: Glycerol dehydrogenase from *Geotrichum candidum*.

4.2.1.3   Enantiotope Selective Oxidation of Prochiral and Meso-Diols

As extensive studies have shown, HLADH is suitable for the oxidation of a wide structural variety of prochiral or *meso*-diols. Cycloalkane-1,2-diols however, are not substrates of HLADH although the corresponding bis(hydroxymethyl) compounds can be subjected to oxidation by this enzyme. Sometimes this particular problem can be solved by turning to another enzyme, for example *cis*-cyclohexane-1,2-diol is oxidized with high enantiotope selectivity in the presence of glucose dehydrogenase [55].

One example of the selectivity of HLADH can be seen in the oxidation of glycerol [100] or of 2-amino-1,3-propanediol [134].

HO—X—OH → HLADH / NAD⁺ → HO—X—CHO   X = OH, NH₂

Table 4.5. Chiral lactones from prochiral or *meso*-diols by HLADH oxidations.

Product		ee(%)	Reference
4.43	R=Me	78	[137,138]
4.44	R=Et	74	[137,138]
4.45	R=Ph	20	[137,138]
4.46	R=Me	>99	[139]
4.47	R=Et	>99	[139]
4.48		>99	[140]
4.49	R=H	>99	[114,140,141]
4.50	R=Me	>99	[114,140,141]
4.51		>99	[114,140,141]
4.52		>99	[114,140,141]
4.53		>99	[139]
4.54		>99	[102,114] [102,114]

Table 4.5. (contd.)

Product		ee(%)	Reference
4.55		>99	[114,140,141]
4.56		>97	[143]
4.57		>97	[143]
4.58		>97	[143]
4.59		>97	[143]
4.60		>97	[143]
4.61		>97	[143]
4.62		>98	[144]
4.63		>98	[144]
4.64		>99	[145]
4.65		>99	[145]
4.66		>99	[145]

When the two hydroxyl groups are in a favorable disposition the products are chiral lactones (Table 4.5). The reaction proceeds via a hemiacetal in two steps, generally with high selectivity. This was initially observed with the oxidation of acyclic diols, where the enantiotope selectivity of the first oxidation step was lower. From the stereoisomeric mixture of hemiacetals, those leading to the predominant lactones are transformed in the second step leaving behind the hemiacetals of opposite configuration [137,139].

The reaction protocol, which involves chemical cofactor regeneration (FMN, O_2 , cf. Table 4.2) [102], cannot be subjected to simple upscaling of the process owing to the difficulties of product inhibition and the relatively poor stability of the enzyme. For large scale operations, an alternative version using enzymic cofactor regeneration and a water/organic solvent two-phase system seems to be more practical [114].

A great majority of the lactones presented in this chapter can also be prepared from products of hydrolase catalyzed enantiotope selective transformation of prochiral or *meso*-diols, or dicarboxylic acid derivatives (cf. Fig. 3.29 in Section 3.3.6.2). The complexities of the oxidative process and the necessity of cofactor regeneration are compensated for by the high enantiomeric purity of the products. Therefore when only a few grams of optically pure product is required HLADH oxidation of *meso*-diols may be preferred, whilst for larger scale preparations a hydrolase catalyzed transformation, ultimately combined with some process to upgrade enantiomeric purity, may be the method of choice.

In the enantiotope selective HLADH oxidation of 3-substituted-1,5-pentanediols, enantiomeric purity declines with increasing bulk of the 3-substituent, a phenomenon demonstrated in the data for 3-methyl- and 3-phenylvalerolactone (4.43 and 4.45 in Table 4.5) for example. In contrast, the high selectivity observed in HLADH catalyzed oxidation of cyclic *meso*-diols was little influenced by ring size and substitution of the parent ring (cf. 4.49-4.63). The enzyme is capable of transforming in an enantiotope selective manner, not only 1,2-diols but 1,3-*cis*-bis(hydroxymethyl) derivatives too and the products (4.64-4.66) arising from such reactions may be enantiomerically quite pure.

Examples of the use of optically pure lactones prepared with the aid of HLADH are the synthesis of the pheromone (+)-grandisol from the alcohol (4.51) [146] or of (–)-sweroside, a compound that plays a role in alkaloid biosynthesis, from the lactone (4.55) [147].

4.2.1.4   Transformations Involving More Than One Type of Selectivity

In oxidoreductase catalyzed transformations several combinations of selectivities can be effected (cf. Section 4.2.1). Some of the more synthetically important and useful possibilities are described here.

i) Regioselective transformation of steroids (Fig. 4.10)

Whilst in oxidations of polyhydroxysteroids with hydroxysteroid dehydrogenase enzymes specific for definite positions of the steroid skeleton it is only possible to effect

regiospecific reactions (Fig. 4.10.a), with polyoxosteroids the same enzymes may catalyze both regio- and diastereoselective reductions of carbonyl groups as part of either the
steroid skeleton or the side chain (Fig. 4.10.b). It is by combination of such operations
that the $3\alpha/3\beta$ inversion of the bile acid shown in Fig. 4.10.a can be implemented, first
by oxidation using 3α-HSDH and then subsequent reduction of the product by 3β-HSDH
[150]:

With apolar ketosteroids, poor solubility of the substrate in water may be a problem.
The solution in this case may be to use water/organic solvent two phase systems [115]
or to work in reverse micelles [116].

ii) Enantiomer and diastereotope selective reduction of racemic ketones (Table 4.6)

When dealing with the reduction of racemic ketones it is advisable to select transformations characterized by high enantiomer and diastereotope selectivity since in this case
both the remaining carbonyl compound and the product alcohol are obtained in high
enantiomeric purity.

Of the alcohol dehydrogenases, most information is available about HLADH, an
enzyme stable enough in its isolated form for preparative use. It reduces racemic 2-
substituted cyclohexanones (4.57, 4.68) with high enantiomer and diastereotope selectivity. Reduction of the analogous tetrahydrothiopyranes (4.69, 4.70) proceeds in a similar
manner, albeit with lower enantiomer selectivity.

The reduction of 3-substituted cyclohexanones (4.71, 4.72) is also highly selective and
this feature is also retained in the reduction of the tetrahydropyran analogs (4.73-76).
With the reduction of 2-substituted tetrahydrothiopyran-4-ones [164], whilst enantiomer
selectivity becomes negligible, diastereotope selectivity remains excellent and thus each
of the diastereomeric alcohols produced is of high optical purity.

When using the bicyclic substrates (4.77-4.79) selectivity is reduced, an observation
that is in accordance with the fact that these compounds do not fit at all onto the lattice
points of the diamond lattice model of the enzyme [54]. In such cases, the cubic section
model may be useful (cf. Section 4.2) although its limitations should be kept in mind.

An interesting feature of HLADH reduction of racemic *trans*-bicyclo[4.3.0]nonanes
(4.80-4.83) is that whilst the reduction of the racemic hydrindan-5-one (4.83) exhibits
poor enantioselectivity, those models in which the methylene group in the five-membered
ring has been replaced by a carbonyl group (4.80), an oxygen (4.81) or a sulphur atom
(4.82) can be transformed with much higher selectivity. Furthermore, some interesting regioselectivity was observed with *trans*-hydrindan-2,5-dione, inasmuch as the 2-carbonyl
group remained intact [156]. Enantiomer selectivity with the corresponding *cis* diastereomers is much lower [156,157].

Racemic *trans*-decalin-2-ones (4.84-85) can be reduced with HLADH enzyme with
excellent selectivity. Reduction of racemic *trans*-2-decalone (4.84) with several different
alcohol dehydrogenases (Fig. 4.11) demonstrated that a given substrate can be transformed with opposite but equally high stereoselectivity, thus reduction of the racemic
substrate with three different enzymes gave rise to three different stereoisomeric alcohols [47,48,165]. HLADH reduction of *cis*-decalones can also be achieved with high
selectivity [references in 52].

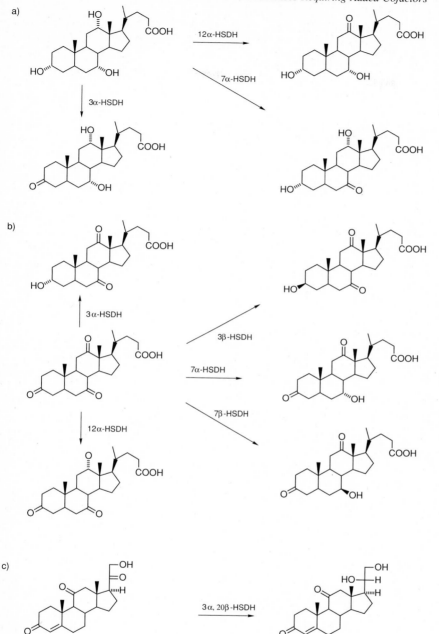

Fig. 4.10.   Regioselective transformations of steroids by hydroxysteroid dehydrogenases: a) oxidations of a bile acid [148,149]; b) regio- and diastereotope selective reductions of a trioxo-bile acid derivative [148]; c) regio- and diastereotope selective reduction in the side chain of a steroid [124]

Table 4.6. Enantiomer and diastereotope selective reduction catalyzed by dehydrogenases.

Substrate[a] (racemic; X=O)			Enzyme	Conv. (%)	Product (4.X a)[a]			Remaining substrate[b] (4.X b,X=0)		Ref.
					X	Y(%)	ee(%)	Y(%)	ee(%)	

	Y	R								
4.67	CH$_2$	Me	HLADH	50	H ''''OH	25	>95	35	50	[52,151]
4.68	CH$_2$	Et	HLADH	50	H ''''OH	37	>95	37	95	[52,151]
4.69	S	Me	HLADH	50	H ''''OH	c	90		66	[152]
4.70	S	Et	HLADH	50	H ''''OH	c	78		58	[152]

	Y	R								
4.71	CH$_2$	Et	HLADH		H ''''OH	12	>95	38	71	[52,151]
4.72	CH$_2$	i-Pr	HLADH	50	H ''''OH	30	77	21	>95	[52,151]
4.73	O	Me	HLADH	50	H ''''OH		>95		85	[153]
4.74	O	Et	HLADH	50	H ''''OH		>95		88	[153]
4.75	O	i-Pr	HLADH	50	H ''''OH		>95		86	[153]
4.76	O	Ph	HLADH	50	H ''''OH		>95		51	[153]
4.77		=X	HLADH	46	H ''''OH	39	64	31	46	[154]

Table 4.6. (contd.)

Substrate[a] (racemic; X=O)			Enzyme	Conv. (%)	Product (4.X a)[a]			Remaining substrate[b] (4.X b, X=O)		Ref.
					X	Y(%)	ee(%)	Y(%)	ee(%)	
4.78			HLADH	25		36	83	58	17	[154]
4.79			HLADH	56			72		74	[155]
	Y	R								
4.80	C=O	H	HLADH			42	94	47	76	[156]
4.81	O	H	HLADH	40		26	>97	48	60	[157]
4.82	S	H	HLADH			33	>97	52	53	[157]
4.83	CH$_2$	H	HLADH	low enantiomer selectivity						[156]
4.84	(CH$_2$)$_2$	H	HLADH			16	>95	31	>95	[52,158]
4.85	(CH$_2$)$_2$	Me	HLADH			35	>95	30	>95	[52,158]
4.86			TbADH	50			>95		>95	[115,159]
				20			>95			[160]
4.87	R^1	R^2								
4.88	Me	Me	HSDH[d]	20			>95			[160]
4.89	H	Cl	HSDH[d]				80			[161]
4.90	Cl	Cl	HSDH[d]				95			[161]

Table 4.6. (contd.)

Substrate[a] (racemic; X=O)		Enzyme	Conv. (%)	Product (4.X a)[a]			Remaining substrate[b] (4.X b,X=0)		Ref.
				X	Y(%)	ee(%)	Y(%)	ee(%)	
4.91		HLADH	47	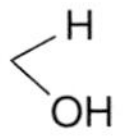	24	90	33	73	[162]
4.92	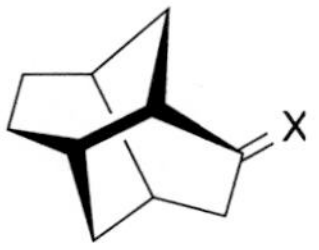	HLADH	43			90		68	[163]
	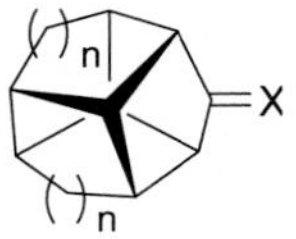								
4.93	n=1	HLADH	52	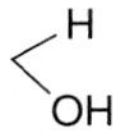	40	76	32	93	[162]
4.94	n=2	HLADH	43	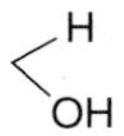	35	96	42	81	[162]

a: Configuration of the faster reacting enantiomer is given in the substrate column; b: The remaining fraction of the substrate has the enantiomeric structure to that shown in the table; c: 5-10% of the diastereomeric alcohol was formed; d: $3\alpha,20\beta$-Hydroxysteroid dehydrogenase.

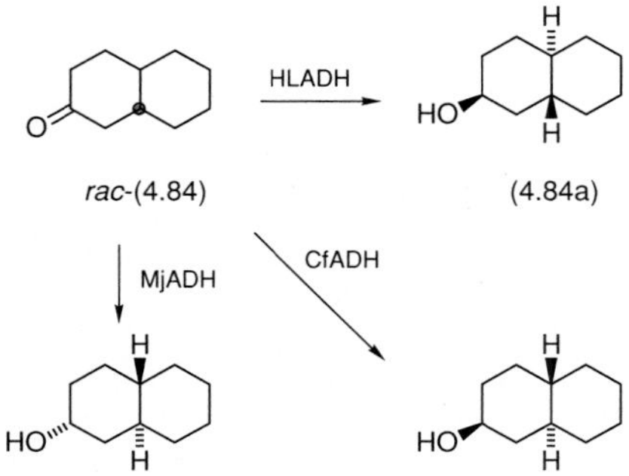

Fig. 4.11.   Alcohol dehydrogenases acting with different selectivities on the same racemic ketone substrate

The reduction of 2-oxo-bicyclo[2.3.0]hept-5-en was studied in connection with syntheses directed towards prostanoids and other natural cyclopentanoids. High enantiomer and diastereotope selectivities were achieved with TbADH enzyme in the model unsubstituted at C-3 (4.86) and with 3α,2β-HSDH enzyme with the 3-substituted compounds.

With the introduction of HLADH it also became possible to reduce the racemic so called girochiral ketones (4.91-94) of C_2 symmetry, and other polycyclic ketones [155,162,163].

In parallel with the HLADH reduction of racemic ketones, oxidation of the corresponding alcohols has also been frequently performed [152,154,155,162,163]. As would be expected, the alcohol that is preferred in the oxidation is that which was the predominant product in the corresponding reductive process.

iii) Reduction of symmetrical ketones

Selective reduction of various ketones of planar, C_i, or C_2 symmetry (see also Section 2.3.2.8) can be realized with HLADH enzyme. Most of the studies in this field were concerned with decalindiones [166-168] (Fig. 4.12) and it was found that with prochiral diketones (Fig. 4.12.a) high enantiotope and diastereotope selectivity predominated. In the unsaturated diketone leading to products in Fig. 4.12.b, whilst the relationship of the oxo groups is homotopic, their faces are enantiotopic. Reduction of these compounds was also highly selective leading to practically enantiomerically pure alcohols.

a)

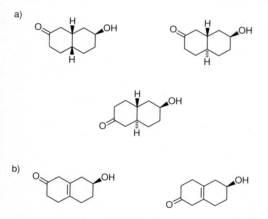

b)

Fig. 4.12. Products from HLADH catalyzed reductions of symmetric decalindiones: a) enantiotope and diastereotope selective reductions; b) enantiotope selective reductions

5 Baker's Yeast

Among microorganisms baker's yeast is very often used as a general biocatalyst. Baker's yeast (*Saccharomyces cerevisiae*) [1] belongs within the class of *Ascomycetes*, and the order of yeasts (*Endomycetales*) to the family of *Saccharomycetae*. It is a eucaryotic, unicellular, spherical organism of 5-10 μm diameter and its vegetative multiplication proceeds in a multilateral manner. In the course of its sexual cycle, vegetative cells coalesce forming a so called ascus enclosing 2-4 spores.

The basic enzyme system of baker's yeast is responsible for the fermentation of glucose to ethanol [1,2]. In the biochemical sequence of alcoholic fermentation, up to the point of pyruvic acid the process is analogous to lactic acid fermentation in higher organisms, however thereafter, under the action of pyruvate decarboxylase, pyruvate is converted to acetaldehyde with the evolution of carbon dioxide. Finally, acetaldehyde is reduced by the alcohol dehydrogenase of baker's yeast to ethanol. The activity of the individual enzymes depends greatly on the method of cultivation. For example, in an intensively aerated culture the production of mitochondrial enzymes important for respiration become prominent. Baker's yeast proliferates only in dilute, 2-3% solutions of sugars; in more concentrated solutions growth is attenuated, although fermentation continues, a fact indicated as shown by the continued evolution of carbon dioxide. Enzymes in baker's yeast are stable, their temperature optimum being 28-35°C. Additionally baker's yeast remains active at low pH values, surviving even pH 2-3 which provides a form of protection from infections by alien microorganisms. The cells are, however, inclined to autolysis induced mainly by lack of oxygen, high temperature and some organic compounds e.g. toluene.

The versatility of baker's yeast as a biocatalyst was first recognized a long time ago [22], but recent demand for stereoselective transformations and the dramatic development of analytical techniques has stimulated a revival in interest and further development of its applications. Nowadays, baker's yeast can be handled by synthetic organic chemists more or less as a simple reagent [3-6,329] (see also reviews on biocatalysis and microbial transformations cited in Sections 1 and 1.1).

The reasons for the extensive use of baker's yeast are as follows:

i) *The active cell mass is extremely cheap and is available in unlimited quantities*. More than 500 000 tons/year of baker's yeast are produced globally [7] and marketed either in the form of cake yeast containing about 70% moisture or as lyophilized active dry yeast with 4-8% moisture content. In the latter form, it has a very long shelf life.

ii) *Handling of baker's yeast is simple*, most of the transformations can be carried out in conventional laboratory or pilot plant equipment using inexpensive additives such as water, saccharose, ethanol, inorganic salts, and eventually other nutrients. Realization of the processes does not require special microbiological or fermentational expertise and the growing of cells, if required, is also not too demanding an operation.

iii) Baker's yeast is neither toxic nor pathogenic and does not require sterile conditions.

iv) Yeast cells contain several enzymes. More than a hundred enzymes have been isolated in a purified form from baker's yeast, of which several are commercially available (cf. ref. [49] in Section 1.1). This facilitates the execution of many different transformations, including those requiring more than one enzyme. Since enzymes act under *in vivo* conditions, regeneration of cofactors, when required, can often be accomplished with the aid of the host enzyme system of the cells, although this is not necessarily the most efficient method [8].

On the other hand, there are a number of less attractive features to baker's yeast catalyzed transformations:

i) Manifestation of more than a single type of enzyme activity may be a disadvantage (cf. Sections 1.2 and 1.2.1).

ii) In reactions performed with living cells *the ratio of cell mass to substrate is usually high* (3-20 to 1, calculated for the dry cell mass) and *the permissible limit of substrate concentration is low.* These factors diminish the overall yield and make the isolation of the desired product from the aqueous medium containing cell mass, nutrients, and metabolites, difficult. When compared to the use of isolated enzymes much more solvent may be necessary for isolation and processing, requiring numerous and varied filtration and extraction operations, and also additional purification of the product.

iii) Reproducibility of the process may often be capricious. It was observed repeatedly that different strains of *S. cerevisiae* performed the transformation of the same substrate with varying selectivities [9-12]. The general sensitivity of biotransformations to experimental conditions (cf. appropriate subsections of Chapter 1 and 2, and also Section 5.1) also applies to baker's yeast [13]. This is, not however critical in terms of its applicability, merely serving as a warning that it is advisable to use yeast of constant quality and to optimize the process for a given substrate.

iv) Baker's yeast cannot be used when either the substrate or the product decompose in water, since the living cells only work in an aqueous medium. Even when using an organic solvent (see later), the cells themselves are in an aqueous micro-environment.

Generally, for the transformation of unnatural substrates, not growing but resting cell cultures are used, preferably under the following conditions:

i) Transformations are generally accomplished with *vigorously fermenting yeast* in a medium containing a substantial amount of sugar [5,6]. Under such conditions the non-buffered medium turns rapidly acidic with pH dropping to 4-5 or even lower.

ii) Occasionally non-natural substrates are transformed in a *non-fermenting medium using a high cell mass/substrate ratio* [14,15]. In this case, an aqueous suspension of baker's yeast is used without any additive whereby the medium stays approximately neutral. Often this method provides improved selectivities over those of fermenting yeast and furthermore, no ethanol is formed facilitating the isolation of the product.

iii) Both fermenting and non-fermenting processes can be realized with *immobilized cells of baker's yeast*. Immobilization has the advantages of the simplicity of product isolation and ultimately increased selectivity. This however, must be balanced against lower productivity due to the diminished activity of immobilized cells and/or a higher degree of product retention. Reaction with fermenting baker's yeast enclosed in a dialysis tube [630] is a special case that should also be mentioned.

iv) Biocatalysis by baker's yeast can be carried out *in organic solvents* with cells encapsulated into reverse micelles [16-18] or by using polymer entrapped yeast cells [19,20,33]. When compared to conventional methods proceeding in the presence of sugar, there have been occasionally reports of enhanced selectivity using this technique. More importantly however, use of an organic medium allows the transformation of substrates that are sparingly soluble in water whilst at the same time considerably reducing the rate of competing undesired chemical reactions (e.g hydrolysis, enolization, racemization).

Of the enzymes present in baker's yeast, the use of oxidoreductases performing carbonyl reductions is most widespread, but hydrolases, lyases, and other enzymes are all also at the user's disposal for synthetic application [3,4,21,22,329].

5.1 Reduction of Carbonyl Compounds with Baker's Yeast

Baker's yeast is useful for the preparation of optically active alcohols by reduction of a variety of carbonyl compounds. Stereoselectivity in reductions with baker's yeast is usually governed by Prelog's rule (Section 4.2), but whilst the sense of selectivity can be predicted with a fair degree of accuracy, its magnitude is much dependent on the constitution of the substrate, the quality of baker's yeast, and also on reaction conditions.

The types of selectivities associated with the reduction of carbonyl compounds with baker's yeast are the same as those that result from the use of alcohol dehydrogenases (cf. Chapter 2 and Section 4.2).

If ultimately poor selectivity results, two factors may be responsible [21,23]:

i) Firstly, it may be due to participation of more than one reductase enzyme with different kinetic parameters and eventually opposite selectivity [21,24-26].

ii) Or secondly, that when offered non-natural substrates none of the enzymes in baker's yeast is completely selective (e.g. the process is catalyzed by a single enzyme but with non-complete selectivity [23]; cf. Section 4.2 and its subsections for examples of biotransformations catalyzed by single oxidoreductases with non-complete stereoselectivity).

Distinguishing between the two situations is simple. In the first case, selectivity is dependent on substrate concentration whilst in the second one it is not [21,24,26]. Conclusive evidence can of course be provided by comparing the selectivities of the isolated enzymes and intact cells [24]. (For data regarding pure enzymes isolated from baker's

yeast see refs. [24,25,27,28,29-31]). In the majority of cases, it turned out that competing enzymes of opposite selectivity was the factor responsible for poor stereoselectivity
[21].

In baker's yeast, reduction of carbonyl compounds is carried out by enzymes requiring
cofactors (NADH, NADPH), thus one condition of their activity is the regeneration of
oxidized cofactors. Most often alcoholic fermentation is exploited for this purpose where
NAD(P)H is regenerated in conjunction with the multienzymic degradation of carbohydrates, the latter being the ultimate electron donor. In the course of this process the
uptake of carbohydrate may be rate limiting [32] and acetaldehyde formed competes for
NADH. Since acetaldehyde is usually a better substrate than the carbonyl compound to
be reduced, formation of one mole of the desired product is accompanied by the production of several hundred moles of ethanol with an appropriate demand for carbohydrates
[14,23].

In transformations accomplished with non-fermenting baker's yeast, the necessary cell
mass is much larger than in the fermenting mode. Here, cofactor regeneration is catered
for by the carbohydrate reserve of yeast and autolysis of the cells.

It has already been discussed in connection with transformations carried out with oxidoreductases, how when using the coupled substrate method, ethanol may be a suitable
electron donor for the regeneration of NAD(P)H[1] when offered to a combined alcohol
dehydrogenase/aldehyde dehydrogenase system (cf. Table 4.1 in Section 4.2). In solutions containing a few percent of ethanol, carbonyl reduction can be achieved under
aerobic conditions [34,35] although curiously, under a nitrogen atmosphere no reduction
takes place [34]. By applying a cell mass/substrate ratio customary in the fermentating
mode, similar results can be achieved [34] and indeed with certain substrates selectivity
may even be better. Alcohols as coupled substrate electron donors can be used in biore-
ductions with immobilized baker's yeast in hexane [331].

Generally, the same methods are available for *optimizing transformations carried out
with the aid of baker's yeast* as are available for other biotransformation processes (cf.
Table 2.2 in Section 2.3.2). Those most frequently used to influence selectivity are as
follows:

i) Modification of pH [36,37], cell mass/substrate ratio [13], substrate concentration
[13,38-40], and sugar concentration [19].

ii) Use of various additives, such as metal salts [41], allylic alcohol, α, β-unsaturated
carbonyl compounds [42,43], ethyl chloroacetate and similar compounds [44].

iii) Modification of the yeast by pretreatment [35] or immobilization [39,41]. Use of dry
yeast instead of row cake yeast [43,45] or of pure strains [9,11,46].

iv) The most frequently used approach is the appropriate modification of the constitution of the substrate. This may not only influence the magnitude but also the sense of
stereoselectivity [26,36,40,47-60].

---

[1] Examples are known where ethanol is used as an electron source for cofactor regeneration in reductions catalyzed by other microorganisms [8,33]. This method may also play a role in processes run in
the presence of sugars, since not only do these produce ethanol as well but most often the substrate
is also added in ethanolic solution.

Clearly then, there are ample possibilities for improving the situation, if a reduction with baker's yeast is showing a good yield but insufficient selectivity.

It should be noted that the outcome of reductions differ considerably depending on whether the substrate is an aldehyde, a ketone, a diketone, or an oxocarboxylic acid. Accordingly, these cases are dealt with individually in the following sections.

5.1.1   Reduction of Racemic Aldehydes

In the reduction of racemic aldehydes with baker's yeast enantiomer selectivity was expected, and, in fact, observed [22] and utilized a long time ago, for example for the resolution of racemic hydroxyaldehydes [61-63].

As Fig. 5.1 illustrates, the method has been extended to the reduction of racemic aldehydes with various types of chirality, although it is apparent that the enantioselectivity of the transformation is not sufficient for synthetic purposes.

Fig. 5.1.a shows the reduction of a racemic aldehyde in the preparation of (S)-citronellol. The product shown in Fig. 5.1.b can also be obtained from a racemic aldehyde, but here, owing to enolization of the substrate, one enantiomer of the product alcohol was isolated in more than 50% yield. Fig. 5.1.c and 5.1.d illustrate the reduction of aldehydes with axial and planar chirality respectively.

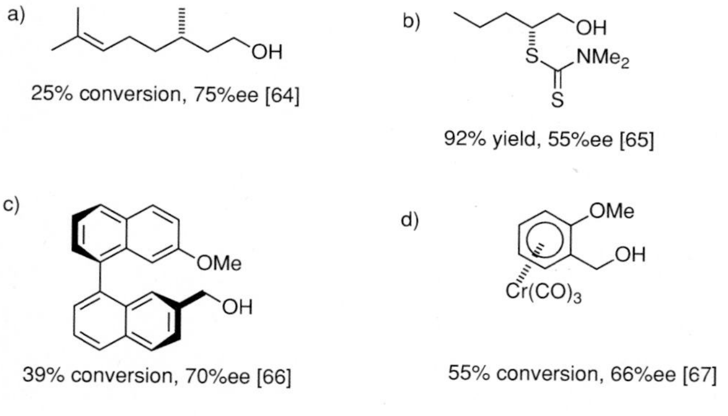

a)
25% conversion, 75%ee [64]

b)
92% yield, 55%ee [65]

c)
39% conversion, 70%ee [66]

d)
55% conversion, 66%ee [67]

Fig. 5.1.   Alcohols prepared by enantiomer selective reductions of racemic aldehydes

5.1.2   Reduction of Ketones

With prochiral ketones the process is enantiotope selective and may, in principle, yield chiral secondary alcohols with total conversion. The racemic ketones however, can be expected to show simultaneous enantiomer and diastereotope selectivity. It is convenient to discuss the two situations independently.

5.1.2.1 Achiral Ketones[2]

The reduction of ketones with baker's yeast has a long tradition [22,68-70]. The first systematic investigations on the stereochemistry of the process were carried out with alkyl ketones [23]. It was established that the predominant product resulting from reduction of methyl and ethyl alkyl ketones was the one predicted by Prelog's rule except in the case of butyl ethyl ketone which gave the (*R*)-alcohol. Sterically hindered carbonyl groups cannot be reduced.

Table 5.1 shows a number of examples of the preparation of chiral alcohols with high selectivity.

Table 5.1. Alcohol products from enantiotope selective baker's yeast reduction of the corresponding achiral ketones.

No.	Product[a]	ee(%)[a]	Reference
	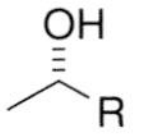 R		
5.1	Et	67	[23]
5.2	Pr	64	[23]
5.3	*i*-Pr	90*	[71]
5.4	Bu	82	[23]
5.5	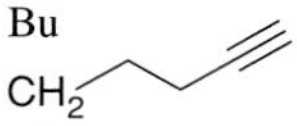	>99	[72]
5.6	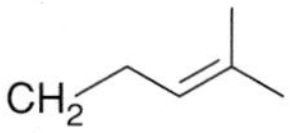	94	[5,68]
5.7	CH=CH-Ph(*E*)	80	[73]
5.8	CF_3	80	[14]
5.9	CH_2Br	96*	[14]
5.10	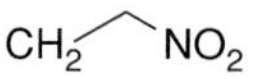	98	[74]
5.11	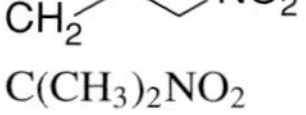	96	[74,75]
5.12	$C(CH_3)_2NO_2$	>95	[76]
5.13	$CH_2OPO(OMe)_2$	74	[77]
5.14	$CH_2SC(S)NMe_2$	>96	[65]
5.15	CH_2SPh	95*	[78]
5.16	CH_2SO_2 Ph	>97	[79,80]
5.17	CH_2OBn	90	[80]
5.18	$(CH_2)_2OBn$	>95	[80]
5.19	Cyclohexyl	95	[14]
5.20	Ph	93	[148,81]
5.21	1-Naphthyl	82	[14]
5.22	$Si(Me)_2Ph$	78	[82]

[2] Additional results on the reduction of keto sulfones [333], hydroxy ketones [334-336], 6-oxoalkyl-1,3-dioxin-4-ones [341], and various bicyclic ketones [337-340] have appeared recently.

Table 5.1. (contd.)

No.	Product[a]	ee(%)[a]	Reference

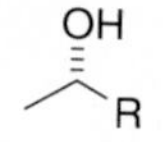

R

5.23	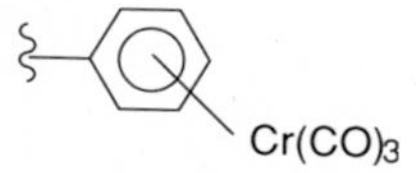	>99 / 73	[48] / [83]
5.24	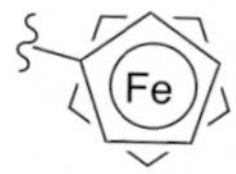	>96	[83]
5.25	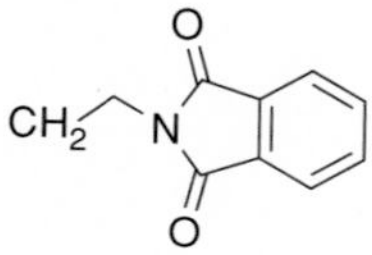	>95	[76]
5.26	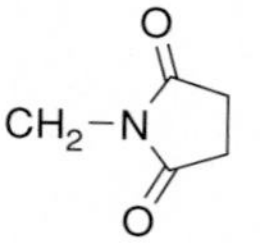	>95*	[76]
5.27	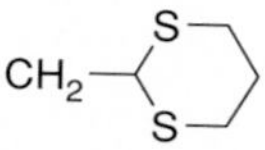	>99	[84,85]
5.28	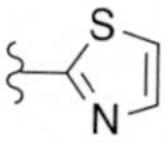	>95	[86]
5.29	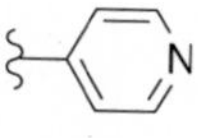	96	[87,88]
5.30	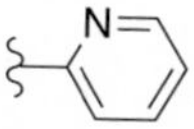	>95	[87]
5.31	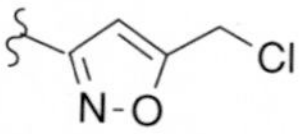	>95[b]	[89]
5.32	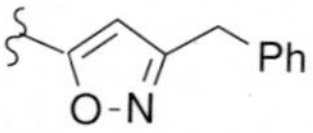	>95[b]	[89]
5.33	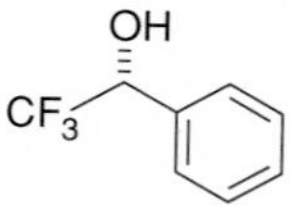	>98	[14,90]
5.34	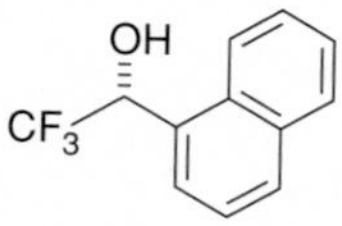	81	[14]

Table 5.1. (contd.)

No.	Product[a]	ee(%)[a]	Reference
5.35	OH CF$_3$ CH(S——)$_2$	55	[91]
5.36	(structure with OH and N-sulfonyl ring, O=, O$_2$)	>95	[76]
5.37	OH, SO$_2$——CH$_3$	63	[79]

	OH HO⎯⎯⎯⎯R R		
5.38	Me	90	[5]
5.39	*i*-Pr	>99	[92]
5.40	*t*-Bu	>99	[92]
5.41	Pentyl	>99	[93]
5.42	Ph	92	[47,94]

5.43	OH AcO⎯⎯⎯Ph	94	[47]

5.44	OH HO⎯⎯⎯SPh	>99	[95]

	OH HO⎯(⎯)$_n$⎯SO$_2$Ph		
5.45	n=1	99	[96]
5.46	n=3	93	[96]
5.47	n=5	72	[96]

5.48	OH HO⎯⎯⎯OBz	>99	[49]
5.49	OH AcO⎯⎯⎯OBn	>99	[49]
5.50	OH N$_3$⎯⎯⎯OAc	>96	[97]
5.51	OH Cl⎯⎯⎯SO$_2$Ph	84[c]	[98]

Table 5.1. (contd.)

No.	Product[a]	ee(%)[a]	Reference

X			
5.52	Me	95	[91]
5.53	MeOCH$_2$OCH$_2$	95	[91]

X			
5.54	Me	>95	[86,99]
5.55	CF$_3$	67	[86]
5.56	Et	>95	[86,99]
5.57	Butyl	>96	[99]
5.58	Hexyl	>95*	[86]
5.59	HOCH$_2$	>95*	[86]
5.60	HO(CH$_2$)$_3$	>96	[99]
5.61	BnOCH$_2$	>95	[86]
5.62	MeOCH$_2$OCH$_2$	>95	[86]

5.63		>96	[99]

5.64		>96	[99]

a: Usual range of chemical yields is between 50-100%. The ee values marked with an asterisk refer to lower (20-50%) chemical yields. b: Absolute configuration is not given c: After recrystallization the product is optically pure.

Selectivity with methyl alkyl, alkenyl, and alkinyl ketones yielding (S)-alcohols is high (e.g. 5.1-5.6) and improves with the size of the substituent, although rates drop with higher molecular mass substrates.

It is evident that the variability of substrates is considerable. Substituents may be halogen atoms[3] (5.8, 5.9, 5.33-5.35, 5.51), a nitro group (5.10-5.12), sulphur containing functional groups (5.14-16), phosphoric esters (5.13), or a free (5.38-5.42) or protected hydroxy group (5.17, 5.18, 5.43). Furthermore, substituents may contain cycloalkyl (5.19, 5.33), aromatic (5.20, 5.21) and various heterocyclic groups (5.25, 5.26, 5.28-32). For synthetic purposes thioacetals (5.27, 5.35, 5.52-5.62), thioketals (5.63-64), and bifunctional compounds with transformable functional groups (5.44-51) proved to be very useful and one even more exotic example is the products arising from the transformation of metallo-organic compounds (5.22-24).

There is clearly a marked difference between the reduction of dialkyl ketones and those containing other substituents, a difference that cannot be ascribed to steric effects alone. For example, it is conspicuous that ketones with bulky groups (e.g. 5.22-24, 5.36, 5.37) can be smoothly reduced and an interesting observation that in the reduction of hydroxymethyl ketones (5.38-42) the directing effect of the polar hydroxy groups overrides steric effects, as is illustrated by the reduction of 2-oxopropanol, for example, which fails to follow Prelog's rule (5.38). Also, in the reduction of the thioketals (5.63-64), a slight modification of substitution results in the inversion of stereoselectivity. A similar inversion of selectivity occurs between hydroxymethyl compounds and their *O*-protected derivatives (the pairs 5.42 and 5.43, and 5.17 and 5.38). All this serves to illuminate the rather limited scope of Prelog's rule.

Different approaches to increasing selectivity can be illustrated by the reduction of acetophenone. Optical purity is substantially increased by working under non-fermenting conditions [14], or by taking a modified substrate with removable groups such as 4-iodoacetophenone [54] or the metallocene (5.23) [48].

There are many examples in syntheses of preparative conversion of prochiral ketones by baker's yeast to products of high optical purity. For example, the acetylenic alcohol (5.5) was an intermediate in the synthesis of brefeldin-A [72] and compound (5.6) is the pheromone (*S*)-sulcatol [101]. The nitro compound (5.10) can also be converted to the above natural products [102] or to other pheromones [332].

1,3-Dithianes proved to be versatile intermediates. Thus (5.27), (the antipode of which is accessible with the aid of other microorganisms [85,103,104]) was used both for the synthesis of the macrolide antibiotic grahamimycin-A [84] and of a number of optically pure pheromones [85,105]. The dithiane (5.56) was converted to the pheromone (+)-*endo*-brevicomin [106].

Alcohols derived from arylthiomethyl or arylsulphonylmethyl ketones have also found several applications. For example, using (*R*)-3-phenylthiopropane-1,2-diol (5.44) both enantiomers of the hexadecan-5-olide could be prepared [95]. (*S*)-2-Phenylsulfonylmethyl-oxirane, a molecule readily accessible from the chloromethyl compound (5.51) is also a valuable intermediate [98]. Finally, the azido-alcohol (5.50) can be converted to optically active β-blockers [97].

---

3 Reduction of methyl-perfluoroalkyl ketones provided chiral alcohols with good enantiomeric purity but unknown configuration [100].

5.1.2.2   Racemic Ketones

In the reduction of racemic ketones one can expect both enantiomer and diastereotope selectivity to predominate. If the racemic ketone substrate contains a labile chiral center, as in 2-phenylthio cycloalkanones [342], reduction may lead nearly completely to one of the possible diastereomeric products in high enantiomeric purity.

i) Table 5.2 lists *reductions proceeding with high enantiomer and diastereotope selectivity.* With a view to the preparation of monocyclic β-lactams, various microorganisms were tested [109] and baker's yeast was found to produce the alcohol (5.68); additionally resting cells were more selective than growing cells. Reduction of compounds (5.69-71) leading ultimately to prostacyclin analogs [110] uncovered several interesting features. Whilst the same process exhibits poor enantiomer selectivity with cyclobutanes condensed with a five membered ring (cf. Section 2.3.2.8 and Fig. 5.2), in the present case enantiomer selectivity is high both in the reduction of the cyclobutanone (5.69) and of the dichloro-analog (5.71). It is surprising that reduction accelerates dramatically with the number of chlorine atoms and that it is the sterically most shielded ketone which is reduced most rapidly.

Table 5.2. Enantiomer and diastereotope selective reduction of racemic ketones by baker's yeast.

No	Substrate[a] (racemic, X=O)		Alcohol product[a] (5.X a)			Remaining substrate[b] (5.X b)		Ref.
			X	Y(%)	ee(%)	Y(%)	ee(%)	

	R¹	R²						
5.65	Cl	H		32	>95	40	>95	[107]
5.66	OAc	H		40	54	19	>95	[107]
5.67	OAc	*t*-Bu		30	>95	40	>95	[107]
5.68				42	90	27	91	[109]

Table 5.2. (contd.)

No	Substrate[a] (racemic, X=O)	Alcohol product[a] (5.X a)			Remaining substrate[b] (5.X b)		Ref.
		X	Y(%)	ee(%)	Y(%)	ee(%)	
5.69	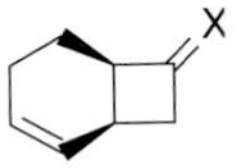		25	92	32	40	[110]
	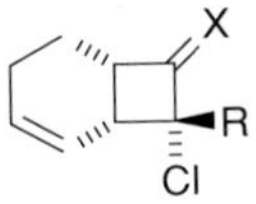						
5.70	R=Cl[c]	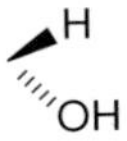	27	>99	14	82	[110]
5.71	R=H	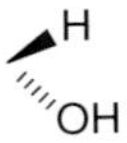	17	56	67	12	[110]
5.72	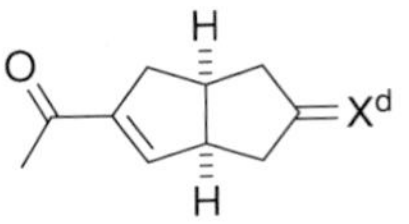	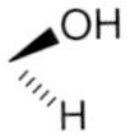	25	>99	22	49	[111]
	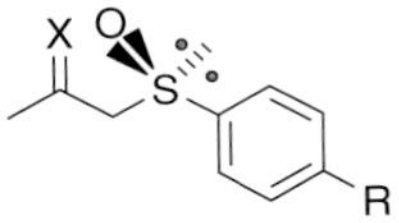						
5.73	R=H	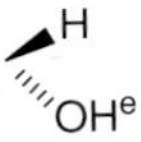	35	>95	30-50	>95	[78,107]
5.74	R=Cl	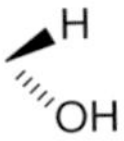			38	>99	[108]

a: The faster reacting enantiomer is shown; b: The remaining substrate is the enantiomer of the substrate shown here; c: This reduction was extremely fast and was accomplished in 45 min [110]; d: In this reduction 10% of α, β-saturated ketones were also obtained; e: In this reduction approximately 10% of racemic diastereomeric alcohol was also obtained.

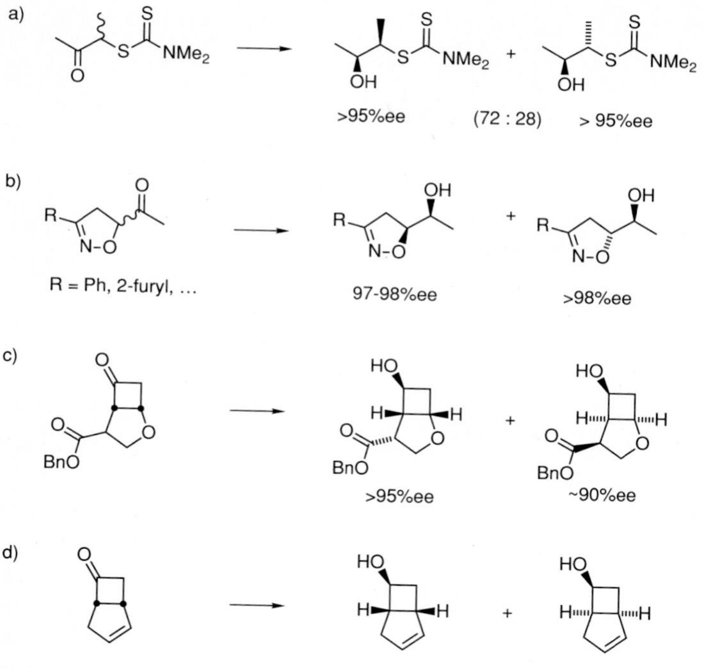

a)

>95%ee   (72 : 28)   > 95%ee

b)

R = Ph, 2-furyl, ...

97-98%ee   >98%ee

c)

>95%ee   ~90%ee

d)

Fig. 5.2.   Reduction of racemic ketones with low enantiomer and high diastereotope selectivity

The optically active sulfoxides (5.73-74) prepared with the aid of baker's yeast are also versatile intermediates, the compound (5.74.b) for example yields pure enantiomers of 2-hydroxycarboxylic esters [108].

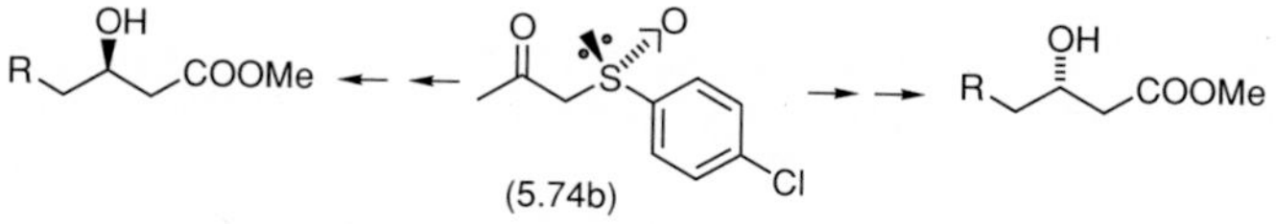

(5.74b)

Baker's yeast may be utilized in the preparation of steroid intermediates [116,117], such as the tricyclic alcohol shown below [116]:

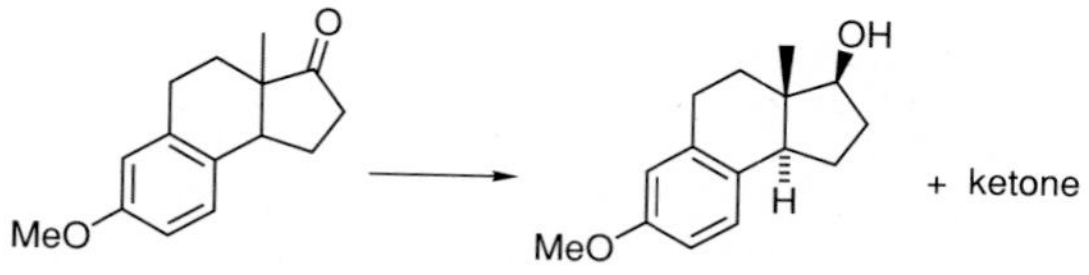

+ ketone

ii) Reductions proceeding with poor enantiomer selectivity but high diastereotope selectivity (Fig. 5.2) may still be very useful since here the two diastereomeric alcohols

formed in high enantiomeric purity can subsequently be separated by conventional methods (cf. Section 2.3.2.8).

Note that in the transformation shown in Fig. 5.2.a the product ratio may be shifted,
due to enolization, in favor of the more rapidly formed diastereomer [65].

Diastereomeric alcohols arising from the reduction of 5-acetyl-2-isoxazolines [112]
(Fig. 5.2.b) are also useful chiral building blocks.

In contrast to the corresponding six membered ring compounds, enantiomer selectivity
of cyclobutanones condensed to a five membered ring is low (Table 5.2, 5.69-71). Figs.
5.2.c [113] and 5.2.d [114,115] illustrate the use of this transformation in the synthesis
of both penicillin analogs and prostaglandins respectively.

iii) Reduction of racemic α-hydroxy and α-acetoxy ketones.

Reduction of α-hydroxy ketones is related to various diols that can be prepared by
means of acyloin condensation from aldehydes with the aid of fermenting yeast (cf. Section 5.5.1). It was observed [118,119] (Fig. 5.3.a) that the sense of diastereoselectivity
always corresponds to that of the acyloin diol, i.e. it is *re* with carbonyls near to the end
of the chain and *si* for a carbonyl group which is flanked by a bulky R substituent and
the hydroxy group.

In transformations studied in connection with the synthesis of *endo*-brevicomin, a
bicyclic pheromone [120] (Fig. 5.3.b), the dihydropyran substrate was found to suffer
first ring cleavage by addition of water to an open chain racemic ketone which is then
reduced, in accordance with the statement above, in a *re* selective manner. Enantiomer
selectivity with this substrate is high, but it is variable with the corresponding acetyl or
butyryl substituted dihydropyrans.

These observations lose their validity with structurally rather different racemic
α-hydroxyketones, e.g. those shown in Fig. 5.3.c, which were studied in relation the synthesis of tetracycline intermediates using baker's yeast [121]. As Prelog's rule predicts,
this reduction is *re* selective, but lacks appreciable enantiomer selectivity so resulting in
a mixture of diastereomers separable by crystallization.

With reduction of open-chain α-hydroxy and α-acetoxy ketones, valuable intermediates become accessible enantiomerically pure form (Table 5.3) [126,127]. As before, *re*
selectivity is high and mostly independent of substitution, whilst enantiomer selectivity
is dependent on the size of substituents R^1-R^4. In the reduction of compounds (5.75-86.a)
that contain few or small substituents, the alcohols (5.75-79.c) are formed in varying
amounts. With ketones to which numerous or bulky substituents are attached, enantiomer
selectivity is favored. After suitable transformations the products from these reductions
can be converted to optically pure 1,3-dioxolanes [126,127]:

(5.75-86b) ⟹ (5.75-86d)   (5.75c, 5.76c, 5.79c) ⟹ (5.75e, 5.76e, 5.79e)

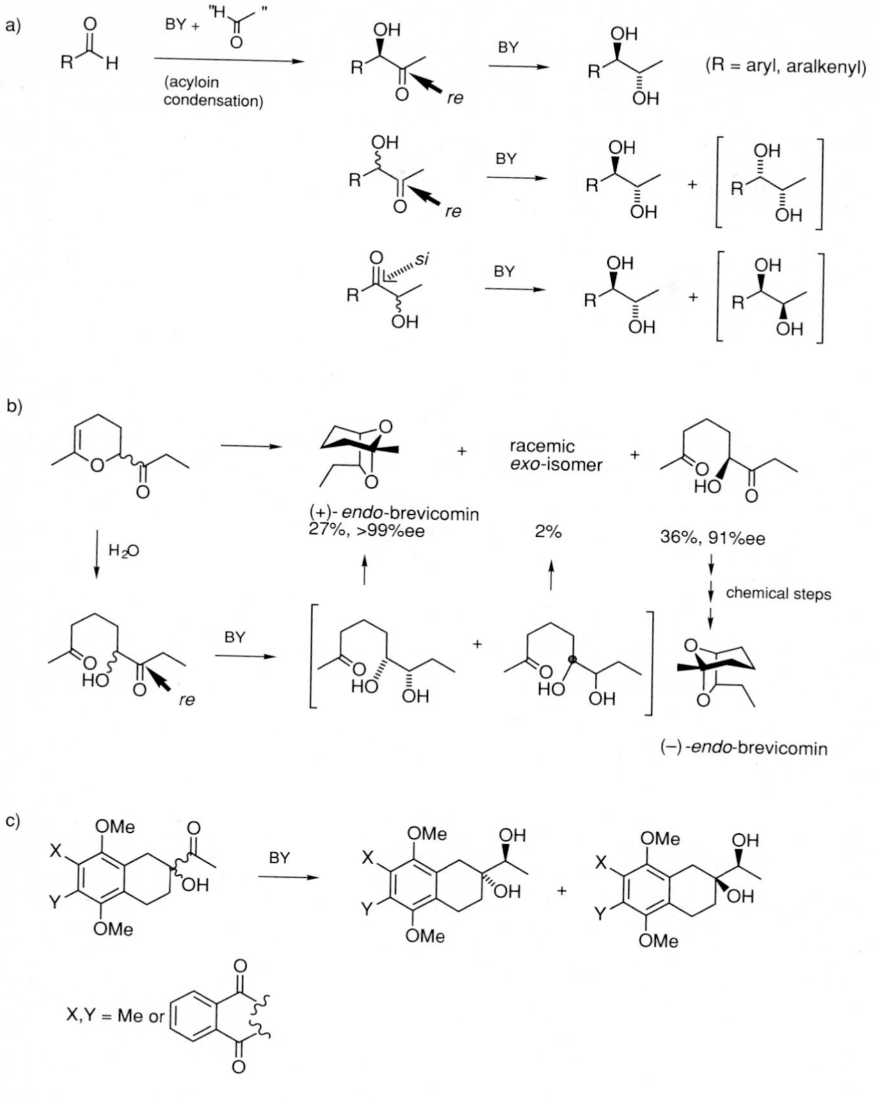

Fig. 5.3.   Reduction of racemic α-hydroxy ketones

The dioxolanes in turn, (also accessible from the corresponding acyloin diols, cf. Section 5.5.1), were used, amongst others, in the synthesis of L-deoxysugars [128], N-trifluoroacetyl-L-acosamine [129], numerous pheromones [127,205], several unusual sugars [122], and other related compounds [130-133].

Some of the diols included in Table 5.3 can also be utilized via other intermediates, one example of which is the synthesis of the pheromone (4S,5R)-sitophilure from the diol (5.80.b) [134].

Table 5.3. Reduction of racemic α-hydroxy- and α-acetoxy-ketones.

a   →[BY]   b (mafor product)   +   c (minor product)

No.	Product(s): b (and c) R¹	R²	R³	R⁴	Yield[a] (%)	de (%)	ee (%)	Ref.
5.75	H	H	H	H	70-80	20	90[b]	[119]
5.76	H	H	Me	H	70-80	20	>95[b]	[119]
5.77	H	H	H	Me	15-20	80	>95	[119]
5.78	H	H	Et	H	15-20	80	>95	[119]
5.79	H	Me	H	H	80	0	>95[b]	[119,122]
5.80	H	Me	H	Me	15	>90	>95	[119,122]
5.81	Ac	H	H	(structure) c)	30	>90	>95	[123]
5.82	Ac	H	H	(structure) d)	20	>90[d]	>95	[124]
5.83	Ac	H	H	(structure)	20	>90	>95	[124]
5.84	Ac	H	H	CH₂—(structure)	20	>90	>95	[125]
5.85	Ac	H	H	CH(structure)	30	>90	67[e]	[125]
5.86	Ac	H	CH₃	CH(structure)	35	>90	50[e]	[125]

a: Combined yields of reduced products; b: The minor diastereomer has the same high optical purity; c: The substrate was the diastereomerically pure racemic ketone. The other diastereomeric pair of enantiomers proved to be no substrate for baker's yeast [123]; d: Substrate was a 1:1 mixture of the two racemic diastereomers. At the chiral centre in R⁴ de=80%; e: After crystallization optically pure products can be obtained.

5.1.3 Reduction of Dioxo Compounds

In the course of the reduction of compounds with two constitutionally different oxo groups, stereoselectivity is often associated with a high degree of regioselectivity. Thus, either just one or both carbonyl groups out of two may be reduced. Since the outcome of this reaction depends greatly on the distance between the two carbonyl groups, it is convenient to classify such reductions according to this feature.

5.1.3.1 1,2-Dioxo Compounds

In the transformation of α-oxoaldehydes an interesting dismutation can take place [135]:

It has long been known that reduction of symmetrical 1,2-diketones in two steps yields optically active diols:

R	Ref.
Me	[136]
Ph	[136]
(furyl)	[137]
(thienyl)	[138]

In terms of synthetic application, it is products arising from the reduction of methyl α-diketones (Table 5.4) that are of primary interest. Results depend greatly on experimental conditions, a feature strongly suggested by the data on reduction of 1-phenyl-1,2-propanedione (5.87.a). With non-fermenting yeast and at low temperature any of the oxo groups can be reduced with high enantiotope selectivity to a monoalcohol, whilst on reduction of the second oxo group a common product, the alcohol (5.78.b), is obtained. Executing the reduction at lower pH with fermenting yeast greatly improves regioselectivity, and thus the alcohol (5.87.b) can be isolated.

The minor pheromone component of grape borer (*Xylotrechus pyrrhodeus*) is the main reduction product of diketone (5.88.a), whilst the principal pheromone of this insect, a *syn*-diol (5.88.e) cannot be prepared this way; it is however, accessible from the same substrate using *Beauveria sulfurescens* [141].

The antifeeding agent against yellow butterfly, 1-hydroxyethyl-pent-2-enolide was prepared from a diol (5.89.d) by yeast reduction [142], whilst the diol (5.90.d) facilitated the synthesis of an optically pure deoxy sugar.

Table 5.4. Reduction of some 1,2-diketones by baker's yeast.

No.	R	Process[a]	b Y(%)	ee(%)	c Y(%)	ee(%)[b]	d Y(%)	ee(%)	d:e ratio	Ref.
5.87	Ph	A	46	94	36	89(R)	7			[139]
		B					89	94		[139,343]
		C	80	>95						[140]
5.88	C$_5$H$_{11}$	B	50	92	22[c]	(S)[c]	5	99		[141]
5.89	CH$_2$SPh	D					56[d]	>99[d]	86:14	[142]
5.90	⟨S,S⟩CH	C	60	93[b]						[143]
		D					82[e]	97[e,f]	95:5	[143]

a: A, non-fermenting yeast, 5 °C; B, non-fermenting yeast, RT, 2 h; C, fermenting yeast, 2 h; D, fermenting yeast, ≥48 h; b: Configuration of the product in brackets; c: Determined by GLC; d: After separation from the other diastereomer by crystallization; e: After separation; f: After one crystallization the product is optically pure.

Unfortunately, certain 1,2-diketones such as 2,3-pentanedione, 2,3-hexanedione and 1,2-cyclohexanedione exhibited no appreciable enantiotope selectivity on reduction by baker's yeast [140].

5.1.3.2 1,3-Dioxo Compounds

The two oxo groups in 1,3-dioxo compounds are favorably situated for the reductive process. An indication that there is assistance from the second oxo group is clear in the observation that after the first step, reduction generally ceases.

i) Reduction of open chain 1,3-diketones (Table 5.5)
From data presented in Table 5.5 it is apparent that if R^1 is a methyl group (*S*)-alcohols (5.91-5.99.b) are formed with high regio- and enantiotope selectivity, whereas when R^1 is ethyl, enantiotope selectivity is poor and in the opposite sense (5.100.b). Alcohols (5.96.b) and (5.98.b) are used in the synthesis of pheromones [147,148].

ii) Reduction of prochiral cyclic 1,3-diketones (Table 5.6)
Reduction of prochiral non-enolizable 2,2-substituted cycloalkane-1,3-diones was first studied in connection with the microbial transformation of steroids using, amongst others,

Table 5.5. Reduction of open chain 1,3-diketones with baker's yeast.

No.	Product (5.x b)[a] R[1]	R[2]	ee(%)[a]	Ref.
5.91	Me	Me	>99	[144-146]
5.92	Me	Et	99	[144-146]
5.93	Me	Bu	>99*	[144]
5.94	Me	i-Pr	92*	[144]
5.95	Me	CH₂ (chain)	96*	[144]
5.96	Me	CH₂ (chain, alkene)	99	[145-147]
5.97	Me	CH₂ (long chain)	97*	[144]
5.98	Me	CH₂ (dioxolane)	98	[148]
5.99	Me	Ph	98	[144,146]
5.100	Et	Et	30[b]	[145]

a: Yields 50-100%, reductions reported to have lower yields (20-50%) are marked with an asterisk; b: Configuration is *R*.

baker's yeast [156] (see also the relevant example in Section 2.1.3). For example, reduction of diketone (5.107.a) yields an intermediate of estradiol-3-methyl ether synthesis [151]. With prochiral diketones, the enantiotopically related carbonyl groups can be attacked from either diastereotopic face of each group. It was found that whilst enantiotope selectivity (i.e. preference for the *pro-R* or *pro-S* carbonyl group) was variable, diastereotope selectivity (i.e. approach from the *re* or *si* face of the carbonyl group) was usually high. Enantiotope selectivity using such prochiral cyclic ketones can be achieved in reductions performed with *Schizzosaccharomyces pombe*, using additives such as allyl alcohol and similar unsaturated alcohols or ketones [157]. This indicates that enantiotope selectivity can be influenced by similar additives or by other methods (cf. Section 5.1).

In reductions carried out with baker's yeast on non-steroidal diketones, diastereoselectivity was generally high and independent of ring size and the substituent R. In contrast, however, enantiotope selectivity was greatly affected by these structural features. Using five-membered cyclic compounds (5.101-106.a) reduction of the *pro-S* (referring to the C_2 prochiral center; cf. equation in Table 5.6) carbonyl group was favored, whereas with the six-membered ring analogs (5.108-115, 5.117.a; 5.116.a is an exception) transformation of the *pro-R* carbonyl group was faster. In the reduction of cyclopentanediones, the degree of enantiotope selectivity was higher and more dependent on substituents whilst the generally lower enantiotope selectivity in the six-membered ring series was less influenced by these factors. Increasing the ring size renders diketones (e.g. 5.118-123.a) ever

Table 5.6. Reduction of prochiral cyclic 1,3-diones with baker's yeast.

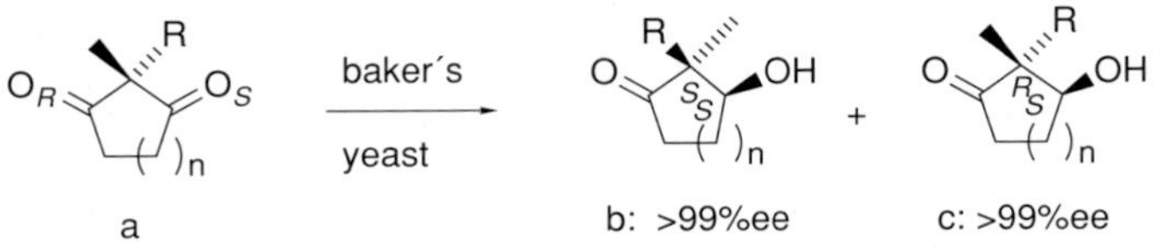

a

b: >99%ee   c: >99%ee

No.	Substrate/Products[a]		Yield (%)	de(%)[a] (dominant product)	Reference
	n	R			
5.101	1	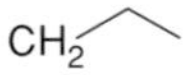 CH₂ (prop-2-enyl)	60	>99(b)	[149,150]
5.102		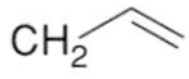 CH₂ (but-2-ynyl/butenyl)	75	80(b)	[149,150]
5.103		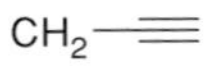 CH₂—≡	60	34(b)	[149,150]
5.104		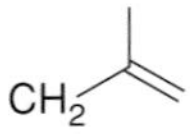 CH₂ (2-methylallyl)	75	>99(b)	[150]
5.105		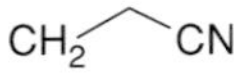 CH₂—CN	71	92(b)	[150]
5.106		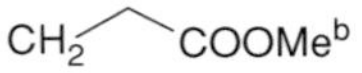 CH₂—COOMe[b]	9[b]	>99(b)	[150]
5.107		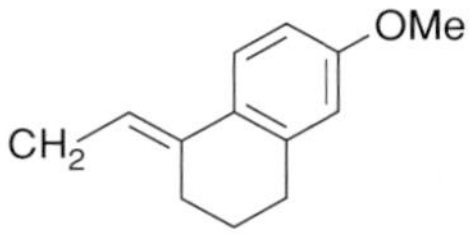 CH₂ (OMe-tetralinylidene)		60(b)	[10]
			90	>90(b)[c]	[150]
5.108	2	CH₃	52		[152]
5.109		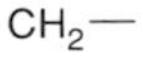 CH₂—		46(c)[d]	[153]
5.110		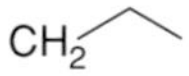 CH₂ (allyl)	80	56(c)	[150,154]
5.111		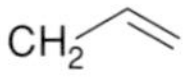 CH₂ (butenyl)	80	10(c)	[150,154]
5.112		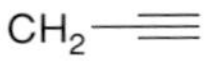 CH₂—≡	75	46(c)	[150,154]
5.113		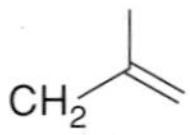 CH₂ (2-methylallyl)	49	20(c)	[150]
5.114		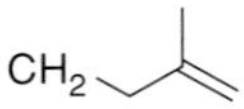 CH₂ (prenyl)	75	52(c)	[154]
5.115		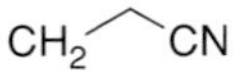 CH₂—CN	49	40(c)	[150]
5.116		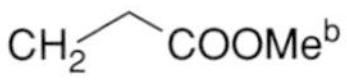 CH₂—COOMe[b]	52[b]	>99(b)	[150]
5.117		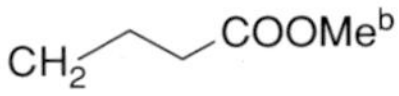 CH₂—COOMe[b]	20[b]	30(c)	[150]

Table 5.6. (contd.)

No.	Substrate/Products[a]		Yield (%)	de(%)[a] (dominant product)	Reference
	n	R			
5.118	3	CH₂ (cyclopropylmethyl)	10	96(c)	[155]
5.119		CH₂ (allyl)	20	>99(b)	[155]
5.120		CH₂—≡	60	42(b)	[155]
5.121		CH₂ (methallyl)	40	10(b)	[155]
5.122	4	CH₂ (allyl)	5	64(b)	[155]
5.123	5	CH₂ (allyl)	0		[155]

a: Ee >98% for all products; b: Products are the corresponding bicyclic lactones; c: *S. cerevisiae* cells were immobilized in polyacrylamide matrix; d: In *Pichia terricola* reduction de's of product b were >99%.

worse substrates although diastereoselectivity remains high throughout. In the case of the structurally closely related 1-substituted bicyclo[2.2.2]octane-2,6-diones, reduction with baker's yeast was also highly enantiotope and diastereotope selective [344, 345].

Many of the products shown in Table 5.6 are useful chiral building blocks. Thus the cyclopentanol (5.102.b) is the key intermediate in the synthesis of coriolin [149,160] and of the antitumor mycotoxin anguidine [158,159]. The (3S)-2,2-dimethyl-3-hydroxy-cyclohexanone (5.108.b) is also a versatile intermediate (Fig. 5.4). The ketol (5.109.b) obtained as the pure stereoisomer with the aid of *Pichia terricola* was used in the synthesis of ethyl homologs of juvenile hormones [153,173,174].

5.1.3.3 Compounds Containing Two Oxo Groups in 1,4 or More Distant Positions

In such compounds reduction of the two oxo groups is practically independent. Nevertheless, if the constitution of the two functional groups is different, a degree of regioselectivity can be expected; this situation can in fact be observed in the preparation of the alcohol used in the synthesis of (S)-β-caroten-2-ol (Fig. 5.5.a) [175].

Amongst symmetrical alkanediones, 2,5-hexanedione [176] and 2,5-heptanedione [34] under reduction with baker's yeast yield the corresponding (S,S)-diols in high enantiomeric and diastereomeric purity (Fig. 5.5.b) although reduction is attenuated with increasing chain length. (2S,6S)-2,6-Hexanediol was converted to 2,5-dimethylpyrrolidine with C₂ symmetry [177].

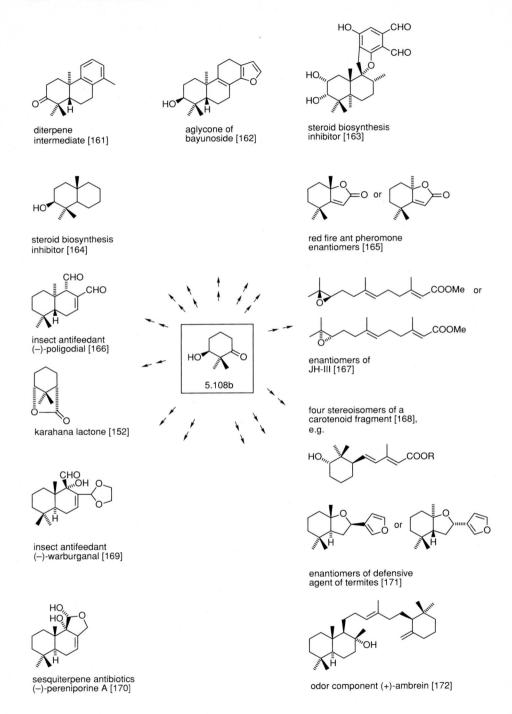

diterpene
intermediate [161]

aglycone of
bayunoside [162]

steroid biosynthesis
inhibitor [163]

steroid biosynthesis
inhibitor [164]

red fire ant pheromone
enantiomers [165]

insect antifeedant
(−)-poligodial [166]

enantiomers of
JH-III [167]

karahana lactone [152]

5.108b

four stereoisomers of a
carotenoid fragment [168],
e.g.

insect antifeedant
(−)-warburganal [169]

enantiomers of defensive
agent of termites [171]

sesquiterpene antibiotics
(−)-pereniporine A [170]

odor component (+)-ambrein [172]

Fig. 5.4.   Synthetic utilization of (*S*)-3-hydroxy-2,2-dimethyl-cyclohexanone (5.108b)

Fig. 5.5.c illustrates the manner in which the enantiomer selectivity of the process was exploited in the reduction of a racemic diketone of C_2 symmetry to prepare the diketone in an optically active form [178].

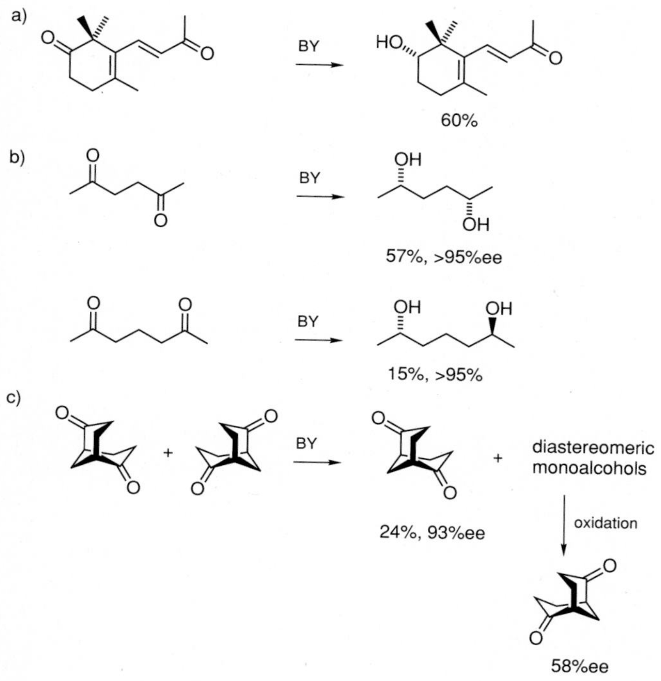

Fig. 5.5.   Examples of different types of useful selectivities in reduction of diketones

5.1.4   Reduction of Oxocarboxylic Acid Derivatives

In the reduction of oxocarboxylic acid derivatives with baker's yeast, the relative position of the two functional groups is imperative and this Section is thus subdivided according to this viewpoint.

5.1.4.1   2-Oxocarboxylic Acid Derivatives

The title compounds are favored substrates for baker's yeast since it contains enzymes with both L- and D-LDH activity, as well as an ADH enzyme which is known to catalyze the reduction of pyruvate to *(R)*-lactate [27].

It has long been known that free 2-oxocarboxylic acids under decarboxylation yield an alcohol one carbon atom shorter. This transformation is therefore of very limited synthetic interest.

Reduction of the corresponding esters, in turn, gives 2-hydroxycarboxylic esters (Table 5.7). Selectivity of the reaction is heavily dependent on the substituents of the substrate and can be predicted by Prelog's rule. With bulky groups in the vicinity of the carbonyl group (e.g. 5.124-131.a), direct reduction yields predominantly the (R)-hydroxy acids. Methyl 2-oxo-3-phenylpropionate (5.132.a) represents an intermediate situation giving an almost racemic product. When the substituents close to the carbonyl group are small, (S)-alcohols are obtained, but in contrast to the high enantiotope selectivity experienced with aryl substituted 2-oxocarboxylic esters, selectivity in this group is greatly reduced and variable. Selectivity could be improved by changing reaction conditions and/or by immobilizing the yeast cells however [19].

The synthetic application of products derived from 2-oxocarboxylic esters is exemplified by the *anti*-chlorohydrin (5.128), from which 14,15-leukotriene A$_4$ can be prepared [182] or by the preparation of (R)-hexahydromandelic acid from the diastereomeric mixture (5.130.b) [185].

5.1.4.2   3-Oxocarboxylic Acid Derivatives

Reduction of 3-oxocarboxylic acids and their esters is one of the most important reactions performed with baker's yeast. Not only do the reactions proceed readily due to acceleration by a β-positioned oxo group, but also because the products are useful, widely applicable synthons. In the present section, reduction of achiral compounds unsubstituted at C-2 is discussed, for which one would expect to see only enantiotope selectivity predominate (Table 5.8). Reduction of the 2-substituted compounds is dealt with in the next chapter.

The first reduction of this kind by baker's yeast, namely that of acetoacetic acid to (S)-hydroxybutyric acid, was described as early as in 1931 [196]. The last twenty years however have witnessed a revival in interest for this reaction, mainly from those with preparative applications in mind. In this family of compounds, Prelog's rule is only really suitable for predicting inversions in the sense of selectivity in the event of a change of substituents.

L-hydroxy acid derivative   D-hydroxy acid derivative

Table 5.7. Reduction of α-keto esters by baker's yeast.

No.	R^1(a)	R^2	Config.	Y^b (%)	de^c (%)	ee (%)	Ref.
5.124	(phenyl)	Me	R	59		>99	[2]
		Et	R	68		>95	[95]
5.125	(furyl)	Et	R	49		>95	[2]
5.126	MeOOC–(furyl)	Me	R	62		>95	[2]
5.127	(dithiane S S)	Me	R	80		64	[181]
5.128	(chain, Cl)	Et	R	67	60	>99^d	[182]
5.129	n=1	Me	R	51	50	>95^d	[183]
5.130	n=2	Et	R	57	80	99^d	[183]
5.131	n=3	Et	R	34	50	<60^d	[183]
5.132	(benzyl)	Me		44		0	[9]
5.133	n=0	Et	R^e			92^e	[2]
		Et	S	47		91	[19]
5.134	n=1	Et	S	42		75	[19]
5.135	n=2	Et	S	36		31	[19]
5.136	n=3	Et	S	25		>97^f	[19]
5.137	n=4	Et	S	31		63^g	[19]
5.138	(PhCH2–O–C(=O)–NH–chain)	Me	S	47		49	[184]

a: In reductions of racemic α-keto esters the configuration of the major diastereomeric product is given here; b: Combined yield of α-hydroxy esters; c: Diastereomeric excess refers to the major diastereomer; d: For both diastereomers; e: With a pure S. cerevisiae strain; f: With non-fermenting yeast; g: With yeast cells entrapped in polyurethane polymer.

Table 5.8. Reduction of achiral 3-oxo-carboxylic acid derivatives with baker's yeast.

No.	R¹	R²	Config.[a] C.I.P. (D/L)	Condition[b]	Yield (%)	ee (%)	Ref.
5.139	n=0	Et	S (L)	A	50-76	82-87[c]	[6]
				B		>98	[39]
				C	70	95	[34,35]
				D	75	99	[44]
				E		>95	[20]
				F	65	99	[41]
5.140	n=1	Me	R (D)	A	60	40	[53,186]
				H	56	96	[42]
5.141	n=1	Et	S (L)	D	63	94	[41]
				F	44	89	[44]
5.142	n=1	octyl	S (L)	A	64	97	[53]
5.143	n=2	Et	R (D)	A	26	96	[187]
5.144	n=4	K	R (D)	A	26	d	[188]
5.145	n=13	K	S (D)	A	40	>98	[189]
5.146		K	S (D)	A	16	98	[190]
5.147		Et	R (D)	A	70	70	[191]
5.148		K	R (D)	A	38	>99	[52]
5.149		K	R (D)	A	59	>99	[52]
5.150	Ph	Me	R (L)	A[e]	70	>90	[9]
5.151	Cl	Me	S (D)	A		62	[40]
			S (D)	B		31	[39]
			S (D)	C		90	[39]
5.152	Cl	Et	S (D)	A		55	[40]
			R (L)	A[e]	60	70	[48]
			R (L)	D	70	80	[44]
5.153	Cl	octyl	R (L)	A		>98	[40]
5.154	Br	Et	S (D)	A	45	>99	[50]
5.155	CF₃	Et	R (L)	A	75	49	[192]
			R (L)	D	60	84	[44]

Table 5.8. (contd.)

No.	R¹	R²	Config.[a] C.I.P. (D/L)	Condition[b]	Yield (%)	ee (%)	Ref.
5.156	CCl₃	Et	S (D)	A	70	84	[192]
5.157	N≡C	octyl	S (L)	A	77	82	[55]
5.158	N₃	Et	R (L)	A	80	80	[50]
5.159	N₃	octyl	R (L)	A	80	>95	[50]
5.160	(benzyl-O-)	Et	S (L) / R (D)	A / A[f]	58-73 / 11	56-71 / 48	[13,193] / [13]
5.161	(t-Bu-O-)	Et	R (L)	A	72	97	[193]
5.162	(benzyl-O-)	Me	g	A	65	30	[194]
5.163	(benzyl-O-)	i-Bu	S (L)	A	70	90	[194]
5.164	MeO	Bu	S (L)	A	48	82	[51]
5.165	HO	pentyl	S (L)	A	59	87	[51]
5.166	(benzyl-O-)	K	R (D)	A	51	>99	[52]
5.167	(Ph-S-)	Me	R (L)	A[e]	42	73	[46]
5.168	(Ph-SO₂-)	Me	S (D)	A	80	98	[54]
5.169	(Ph-S-)	K	S (D)	A	8	>99	[53]
5.170	(Ph-S-)	Me	S (D)	A	88	42	[53]
5.171	(dithiane)	Me	S (D)	B	28	78	[195]

a: D and L descriptors refer to the enantiomeric products b and c respectively; b: Conditions: A: traditional fermenting yeast; B: non-fermenting yeast; C: yeast cells in 5% ethanolic solution under aerobic conditions; D: B + ClCH₂COOEt; E: polymer entrapped yeast cells in organic solvent containing media; F: immobilized yeast cells in MgCl₂ containing solution; G: yeast cells immobilized by polyurethane entrapment; H: A + allyl alcohol; c: Optically pure hydroxyacid can be obtained from this product (see text); d: High enantiomeric purity; e: With a pure *S. cerevisae* strain; f: At high substrate concentration and low yeast/substrate ratio; g: Not reported.

A decrease in the size of R^1 and increase in the size of the esterifying group (R^2) generally enhances L-selectivity (in order to be independent of eventual heavy atom substitution in R^1, D and L descriptors are more convenient here), whilst a bulkier R^1 shifts reduction in favor of the D enantiomer. Accordingly, L selectivity was experienced with the 3-oxobutanoate (5.139.a) containing a small R^1 group and with its trifluoromethyl analog (5.135.a), whilst on reduction of the methyl or ethyl esters of analogs with larger R^1 substituents (e.g. 5.140-149.a and 5.151-154.a) under conventional conditions (i.e. in the presence of sugar and fermenting yeast) D selectivity was experienced. (The same trend was observed on reduction of C-4 to C-18 3-oxoalkanoic acids [346].) It is however evident from Table 5.8, that often, for example with *tert*-butoxy- (5.161.a) or benzyloxymethyl- (5.160.a) substituted β-ketoesters, despite the presence of relatively bulky R^1 groups, that L products still predominate. It has also been observed that enantiomeric purity is greatly dependent on reaction conditions, and not least on the quality of yeast. Occasionally, for instance with the benzyloxymethyl or chloromethyl substituted compounds (5.160.a) and (5.152.a) respectively, inversion of selectivity has been observed on changing conditions.

These findings concur with the fact that in baker's yeast, reduction of a single substrate may be catalyzed by more than one enzyme, and eventually by enzymes with opposite stereoselectivity [21]. Thus net selectivity is a resultant of competing processes. This assumption was later supported by the isolation of three enzymes from baker's yeast, two of which proved to be D enzymes and one an L enzyme towards alkyl 4-chloro-3-oxobutanoates used as model substrates [24]. In a similar study on ethyl 3-oxobutanoate (5.139.a) it was found that with pure enzymes isolated from baker's yeast, *in vitro* reduction to both the pure ethyl (*R*)- or (*S*)-3-hydroxybutanoate could be accomplished [31].

The factors influencing enantiotope selectivity with 3-oxocarboxylic acid derivatives can be summarized as follows:

L selectivity is promoted by selecting a more bulky esterifying alcohol [40,50,51,53,55, 194] , or by inhibition of the D enzymes, for example with chloroacetic acid esters [44] or certain salts such as magnesium chloride [19].

D selectivity is promoted by increasing the size of R^1 by attaching to it a removable group [53] and/or by decreasing the bulk of the esterifying alcohol, ultimately using the potassium salt of the free acid [52]. This latter measure involves not only a steric effect, but also more importantly a change from a neutral substrate to an ionic one. The free acid and the corresponding ester may participate in quite different enzymic processes. This is further indicated by the fact that it is possible to reduce the acid (5.166.a) with baker's yeast, but not its esters [52]. L Enzymes can be inhibited with allylic alcohol or α,β-unsaturated ketones [42].

The effect of immobilization of yeast cells on selectivity is dependent on the nature of both the substrate and the supporting polymer [39].

Several methods have been tested for improving the enantiomeric purity of 3-hydroxycarboxylic acid derivatives, mostly using ethyl (*S*)-3-hydroxybutanoate as the model substance. This compound can be prepared in high optical purity by recrystallization of its 3,5-dinitrobenzoate [6] or amide [197] or by lipase catalyzed enantiomer selective acylation when it remains unchanged [198]. Recrystalization proved to be useful also in the purification of 3,5-dinitrobenzoates of methyl (3R)- or (3S)-3-hydroxyvalerates [53,199], of the halogenated hydroxyesters (5.155.b) and (5.156.b) [192] or of the acid obtained from the ester (5.167.b) [192].

Among 3-hydroxycarboxylic acid esters, those of the pure enantiomers of 3-hydroxybutyric and -valeric acid are most important for syntheses (Fig.5.6).

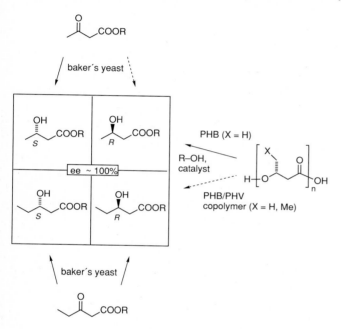

Fig. 5.6.   Possible sources of 3-hydroxy-butanoate or valerate enantiomers of high enantiomeric purity

Ethyl (S)-3-hydroxybutanoate (5.129.b), methyl (R)-3-hydroxypentanoate (5.140.b) and octyl (S)-3-hydroxypentanoate (5.142.b) can be prepared in over 95% optical purity using baker's yeast (Table 5.8) the residual antipodal impurity may be removed by the methods mentioned above.

Alkyl (R)-3-hydroxybutanoates can be obtained enantiomerically pure by alcoholysis of a poly(hydroxybutanoate) (PHB) produced in large quantities by certain microorganisms such as Alcaligenes eutrophus [200,201] or Zooglea ramingera [202] and available commercially. A PHV/PHB copolymer produced by Alcaligenes eutrophus is also avaiable as a commercial product and contains more than 30% 3-hydroxyvalerate. After ethanolysis of the copolymer, pure (3R)-3-hydroxyvalerate can be separated from the mixture by distillation [201]. Reduction of methyl 3-oxovalerate (5.140.a) with baker's yeast [42] or with Hansenula polymorpha [203], and in addition enantiotope selective hydroxylation of valeric acid at C-3 with a mutant strain of Candida rugosa [204] all yield (R)-3-hydroxyvalerates of high enantiomeric purity.

The above products have found several synthetic applications. Both enantiomers of alkyl 3-hydroxybutanoates have been used in the synthesis of various pheromones [205]. Fig. 5.7 shows the synthesis of some pheromones and other compounds based on optically pure 3-hydroxybutanoates that were not included in the preceding reference.

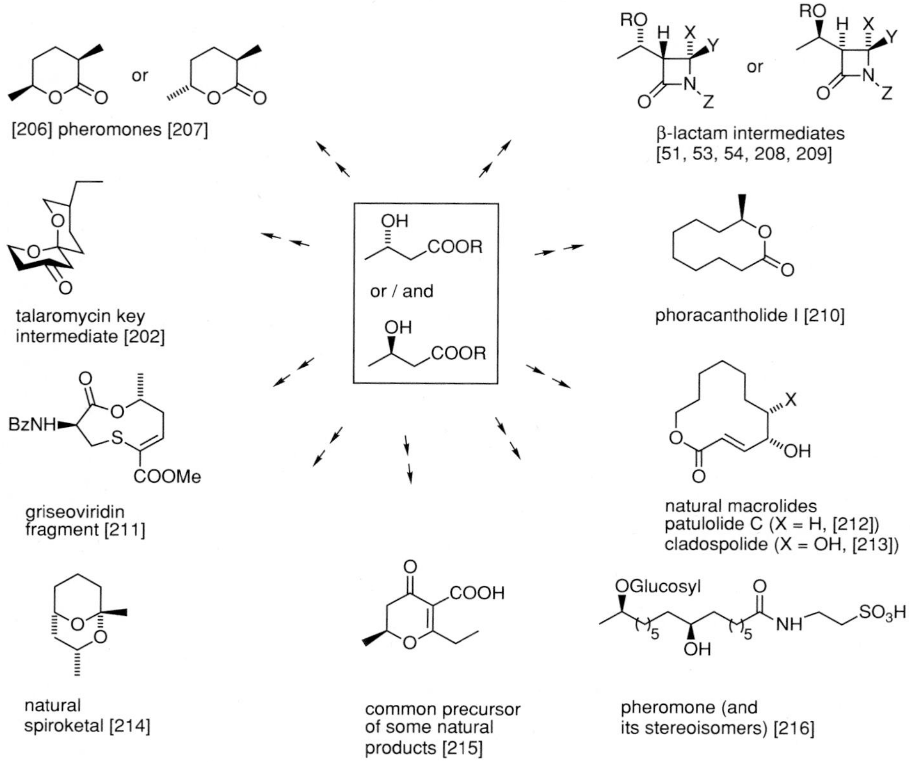

[206] pheromones [207]

β-lactam intermediates
[51, 53, 54, 208, 209]

talaromycin key
intermediate [202]

phoracantholide I [210]

griseoviridin
fragment [211]

natural macrolides
patulolide C (X = H, [212])
cladospolide (X = OH, [213])

natural
spiroketal [214]

common precursor
of some natural
products [215]

pheromone (and
its stereoisomers) [216]

Fig. 5.7.   Synthetic potential of 3-hydroxy-butanoate enantiomers

3-Hydroxyvalerates are also valuable intermediates in pheromone synthesis [205]. Additional examples of syntheses starting from various other chiral synthons included in Table 5.8 are the preparation of an avermectin B subunit from 3-hydroxy-4-methylhexanoic acid (5.146.b) [190], the synthesis of goniothalamin [191] and of the lactone moiety of the anticholesteremic agent compactin [191,217] from ethyl 3-hydroxyhexanoate (5.147.b):

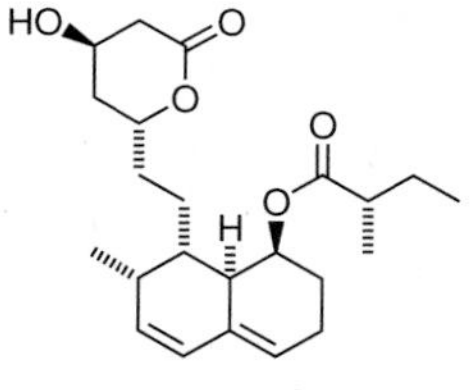

compactin

The homologous hydroxy acid (5.148.b) also played a role in compactin syntheses. (S)-citronellol can be prepared from the hydroxy acid (5.149.b) [218] or from ethyl (S)-3-hydroxybutanoate [219] with high enantiomeric purity, and from the nitril (5.157.b) (R)-α-lipoic acid, an intermediate in lipid metabolism, was obtained [55].

The azido compound (5.159.b) was used to synthesize the enantiomers of carnitine [50]. L-carnitine can, in addition, be prepared from the chloro-compound (5.153.b) [40], which may also be converted to various 1,3-diol intermediates [220].

5.1.4.3   2-Substituted 3-Oxocarboxylic Acid Derivatives

The outcome of baker's yeast reduction of 3-oxocarboxylic acid derivatives that are monosubstituted or substituted by two different groups at C-2 is more complex since in this case, the substrate becomes chiral and both enantiomer and diastereotope selectivities are possible.

With the 2,2-disubstituted derivatives (Fig.5.8) the eventual chiral centre is stable. Reduction of ethyl 2-allyl-2-methyl-3-oxobutanoate [221] or of 1-allyl-1-ethoxycarbonyl-2-oxocyclohexane [222], both of which belong to this group, is highly enantiomer and diastereotope selective (Fig. 5.8.a). Similarly, both of the pure enantiomers of 2-fluoro-2-methyl-3-oxocarboxylic acid ethyl esters undergo reduction with high diastereotope selectivity (Fig. 5.8.b) [223].

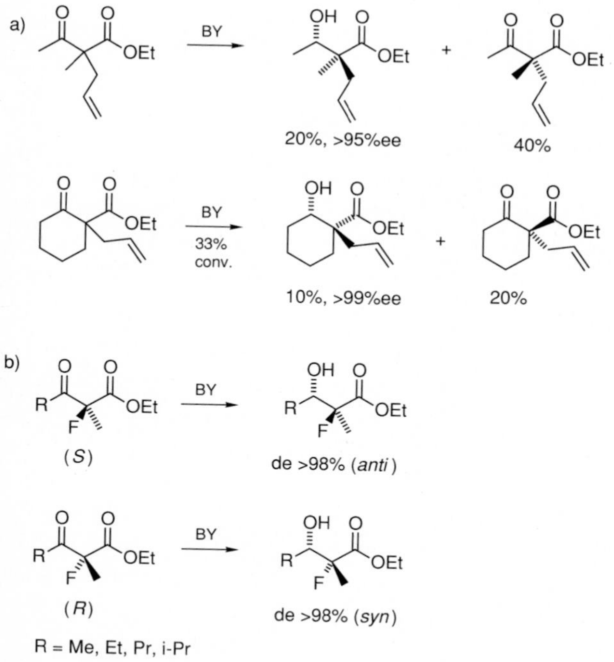

Fig. 5.8.   Reduction of 2,2-disubstituted-3-oxo-esters

Owing to a strong tendency to enolization, the centre of chirality in substrates with a single substituent at C-2 is unstable and the enantiomers are interconverted via the enolate (Fig. 5.9). The final product depends both on enantiomer and diastereotope selectivity and on the rate of enolization. If both modes of stereoselectivity are high and the rate of enolization is comparable to the rate of reduction then, in principle, the total amount of substrate can be converted to a single stereoisomer.

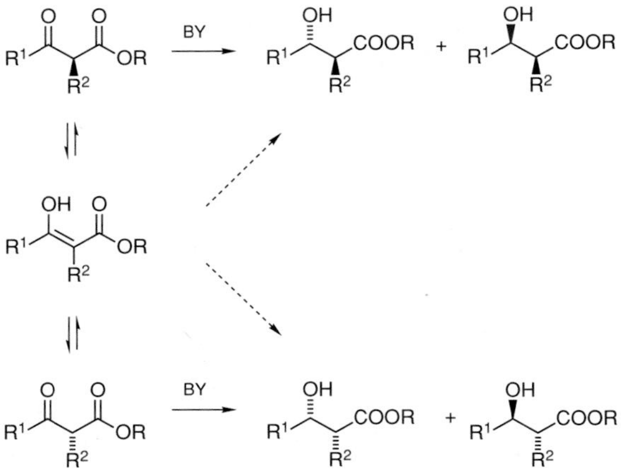

Fig. 5.9.   Equilibrium between the enantiomers of 2-substituted-3-oxo-carboxylic acid derivatives in yeast reductions

If enolization is slower than reduction, enantiomer and diastereotope selectivity are decisive. This is illustrated in Fig. 5.10. Performing the reduction of ethyl 2-benzyloxy-3-oxobutyrate in an acidic medium, in which enolization is fast, the *anti*-alcohol formed from the more rapidly reduced enantiomer is the predominant product. At neutral pH, in turn, when enolization becomes slow, high diastereotope selectivity in the system prevails and optically pure diastereomers are formed in a 1:1 ratio from the racemic substrate [36].

In principle, the possibility that some of the product is formed by reduction of the carbon-carbon double bond of the enolate cannot be excluded, but the facts discussed above and the results from studies on the reduction of acyl enolates [59] tend to make this hypothesis improbable (Fig. 5.11). Reduction of acyl enolates gives rise to mixtures of *syn* and *anti* diastereomers, which arise either as a result of carbonyl reduction following hydrolysis of the substrate and/or by hydrolysis following reduction of the C=C bond of the enolate. The latter possibility can be excluded since 2-substituted-3-acyloxycarboxylic esters do not hydrolyze in the presence of baker's yeast under such conditions [59]. In comparison to the reduction of the corresponding 3-oxoesters, a slight increase in the diastereomeric excess of the *syn* product was recorded.

The outcome of the reduction of 2-substituted-3-oxo esters is much dependent on the nature of the substituents R^1 and R^2 (Fig. 5.8).

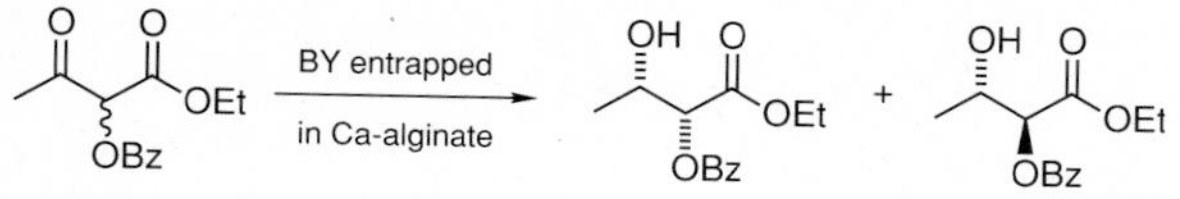

pH	syn		anti
acidic	6	:	94
	7	51	:	49

Fig. 5.10.   Effect of rate of enolization on product distribution

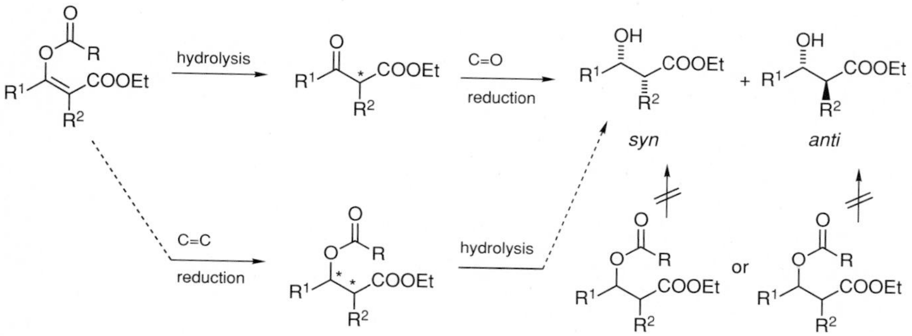

Fig. 5.11.   Baker's yeast mediated transformation of acyl enolates of 2-substituted-3-oxo-esters

i) With α-formyl esters (Table 5.9) under the usual experimental conditions, only enantiomer selectivity and the consequences of enolization can be observed. Results very much depend both on the constitution of the substrate and on the reaction conditions. The role of the esterifying alcohol is illustrated by a comparison of various esters of α-formylpropanoic acid [57]. Reduction of the *tert*-butyl ester (5.173.a) shows much better enantiomeric purity than that of the ethyl ester. The influence of reaction conditions can be seen when the enantiomeric purity of products obtained from ethyl α-formylbutyrate are compared (5.174.a).

Table 5.9. Enantiomer selective reduction of α-formyl esters with baker's yeast.

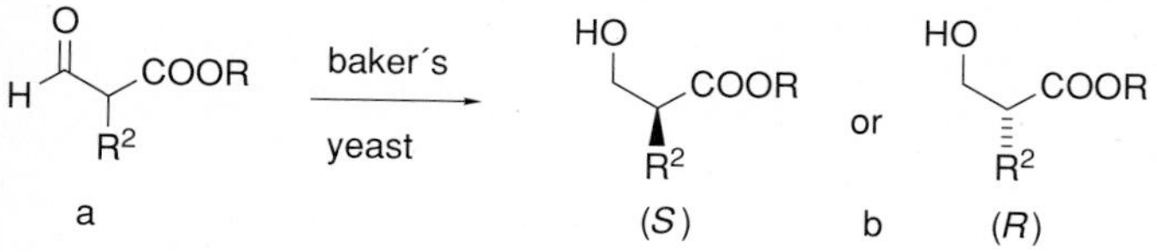

No	R^2	R	Yield(%)	Config.	ee(%)	Reference
5.172	Me	Et		R	63	[224]
5.173		CH$_2$C(CH$_3$)$_3$	78	R	90	[57]
5.174	Et	Et	88[a]	S	50 (91)[a]	[35]
5.175	Ph	Et		unknown	83	[224]

a: In 5% ethanolic solution with aeration.

ii) Reduction of open chain 2-substituted 3-oxoesters (Table 5.10)[4] has been studied in much detail, and as such a number of general conclusions can be drawn.

Table 5.10. Reduction of 2-substituted 3-oxo-esters by fermenting baker' s yeast.

Scheme: R^1–C(=O)–CH(R^2)–C(=O)–OR (a) → baker's yeast → R^1–CH(OH)–CH(R^2)–C(=O)–OR (b, syn) + R^1–CH(OH)–CH(R^2)–C(=O)–OR (c, anti)

No.	R^1	R^2	R	Yield (%)	de(%) (syn/anti)	ee[a] (%)	ee[a] (%)	Ref.
5.176	Me	Me	Me	71	62(s)			[56]
5.177			CH$_2$Ph	44	34(s)		84	[225]
5.178			octyl	82	90(s)			[56]
5.179			CH$_2$C(CH$_3$)$_3$	57	92(s)			[58]
5.180	Me	Et	octyl	84	52(s)			[58]
5.181	Me	Pr	CH$_2$C(CH$_3$)$_3$	68	60(s)			[58]
5.182			t-Bu	76	50(a)			[58]
5.183	Me	CH$_2$CH=CH$_2$	t-Bu	36	88(a)			[58]
5.184	Me	CH$_2$C≡CH	Bu	97	38(s)			[58]
5.185			t-Bu	68	66(a)			[58]
5.186	Me	CH$_2$CH=CMe$_2$	Et	29	n.d.			[197]
5.187	Me	CH$_2$Ph	Et	84	33(a)			[226]
5.188	Me	OH	Et	68	62(a)		90	[227]
5.189	Bn	OH	Et	60	60(a)		70	[227]
5.190	Me	OAc	Et	53	64(a)[b]			[36]
5.191	Me	SMe	Me	72	44(s)			[228]
5.192	Me	SPh	Et	49	46(s)			[228]
5.193	Et	SMe	Et	44	6(a)			[228]
5.194	BnOCH$_2$CH$_2$	SMe	Et	30	36(a)			[228]
5.195	Me	(1,3-dithian-2-yl)	Me	42	92(s)			[229]
5.196	MeOOC	Me	Me	57	6(s)	65	20	[230]
5.197	BnOCH$_2$	Me	Me	50	80(a)	68	6	[231]
5.198			i-Pr	64	52(a)		50	[231]

a: If not stated otherwise ee >95%; b: Reduction was carried out with Ca-alginate entrapped yeast cells. The acetate of *anti* product was partly hydrolysed to diol. Amount of diol was taken into account in de value.

Diastereotope selectivity of reductions is usually high and the predominant configuration conforms to the restrictions of Prelog's rule. Enantiomer selectivity, in turn, depends heavily on reaction conditions and on the nature of substituents. Inconsistency of enantiomer selectivity, also results, in this case, as a consequence of the fact that reduction may be catalyzed by several enzymes of different selectivity. Fig. 5.12 illustrates the

---

[4]   For the reduction of ethyl-methyl 2-acetyl-pentandioate see ref. [347].

reduction of ethyl 2-acetyl-4-pentenoate with fermenting baker's yeast and with a purified enzyme isolated from the same source. It is apparent that the intact cell contains
at least one additional enzyme of opposite stereoselectivity for which this ketoester is
a substrate, since reduction with the isolated enzyme yielded a single product of *syn*
relative configuration, whilst fermenting yeast produced the *anti* isomer in excess.

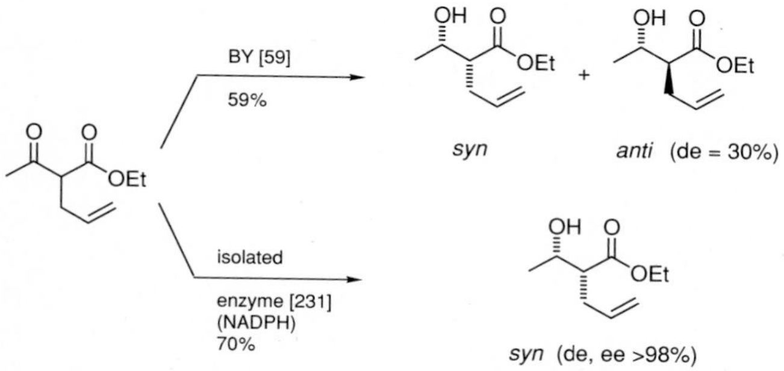

Fig. 5.12.   Comparison of the reductions of ethyl 2-allyl-3-oxo-butanoate by intact yeast cells and an
isolated enzyme

Reduction of other 2-substituted-3-oxocarboxylic acid derivatives, e.g. of thioesters,
thiocarboxylic acid thioesters [233] or nitriles [232] proceeds in a similar manner, i.e.
with high *re* selectivity and variable enantiomer selectivity to give a mixture of (3S)-*syn*-
and (3S)-*anti*-alcohols.

Reduction of the potassium salts of the free acids is, however, considerably different
from that of the corresponding esters [234] providing as the only product (R)-*anti*-
hydroxycarboxylic acids.

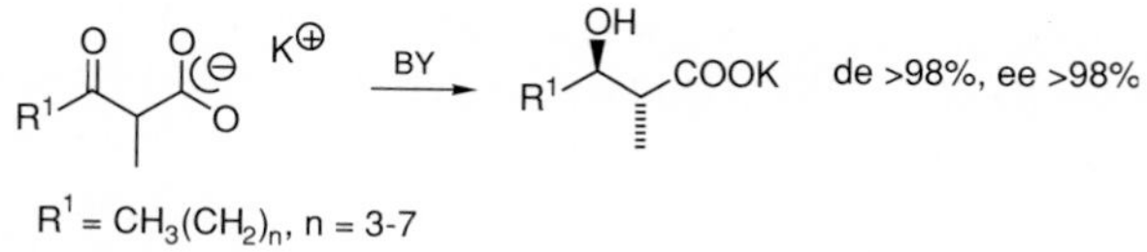

Unfortunately, substrate tolerance of this transformation is narrow, C-2 cannot be
substituted with a group larger than methyl and R^1 can have neither a longer nor a
shorter chain than that shown. This unusual specificity associated with high selectivity
may be interpreted by assuming that these carboxylates are only substrates for a single
enzyme.

Data in Table 5.10 suggest that diastereoselectivity is primarily influenced by the R^1
substituent. Ketoesters in which R^1 is a small group like methyl or ethyl (5.176-188,
5.190-193.a) yield products with very high enantiomeric purity. On increasing the size
of R^1, as is the case for example with esters (5.196.a) and (5.197.a), diastereotope
selectivity diminishes, although this effect can be offset to a certain degree by choosing
a bulky esterifying alcohol (e.g. 5.198.a). Often esters with bulky R^1 groups (e.g. when
R^1=phenyl [235,236]) cannot be reduced at all with baker's yeast, or if, only in very

poor yield (e.g. R^1= PhCH=CH- [237]). (Note that good results can be achieved with the same substrates using another microorganism).

The nature of R^2 and of the esterifying alcohol also has a profound effect on enantiomer selectivity [58]. Changing the ester group can invert the sense of selectivity as in the case of the esters (5.181.a), (5.182.a) and (5.184.a), (5.185.a) respectively.

The utilization of alcohols obtained from 2-substituted 3-oxoesters can be illustrated by a synthesis starting from the ester (5.176.b) and aimed at the elucidation of the absolute configuration of the pheromone (2R,3R)-stegobinone [238] or by the synthesis of the pheromone (S)-sulcatol from the ester (5.186) [197].

iii) Reduction of cyclic β-oxocarboxylic esters (Table 5.11) with baker's yeast takes place with excellent diastereotope and enantiotomer selectivity. Older experiments, carried out predominantly in the fermenting mode on 2-oxocyclopentane and cyclohexane esters, gave somewhat capricious results showing a tendency to *syn* predominance [60,244] and enantiomeric purity of the main product [9,222] in the range 70-95%. The results of reduction of the ethyl esters (5.199.a) and (5.201.a) suggest that reduction with non-fermenting yeast gives higher enantiomer and diastereotope selectivities than reduction in the fermenting mode [15]. Reduction of thioester (e.g. 5.200.a) and thiocarboxylic acid thioester analogs (5.202.a) yielded better selectivities in both respects. Baker's yeast is also quite capable of reducing heterocyclic (e.g. 5.203.a-206.a) and bicyclic β-oxocarboxylic esters (e.g. 5.207.a-5.210.a) too.

The cyclic hydroxyesters prepared by the methods above are also versatile intermediates. Esters of (2S)-*syn*-2-hydroxy-cyclopentane and -cyclohexane carboxylic acids (5.199.b and 5.201.b) were converted by alkylation to enantiomerically pure synthons [245]. From the thioester (5.201.b), the pheromone (−)-frontalin was synthesized [60]. Sulphur containing heterocyclic hydroxyesters such as (5.203.b) and (5.205.b) are convenient intermediates, since their reductive ring opening is facile. After recrystallization, the acid obtained from the ester (5.205.b) is optically pure and was used for the synthesis of anhydroserricornine [241].

Reductive ring opening of the thiopyran ester (5.204.b) leading to methyl (S)-3-hydroxyvalerate (5.140.b) with high enantiomeric purity was of interest since it can be used for the synthesis of fragments leading to macrolide antibiotics [240]. The bicyclic 1,3-dioxolane (5.207.b) was a starting material in the synthesis of (−)-phaseic acid, a compound with plant hormone activity [246] and of a compound that is important for stimulating microbial sporulation [243].

iv) The reduction of β-oxoesters containing stable centre(s) of chirality in addition to the unstable one at C-2, is an even more complex process (Fig. 5.13) and stereodirection by the stable centre predominates. Due to facile enolization, such compounds are mixtures of two diastereomers, epimeric at C-2. Reduction therefore involves diastereomer and diastereotope selectivity, and when the substrate is a racemate, also enantiomer selectivity. This situation is further complicated by the interconversion of diastereomeric substrates by enolization.

Table 5.11. Reduction of cyclic β-oxo esters and some tioanalogs with baker's yeast.

No.	Major product (b, syn)	Yield[a] (%)	de[b] (%)	ee[b] (%)	Ref.
5.199	Y=OEt	80(44[a])			[15,348]
5.200	Y=SEt	88			[60]
5.201		85[a]			[15]
5.202		27			[233]
5.203		n.d.		85	[239]
5.204		62			[238,239]
5.205		71		85	[239,241]
5.206		65[a]	73		[15]
5.207		86[a]			[15]
		70			[242,243]
5.208		12[a]		72	[15]

Table 5.11. (contd.)

No.	Major product (b, *syn*)	Yield[a] (%)	de[b] (%)	ee[b] (%)	Ref.
5.209		67		75	[15]
5.210		63		70	[15]

a: Yields marked refer to products obtained by non-fermenting yeast. b: If not stated otherwise values exceed 95%.

Thus, reduction of the diester in Fig. 5.13.a exhibits poor diastereomer selectivity, but shows high enantiomer and diastereotope selectivity in the case of one of the racemates yielding the minor isomer in high enantiomeric purity. This transformation was used in the synthesis of the fungal toxins talaromycin A and B [247].

In the reduction of the tricyclic β-oxo ester in Fig. 5.13.b, all three types of selectivities are outstanding to the extent that a single stereoisomeric alcohol is obtained optically pure [248-250]. The remaining ketone of 60-85% enantiomeric purity can also be used and serves in the syntheses of carbacyclin [248] and pentalenolactone E methyl ester [250].

Reduction of the corresponding methyl ester shown in Fig. 5.13.c can be performed with similar selectivity, but interestingly the unreduced ketone concurrently underwent enzymic hydrolysis followed by decarboxylation [249].

Microbial reduction of the bicyclic β-oxo-carboxylic acid methyl ester in 5.13.d was undertaken with the synthesis of carbacyclin intermediates in mind [251]. Reduction was selective in every respect, yielding a single stereoisomer in which unexpectedly the (*S*)-hydroxy and methoxycarbonyl groups were disposed *trans* to one another.

The cyclopentenone carboxylic ester in Fig. 5.13.e does not contain an enolizable functionality and is reduced by baker's yeast with high enantiomer and diastereotope selectivity [252].

5.1.4.4   Carboxylic Acid Derivatives Containing the Oxo Group in or More Remote Positions

4- and 5-oxocarboxylic acids were being reduced with baker's yeast in the late fifties, whereby *R* configured γ- and δ-lactones were obtained with high optical purity [253,254]

(Fig. 5.14.a). Of these lactones, (5.214.d) and (5.214.h) are manufactured on an industrial scale in good yield and high optical purity using several tons of yeast per batch [255] and substrate concentrations as high as 80 g/l [253]. High enantiotope selectivity probably results from the fact that these substrates are only accepted by a single enzyme system in baker's yeast. In fact, an NADPH dependent enzyme, stimulated by the addition of ATP, Mg and CoA catalyzes the reduction of the ketoacid (5.213.e) [254].

An extensive study on the scope of this reaction revealed [256] that only the ketoacids (5.211.b-m) and (5.213.b-m) are suitable substrates (Fig. 5.14.a). Reduction of the keto-acids (5.211.b) and (5.213.b) is extremely slow and the short chain ketoacids (5.123.a) and (5.213.a) are not substrates of the system at all.

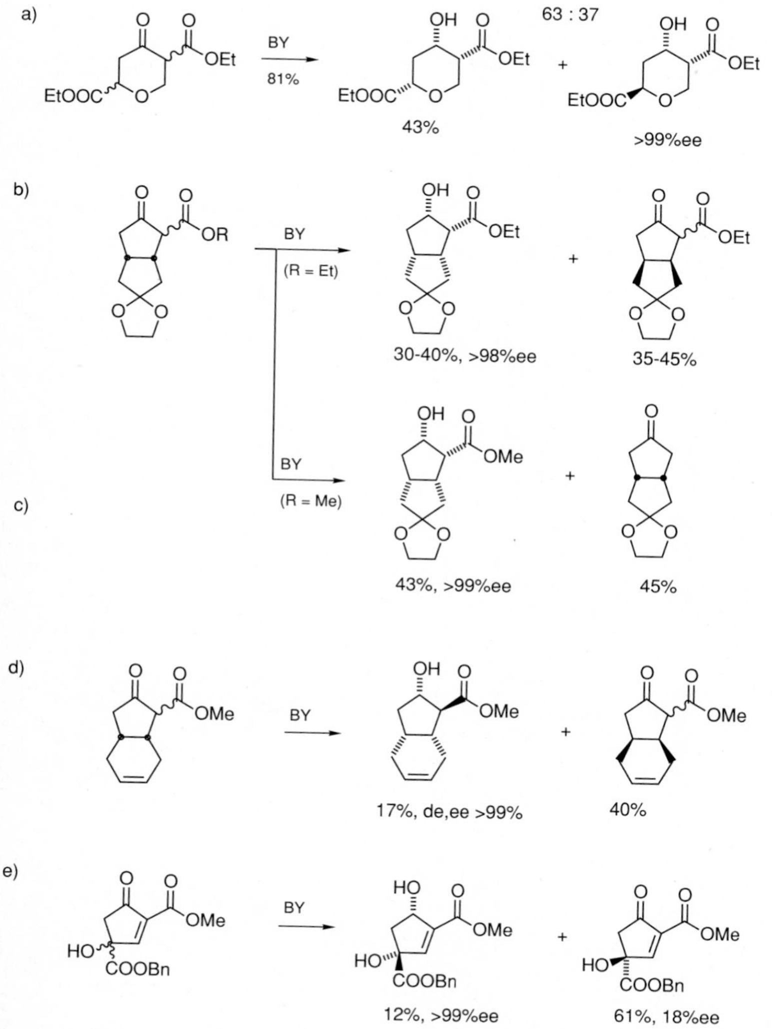

Fig. 5.13.   Reduction of cyclic β-keto esters having additional and stable chiral center(s)

Reduction of the esters of 5-oxocarboxylic acids (Fig. 5.14.b) yields the corresponding 5-substituted valerolactones [256]. In this process first the ester group is hydrolyzed enzymatically which is followed by reduction of the ketoacid as explained above.

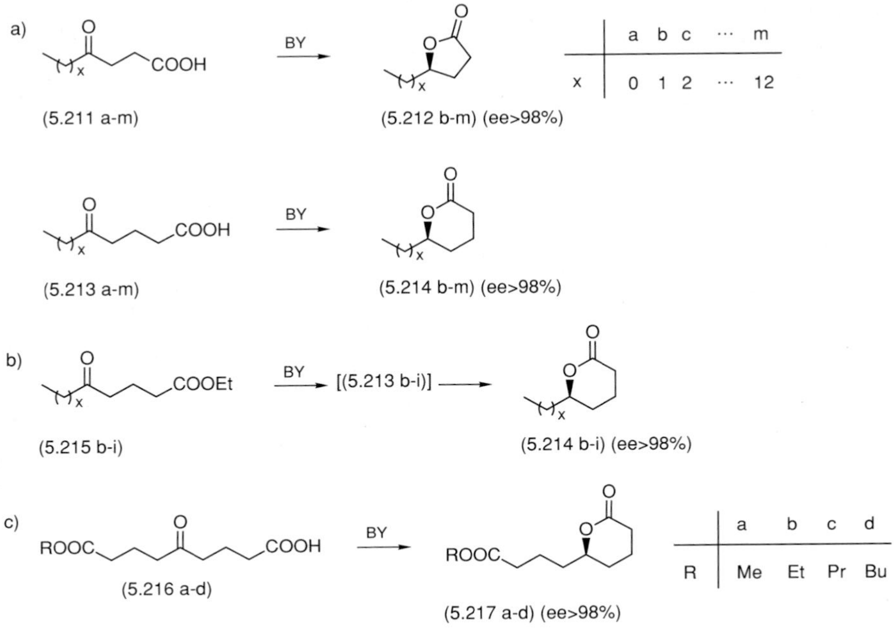
a)
COOH
BY
x
a b c ··· m
x 0 1 2 ··· 12
(5.211 a-m)
(5.212 b-m) (ee>98%)
COOH
BY
(5.213 a-m)
(5.214 b-m) (ee>98%)
b)
COOEt
BY
[(5.213 b-i)]
(5.215 b-i)
(5.214 b-i) (ee>98%)
c)
ROOC
COOH
BY
ROOC
(5.216 a-d)
(5.217 a-d) (ee>98%)
a b c d
R Me Et Pr Bu

Fig. 5.14. Transformation of 4- or 5-oxo-carboxylic acid derivatives to the corresponding γ- or δ-lactones via reductions of the free carboxylates

The dominant role of the free carboxylic acid group is also clearly evident in the reduction of 5-oxononane-1,9-dicarboxylic acid monoesters (5.216.a-d in Fig. 5.14.c) [257], since here also (*R*)-lactones (5.217.a-d) are formed with high enantiomeric purity.

Several important natural products are found amongst γ- and δ-lactones. Reduction with baker's yeast provides a simple route to (*R*)-4-dodecanolide (5.212.i) [258,259] and to (*R*)-5-hexadecanolide (5.214.i) [258-260] both of which are pheromones of various insects.

Another interesting example is the reduction of 4-oxocarboxylic acids substituted with a dibenzofuran unit [261] which provides in high yield the corresponding (–)-hydroxycarboxylic acid.

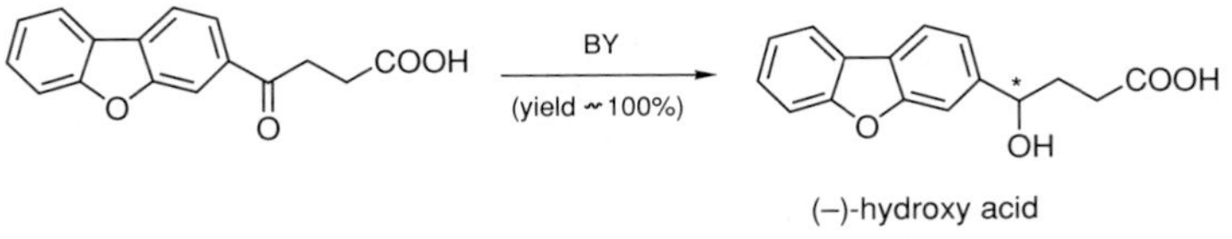
COOH
BY
(yield ~ 100%)
COOH
OH
(–)-hydroxy acid

From results in Fig. 5.15.a concerning the reduction of a number of esters containing the oxo group in the γ or even more remote positions, it becomes apparent that control by an ester group is much less effective than that by the free carboxyl group in the above acids.[5] Stereoselection in this class is usually in accordance with Prelog's rule, especially since the products (5.218-19) are of S configuration (such transformations probably involve another enzyme system different from those reducing 3- and 4-oxocarboxylic acids [13]).

With the synthesis of bifunctional chiral compounds in mind, a study of the reduction by baker's yeast of a number of 3-oxoglutarate and 3-oxoadipate derivatives was undertaken [208] and compounds such as the hydroxyadipates (5.220.a,b) were obtained.

The 2-(1,3-dithianyl)-carbinol (5.221) was synthesized to act as an intermediate in leukotriene synthesis [231], and the hydroxyester (5.222), in turn, was ultimately converted to (R)-lipoic acid, a blood sugar reducer cofactor for α-ketoacid dehydrogenases [262,350].

The difficulties inherent in the synthesis of natural macrocycles can be solved by means of immobilized yeast reduction of long chain ketoacids (Fig. 5.15b). Immobilized yeast can be recycled several times in the preparation of the hydroxycarboxylic acid (5.223) [263,264] which is used for the synthesis of (S)-phoracantholide I, and of the hydroxy-alkynoic acid (5.224) [265,351] which leads, after reduction and cyclization, to a pheromone with a 14-membered ring.

5.2 Reduction of Carbon-Carbon Double Bonds with Baker's Yeast

It has been known for many years that the olefinic bond in α,β-unsaturated carbonyl compounds can be reduced with baker's yeast [22,180,266] and it had already been reported in the thirties that isolated double bonds cannot be reduced this way. For example, it had been clearly shown that the olefinic bond of 3-hexen-2-one is reduced by yeast while that of 2-hexen-5-one is not [266].

Over the last 25 years, this transformation became the subject of more detailed studies. The double bond in allyl alcohols could also be reduced, and the hidden enantiotope selectivity in the reduction of disubstituted (E)-olefins was detected using labelled cinnamic alcohols [267].

---

[5] Recently reduction of alkyl levulinates in good yield and with high enantiotope selectivity has been reported [349].

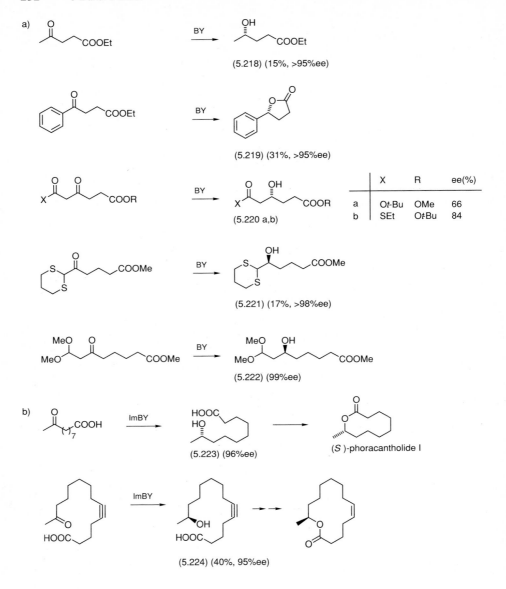

a)

(5.218) (15%, >95%ee)

(5.219) (31%, >95%ee)

(5.220 a,b)

	X	R	ee(%)
a	O*t*-Bu	OMe	66
b	SEt	O*t*Bu	84

(5.221) (17%, >98%ee)

(5.222) (99%ee)

b)

(5.223) (96%ee)

(*S*)-phoracantholide I

(5.224) (40%, 95%ee)

Fig. 5.15.   Reduction of various ketoacid derivatives (ImBY: immobilized baker's yeast)

Reduction of 1,1,2,3-d$_4$-cinnamic alcohol, for example, gave (1*S*,2*R*,3*R*)-d$_3$-3-phenylpropan-1-ol. This product may be formed by enantiotope selective oxidation in the first step, followed by enantiotope selective *anti* addition of hydrogen, and then finally enantiotope selective reduction of the aldehyde group.

A generalized scheme for the transformation of α, β-unsaturated carbonyl compounds or for the corresponding allyl alcohols is depicted in Fig. 5.16 (see also ref. [268]). The

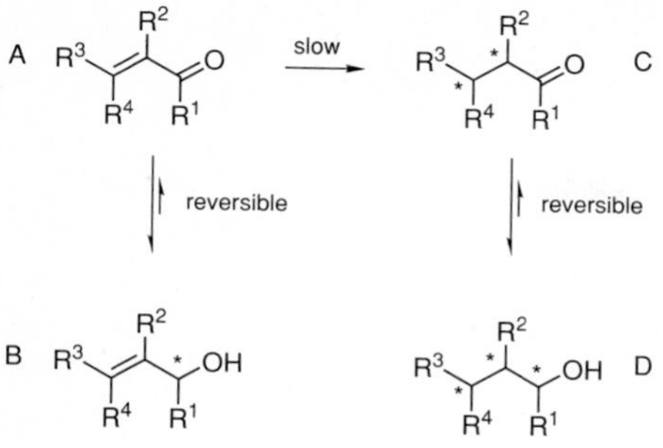

Fig. 5.16.   General scheme for the C=C reduction of α,β-unsaturated aldehydes or ketones and the corresponding allyl alcohols

unsaturated carbonyl compound A yields the allyl alcohol B in a fast, but reversible alcohol dehydrogenase catalyzed process. If $R^1 \neq H$, the allyl alcohol is chiral. Reduction of the carbon-carbon double bond to the saturated product C is generally slower and irreversible. In product C both atoms marked with an asterisk can be chiral. Tetra-substituted compounds are usually not accepted as substrates. The saturated carbonyl compound C can then yield in another fast, alcohol dehydrogenase catalyzed reaction the saturated alcohols D, which in turn may contain a maximum of three chiral centers, i.e. incomplete stereoselectivity may give rise to a mixture of eight stereoisomers.

The fact that in this consecutive series of transformations, generally the reduction of the olefinic bond (A $\rightarrow$ C) is the slowest step allows one to draw the following conclusions:

– The final result is independent of whether the substrate is A or B.
– If the substrate is A, predominantly allyl alcohol B is formed first and this is then transformed slowly into the saturated products C and D. Neglecting side reactions, higher yields of C and D at the expense of B can be expected on increasing the reaction time.
– Carbonyl reduction (C$\rightarrow$D) is most often faster than C=C reduction (A$\rightarrow$C) and thus the equilibrium is generally shifted in favor of the reduced product. This means that the main products are generally B and D.
– Often, depending on conversion and the degree of selectivity several products can be isolated. The following discussion deals solely with products arising from C=C bond reduction and does not address the products of other processes that might arise.

The points outlined above are, of course, not of unconditional validity; with some substrates relative rates may be quite different and the double bond can also be activated by functional groups other than carbonyl, for example neighboring nitro, carboxyl and halogen groups will suffice. In comparison to carbonyl reductions, those of activated double bonds are much slower and this must be taken into account in synthetic applications where much higher than usual yeast/substrate ratios (up to 300:1) may be necessary. As a consequence, special extraction procedures for product isolation and large volumes of solvents may be required.

Olefin reductions can be classified according to the position of substituents, a feature which also has a decisive effect on the steric course of the process.

i) Trisubstituted carbon-carbon double bonds activated at the disubstituted carbon (Table 5.12).

If the activating group is an aldehyde or an aldehyde equivalent, products with a single chiral center are formed. This transformation has only been studied with substrates in which R^2 = Me (5.225-238.a) (see Fig. 5.16). Data in Table 5.12 illustrate that with a wide selection of R^3 groups, reduction shows high enantiotope selectivity. The effect of R^3 is manifest more in the rate and thereby in the chemical yield of the reaction. Reduction of the double bond in allyl alcohols is prevented by acetylation of the alcohol function (5.234.b) and, in the absence of additional activation (e.g. 5.235.b), the double bond of α,β-unsaturated esters also resists reduction. Similar observations were made with other unsaturated esters too [269].

Several of the products enumerated in Table 5.12 have found some form of synthetic application. The unsaturated alcohols (5.230.b) [271], (5.232.b) [273], (5.235.b) [275] and (5.236.b) [276] were used in tocopherol side chain syntheses, the furylethanol (5.228.b) for the synthesis of the pheromone (3S)-4-hydroxy-3-methylbutyric acid [268], and the butyric ester (5.236.b) was incorporated into (25S)-26-hydroxycholesterol [278].

ii) Trisubstituted carbon-carbon double bonds activated at the monosubstituted carbon (Fig. 5.17)

In Fig. 5.17.a the reduction of 3-deuterocinnamaldehyde is depicted yielding (3R)-3-deutero-3-phenylpropanol [267]. Since the van der Waals radius of deuterium and hydrogen are practically the same, this transformation truly represents the reduction of a disubstituted double bond. Results shown in parts b) and c) of the figure concerning compounds with bulky substituents at C-2 can be interpreted as follows:

In substrate S_E, the bulky substituent (L) and the activating group (X) are on opposite sides of the double bond and the entering hydrogen approaches from the top, giving rise to product P. If the configuration of the double bond is opposite, but the direction of the attack of hydrogen remains the same, the enantiomer of the previous product (*ent*-P) is formed. The final enantiomeric composition of the product may be influenced by an E/Z isomerization of the substrate S_Z to the more stable isomer (S_E) concurrent with the enzymic process (k_1 and k_2). Thus, neglecting other factors at this point, the enantiomeric purity of the product depends on the ratio of three rate constants, k_1, k_2, and k_3. This phenomenon has been described for the reduction of (E)- and (Z)-geraniate by *Clostridium* sp. [8]. Performing the reaction with intact cells, whilst both isomers gave (R)-citronellol, the E isomer was generated with much higher enantiomeric purity than the Z isomer. Using a purified enzyme for the same purpose, (E)-geraniate gave (R)-citronellol, (Z)-geraniate, in turn, (S)-citronellol, both products being optically pure. Reduction of geraniol ((E)-5.241) and nerol ((Z)-5.242) or of the corresponding aldehyde to (R)-citronellol proceeds in a totally analogous way [279] (Fig. 5.17.b and c).

Table 5.12. Reduction of some α-methyl substituted (E)-α, β-unsaturated aldehydes or aldehyde equivalents with baker's yeast.

No.	Substrate R³	X	Product Y(%)	ee(%)	Ref.
5.225	Me	CHO	28	95	[269]
5.226	CH₂=CH	CH₂OH	30	>98	[270]
5.227		CH₂OH	a	a	[266]
5.228		CH₂OH	72	>99	[267]
5.229		CH₂OH	a	>90	[4]
5.230		CH₂OH	60	80	[4,271]
5.231	HO	CHO	40	>98	[272]
5.232	AcO	CHO	35	>98	[272-274]
5.233	AcO	CHO	35	>98	[272]
5.234	AcO	CHO	20	>98	[274]
5.235	MeOOC	CHO	35	>98	[274,275]
5.236	EtOOC	CH(OMe)₂	49	>98	[269,276]
5.237		CHO	35[b]	86	[277]
5.238		CHO	48[b]	>99	[277]

a: Data are not reported; b: Isolated as the acid after more than two weeks under aerobic conditions.

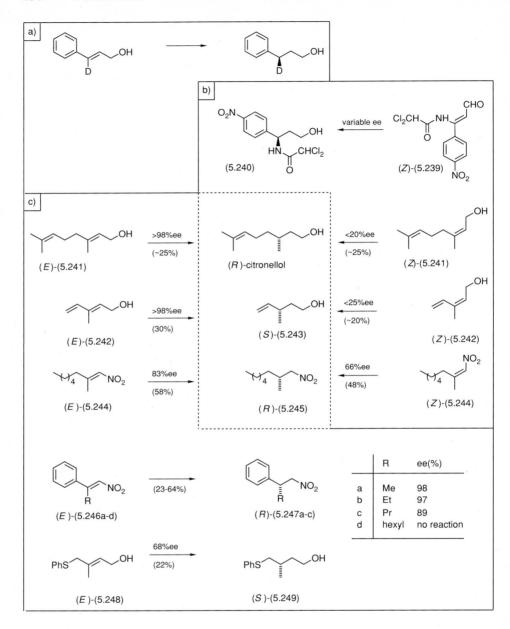

Fig. 5.17.   Reductions of activated trisubstituted olefins bearing two substituents in β-position

Reduction of the allyl alcohols ((*E*)- and (Z)-5.243) to ((*S*)-5.243) [270], as well as of the nitroolefins ((*E*)- and (Z)-5.244) to the nitroalkane ((*R*)-5.245) [280] can be interpreted along the same lines. If we assume that in ((Z)-5.239) the acetamido function is the larger group, then the same rationalization holds for its transformation to the alcohol

(5.240) [281]. The fact that enantiomeric purity strongly depends on reaction conditions can be explained by a change in the relative rates of isomerization and reduction. In the case of the reduction products ((R)-5.241.a-c) obtained from the nitroolefins ((E)-5.246) it was observed that optical purity deteriorated when the difference in size between substituents diminished [280]. The results of reduction of *p*-halogenophenyl substituted nitro compounds (not included here) indicate that electronic effects may also play an important role in the stereoselectivity of this reduction [280]. Conversion of the sulphide ((E)-5.248) to the alcohol ((S)-5.249) is also in accordance with the concept outlined above [277].

With regard to practical application, amongst the reduction products (R)-citronellol is a key intermediate in tocopherol syntheses [273,275].

iii) Reduction of activated carbon-carbon double bonds in other open chain compounds

A carboxylic ester group alone is insufficient for activating an adjacent olefinic bond, and for this reason α,β-unsaturated esters cannot be reduced by baker's yeast (cf. part *i*) in this section). Nevertheless there are several examples of reductions starting from such substrates (Fig. 5.18) and in fact, for the transformations in Fig. 5.18 it has been shown that reduction of the esters (5.250) proceeded via the free acids [282]. A correlation between substrate and product configuration was observed, thus after reesterification, ((E)-5.250) gives rise to the saturated esters (5.251), whilst reesterification of the *Z* stereoisomers yields the enantiomers (*ent*-5.251). The lower optical purity of the latter was shown, however, not to be due to *Z/E* isomerization [282].

Reduction of the fluorinated esters shown in Fig. 5.18.b can proceed with high enantiotope selectivity in an as yet unknown sense, and was proven to involve the parent ester. Here however, poor activation by the ester group is supported by that of the perfluoroalkyl groups.

In Fig. 5.19 some more examples of double bond reduction have been collected. Fig. 5.19.a shows the oxidoreduction of a 2-substituted allyl alcohol [277]. During aerobic fermentation over 3 weeks, oxidation of the saturated product yielded the corresponding carboxylic acid.

Fig. 5.19.b illustrates the reduction of racemic α-methylene-β-hydroxyketones [283]. The process is diastereotope selective but is practically non-selective towards enantiomers and the carbonyl group resists reduction (cf. Section 5.1.3.2). Accordingly, *anti*- and *syn*-α-methyl-β-hydroxyketones are formed in nearly equal amounts and in high optical purity.

Fig. 5.19.c shows reductions of α-chloro-α,β-unsaturated ketones [284]. Initially the olefinic bond is reduced with incomplete enantiotope selectivity which, after brief fermentation, permits the isolation of isomeric α-chloroketones with varying optical purities. After prolonged treatment the enantiomeric saturated α-chloroketones are further reduced with poor enantiomer but high diastereotope selectivity to yield a mixture of enantiomerically pure *anti*- and *syn*-chlorohydrins.

Surprisingly, in the reduction of 15-keto-PGF$_{2\alpha}$ with baker's yeast the double bond was saturated and the ketone group remained intact (Fig. 5.19.d) [285].

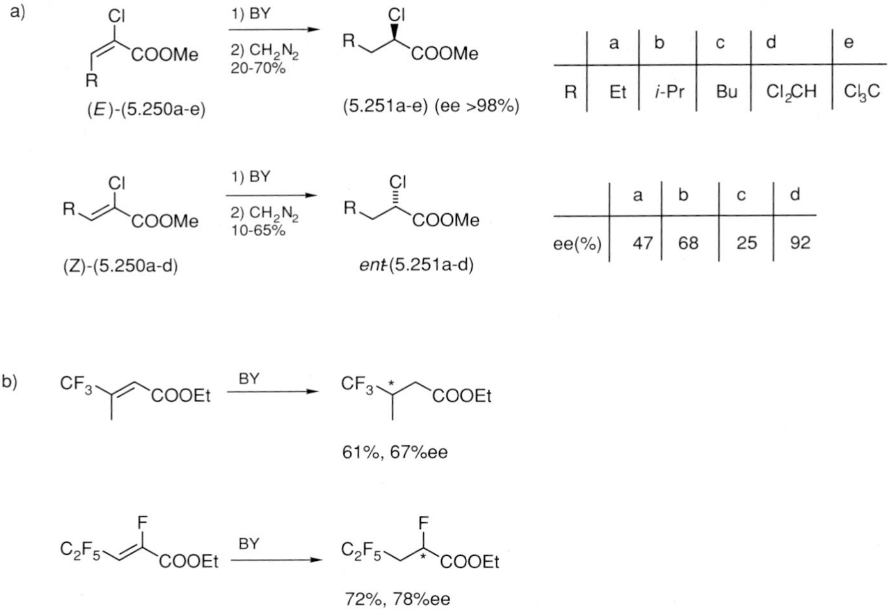
a)
1) BY
2) CH_2N_2
20-70%
(*E*)-(5.250a-e)
(5.251a-e) (ee >98%)
1) BY
2) CH_2N_2
10-65%
(Z)-(5.250a-d)
ent-(5.251a-d)
b)
BY
61%, 67%ee
BY
72%, 78%ee

	a	b	c	d	e
R	Et	*i*-Pr	Bu	Cl_2CH	Cl_3C

	a	b	c	d
ee(%)	47	68	25	92

Fig. 5.18. Products arising from the reduction of C=C bonds in some halogen or perfluoroalkyl substituted α, β-unsaturated ester substrates

The reduction of (2*E*,6*E*)-8-hydroxy-2,6-octadienal containing two activated olefinic bonds (Fig. 5.19.e) deserves some attention since both double bonds are reduced providing pure (2*S*,6*R*)-2,6-dimethyl-octane-1,8-diol [272].

iv) Reduction of activated cyclic olefins (Fig. 5.20)

The reduction of the cyclohexenedione shown in Fig. 5.20.a is remarkably proficient due to an exceptionally favorable productivity number i.e. with a yeast/substrate ratio less than 1:0 and yields the product (5*S*)-2,2,5-trimethyl-1,4-cyclohexandione [286] which is used in the synthesis of carotenoids. The transformation of the cyclic dienol diacetate in Fig. 5.20.b to optically pure (5*R*)-3-acetoxy-5-hydroxy-4,4-dimethyl-2-cyclohexenone was also studied in connection with carotenoid syntheses [287].

Fig. 5.20.c depicts the conversion of the racemic "Woodward lactone" to the saturated ketone, a process also catalyzed by baker's yeast [288].

Lastly it should be mentioned that allenes [289], of which the alcohol shown below is an example, can also be reduced with baker's yeast.

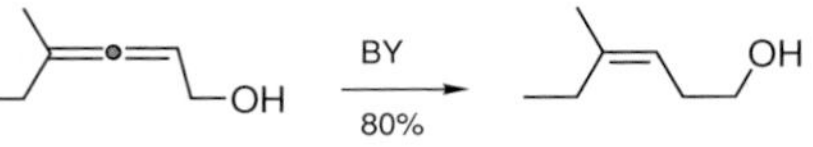
OH
BY
80%
OH

a)

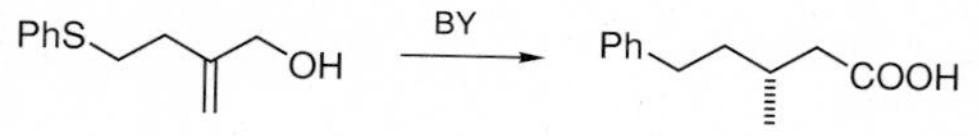

39%, 96%ee

b)

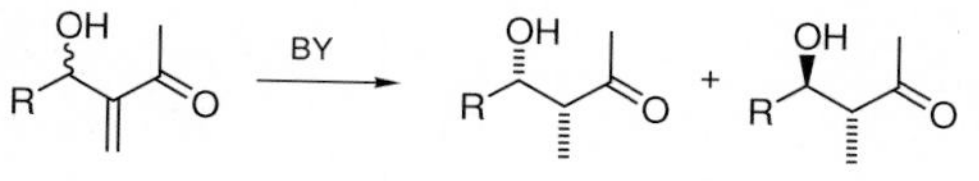

R	Yield (%)	ee(%)	Isomeric ratio	ee(%)
C_2H_5	61	>98	53 : 47	>98
C_5H_{11}	72	>98	36 : 64	69

c)

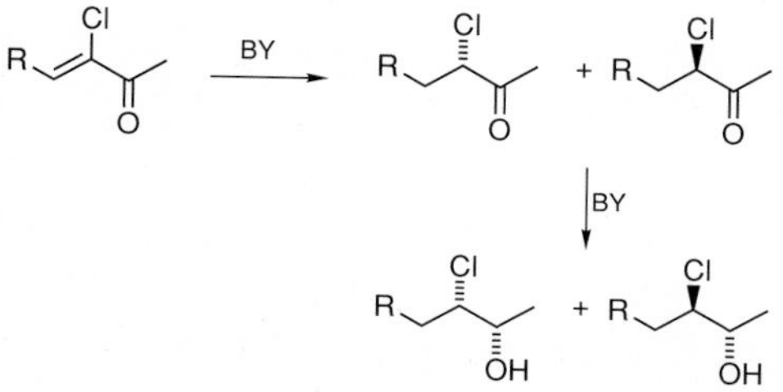

R	ee(%)	syn : anti ratio	ee(%)
C_2H_5	>98	72 : 28	>98
C_8H_{17}	>98	95 : 5	>98

d)

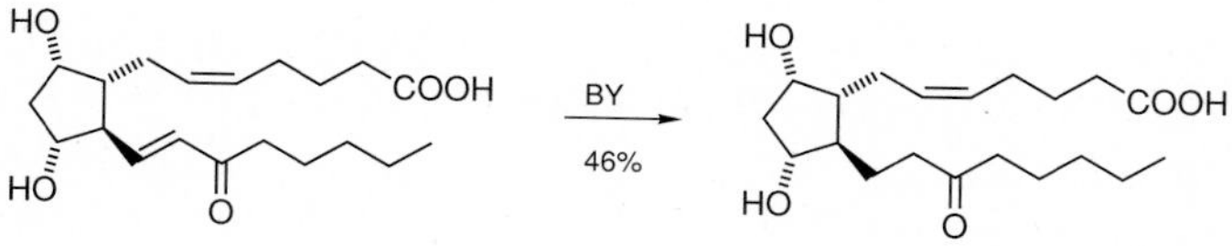

46%

e)

40%

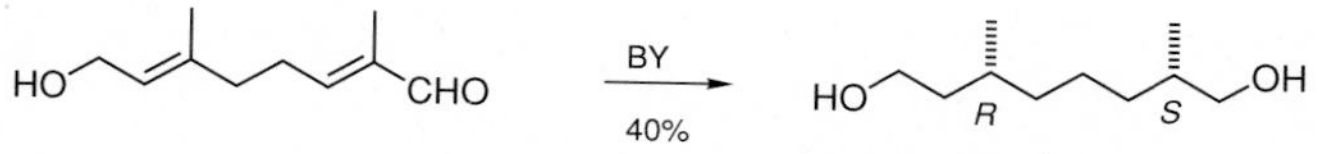

de,ee >96%

Fig. 5.19.   Reductions of C=C bonds in miscellaneous unsaturated compounds

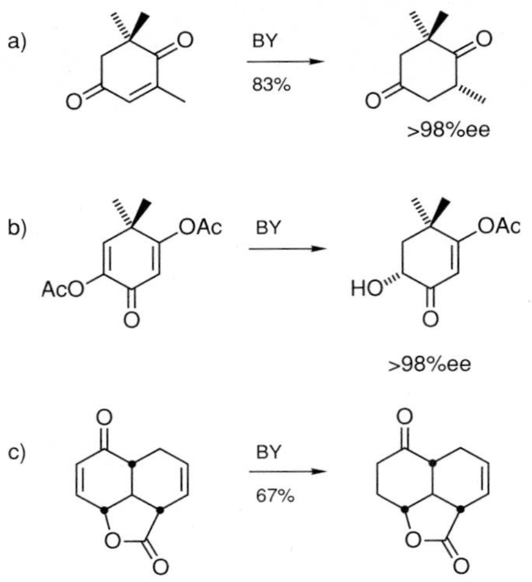

a)

BY
83%

>98%ee

b)

BY

>98%ee

c)

BY
67%

Fig. 5.20.   Reductions of cyclic α, β-unsaturated ketones

5.3 Other Oxidoreductions with Baker's Yeast

Baker's yeast is capable of catalyzing many other oxidoreductive transformations besides those already discussed in the foregoing sections. Early investigations disclosed for example, that quinones may be reduced to hydroquinones [290] or that various nitrogen-oxygen bonds can be reductively cleaved with baker's yeast. In this way aromatic nitro [291,292] and nitroso compounds, and indeed hydroxylamines [293], can be converted to amines.

Recently baker's yeast reduction of nitroacetophenones to aminophenyl-ethanols was reported [81]. In some cases the reaction was chemoselective and the corresponding aminoacetophenones were isolated [294]. The same paper also described that with aromatic nitro compounds, reduction of the nitro group was inhibited by electron donating substituents, but promoted by electron attracting substituents [294].

Baker's yeast can also reduce carbon-nitrogen double bonds, as shown by the conversion of oximes to amines of moderate enantiomeric purity [352].

With a growing culture of baker's yeast regio- and stereoselective dehydrogenation of the dithia analogue of stearinic acid could be accomplished [295]:

Similarly thiastearinic acids were converted to thiaoleic acids [296].

Enantiotope selective sulfoxidation [353] and oxidative conversion of thiocarbamates and thioureas to the corresponding carbamates and ureas [354] are examples of oxidations of sulfur-containing compounds by baker's yeast.

5.4 Hydrolysis with Baker's Yeast

Baker's yeast contains several hydrolytic enzymes and as such it can be used for enzymic hydrolysis too [1,297] (carboxypeptidase Y from baker's yeast, for instance, is actually a commercial product).

In the previous chapters several examples of hydrolyses with baker's yeast have been quoted e.g. the mild hydrolysis of prostanoid esters (cf. Section 2.3.2.1).

Fig. 5.21 lists some additional examples of the efficient preparative utilization of baker's yeast as an enantiomer selective hydrolyzing agent.

Hydrolysis of ethyl α-N-acetamidocarboxylates with fermenting baker's yeast, when stopped at about 50% conversion gave the esters ((R)-5.252.a-e) in high enantiomeric purity and about 40% chemical yield [298] (Fig. 5.21.a). This hydrolysis was shown to be catalyzed by an enzyme with a substrate specificity rather similar to that of α-chymotrypsin (quite possibly the above mentioned carboxypeptidase Y). A mutant strain lacking proteinases failed to perform any hydrolysis [298]. Analogous hydrolyses with the methyl esters of racemic N-acetylalanine, N-acetyl-phenylalanine and L-acetamidophenylacetic acid have also reported [16], using gel entrapped yeast cells in reverse micelles. The (S)-acids were isolated in 96-98% optical purity and about 30% yield, whilst 20% of the unchanged (R)-esters were recovered in an optically pure form.

For the hydrolysis of the alkinyl substituted carbinol acetates shown in Fig. 5.21.b non-fermenting lyophilized baker's yeast was applied [45]. (Similar hydrolyses were also carried out by *Bacillus subtilis* [300]). When the hydrolysis is interrupted at 40% conversion, the alcohols (5.254.a-f) are obtained in better than 90% optical purity and unchanged substrates could be recovered in high optical purity when conversion was carried to 60%. Selectivity was found to be similar when hydrolysis of the racemic acetate (5.253.c) was accomplished in a reversed micelle system [18].

Fig. 5.21.c illustrates the enantiomer selective hydrolysis of racemic 2-acetylpantolactone with non-fermenting lyophilized baker's yeast [301]. Enzymically, such transformations could only be achieved with *Aspergillus* lipase, whilst other common hydrolases failed to give any reasonable results at all [301].

Fig. 5.21 presents the hydrolysis of heteroaryl methyl carbinol acetates. The impact of electronic effects on enantiomer selectivity is striking, to the extent in the case of 2-, 3-, or 4-substituted pyridine derivatives of even inverting the sense of selectivity [87].

a)

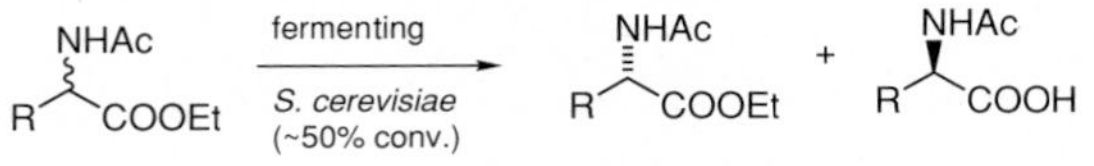

(R)-(5.252a-e)   (not isolated)

	a	b	c	d	e
R	Me	Et	i-Bu	CH₂Ph	(CH₂)₂COOEt
ee(%)	>99	>96	92	>97	89

b)

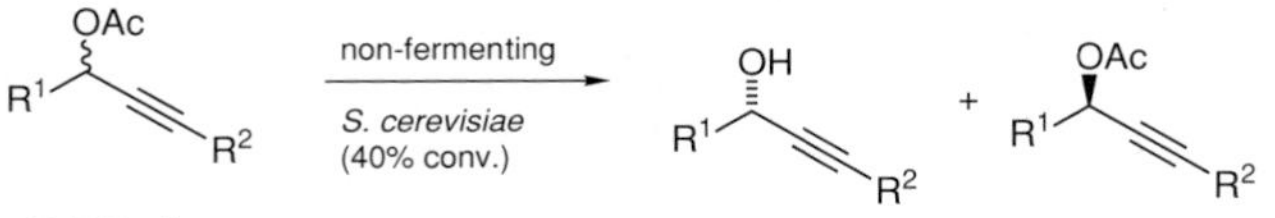

(5.253a-f)

(5.254a-f) (>90%ee)

	a	b	c	d	e	f
R¹	i-Bu	i-Bu	pentyl	hexyl	allyl	CH₂COOEt
R²	H	Me	H	H	H	H

c)

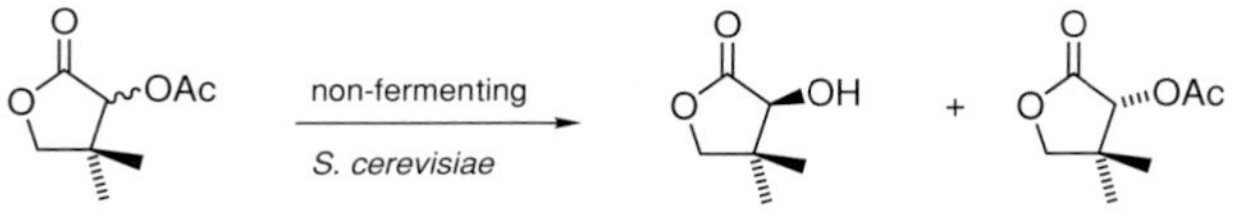

Conv. (%)	ee(%)	ee(%)
40	86	
60		95

d)

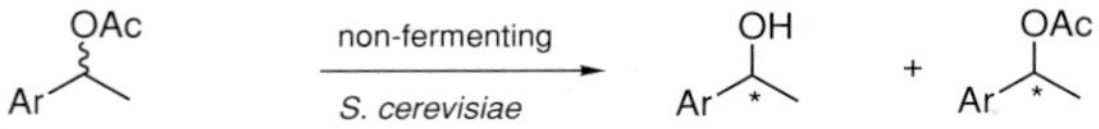

Ar	Config.	ee(%)
2-pyridyl	R	96
3-pyridyl	R	67
4-pyridyl	S	96
2-furyl	R	30
2-tiophenyl	R	54

Fig. 5.21.   Enantiomer selective hydrolyses by baker's yeast

5.5 Exploitation of the Lyase Activity of Baker's Yeast

Considering the versatile enzymic armory of baker's yeast its lyase activity has seldom been used. Decarboxylations catalyzed by baker's yeast have repeatedly been mentioned in this book whilst a further interesting example is the decarboxylation of cinnamic aldehyde to styrene with a mutant of *S. cerevisiae* [302].

Other applications of synthetic relevance are listed in Fig. 5.22.

Compounds containing an activated double bond such as methyl vinyl ketone or ethyl acrylate, can be converted by fermentation in the presence of trifluoroethanol to fluorine containing products (Fig. 5.22.a) [303].

The reduction of 4-hydroxycrotonic aldehydes with a benzyl or benzoyl protected hydroxy group (Fig. 5.22.b) is accompanied by a significant addition of water at the β-position yielding optically pure monoprotected (*S*)-1,2,4-trihydroxybutanes [304,305].

As an extension to the above reaction, the same procedure was carried out in the presence of benzyl mercaptane [305], but only non-enzymic addition took place. Optically pure (*R*)-glyceric acid thiobenzylester was however isolated from the reaction mixture, albeit in low yield (Fig. 5.22.c).

Perhaps the most important synthetic application of the lyase activity of baker's yeast is in acyloin condensation, a process discussed in the following section.

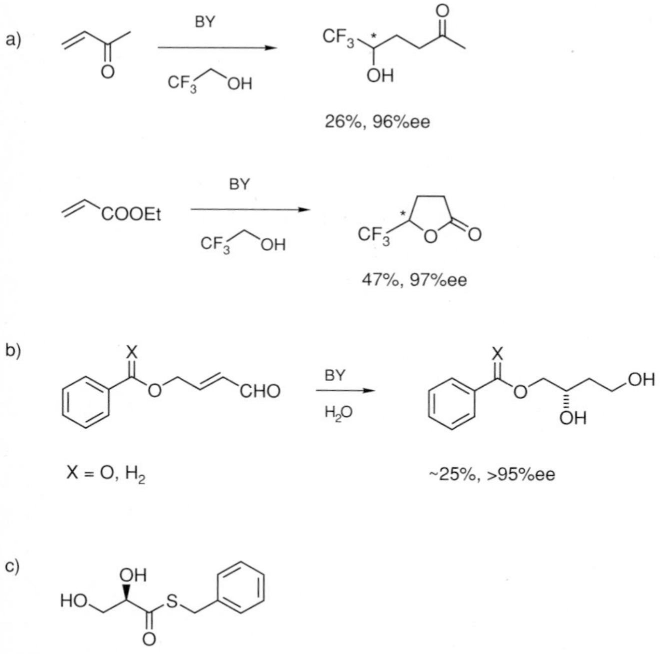

a)

26%, 96%ee

47%, 97%ee

b)

X = O, H$_2$

~25%, >95%ee

c)

Fig. 5.22.   Reactions exploiting the lyase activity of baker's yeast

5.5.1   Acyloin Condensation

As was alluded to in Chapter 1, acyloin condensation of benzaldehyde for the preparation of L-ephedrine was one of the pioneering achievements of preparative biocatalysis. The existence of numerous patents indicate the extent of the practical importance of this procedure (e.g. refs. [314,315,320]).

The formation of several products can be detected when aromatic aldehydes are fermented with baker's yeast (Fig. 5.23). In the first step in a C-2 carboligation process, the aldehydes give rise to (R)-alcohols (5.255) in high enantiomeric purity or can be reduced to the alcohols (5.258). In a secondary redox process, the condensed primary products (5.255) are converted to diols (5.256) [306,307] or occasionally to diones (5.257) [308].

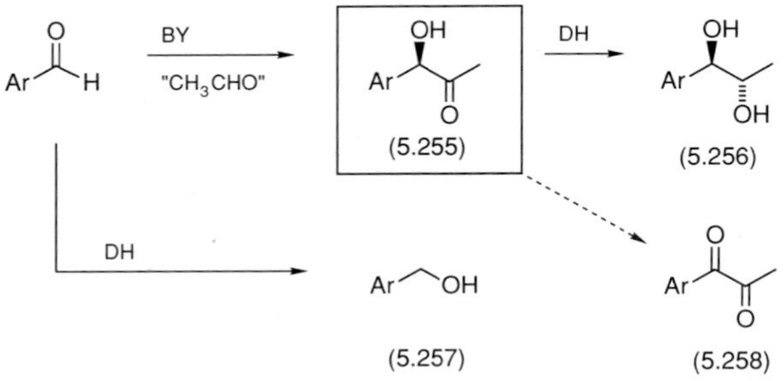

Fig. 5.23.   Transformation of aromatic aldehydes by baker's yeast

The mechanism of the condensation step has been thoroughly studied by many authors [309-312] and it was shown that the presence of pyruvate or a suitable source of carbohydrate is necessary to allow the formation of (5.255) [313]. The process is catalyzed by a thiamine pyrophosphate dependent pyruvate decarboxylase enzyme complex. In the presence of pyruvate, condensation will also take place with carbohydrate free yeast, and in addition, 2-oxobutyrate or 2-oxovalerate are substrates of the enzyme system too. (This is not the case with other, sterically more demanding 2-oxoacids [312]). In the presence of pyruvate, benzaldehyde and other α, β-unsaturated aldehydes are transformed to the corresponding homologs (5.256) [312].

Product ratios in reactions involving aromatic aldehydes are heavily dependent on reaction conditions and indeed the formation of ketols (5.255) at the expense of alcohols (5.256) is favored by adding acetaldehyde [308,314,315]. Substrate concentration is also not unimportant [294] and at a concentration over 2 g/l benzaldehyde appears to have a toxic effect on yeast cells [316].

Ketol/diol ratio is strongly influenced by pH, whereby in the acidic range ketols predominate, whilst at pH 8-9 diols are the primary products [307]. Note that reduction of the ketols (5.255) is not completely diastereotope selective and, in addition to the diols (5.256), a few percent of their *syn* isomer can also be detected [307,317].

Baker's yeast catalyzed acyloin condensation is amenable to many types of aldehydes, reactions with 2- and 4-methylbenzaldehyde [318], with 2- and 4-chlorobenzaldehyde [319] or with vanilline [320] having been known for many years. These results have recently been confirmed by the conversion of substituted benzaldehydes with yeast fermentation to either ketols [321] or diols [307] as the main product. Phenylacetaldehyde [314,315], 1-naphthaldehyde and 2-furylaldehyde [315] also yield optically active ketols.

From a synthetic point of view it is important that various α, β-unsaturated aldehydes can also be substrates in condensations (Fig. 5.24). Apart from the allyl alcohols and saturated alcohols produced by competing reductive processes, the primary product is the ketol (5.260), which has very high optical purity [317]. With the diols (5.261.a-e) shown in the figure, these are the main acyloin type products which on reduction with a high degree of *re* selectivity, yield the corresponding optically almost pure diols (5.261).

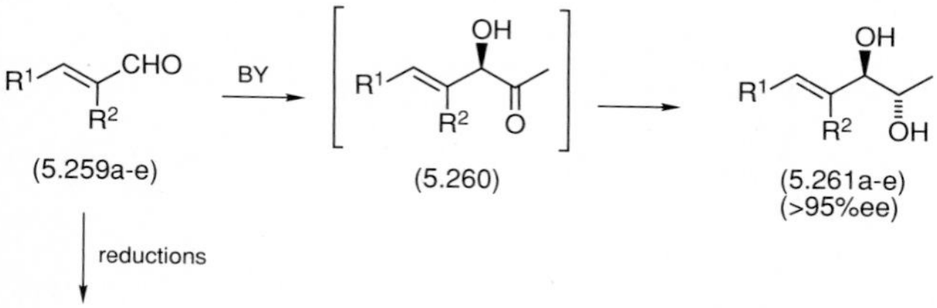

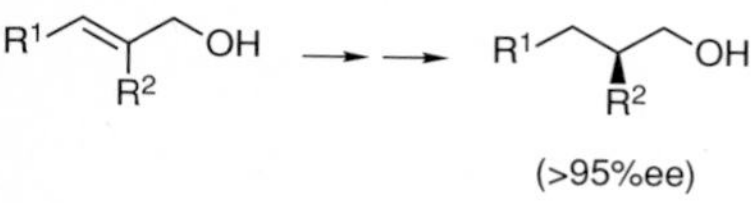

5.261	R^1	R^2	Y(%)	Ref.
a	Ph	H	8	[306]
b	Ph	Me	30	[306]
c	Ph	Br	50	[306]
d	2-furyl	Me	20	[322]
e	EtOOC	Me	35	[312]

Fig. 5.24.   Transformations α, β-unsaturated aldehydes by baker's yeast fermentation to products of high enantiomeric purity

The diols (5.261) are very useful synthetic intermediates [126,127], the role of the acetonides that can be prepared from them have already been mentioned (cf. Section 5.1.2.2). We can now add that the diol (5.261.a) may be converted to both enantiomers of a pheromone like 4-hexanolide [328] whereas (5.261.d) is a versatile chiral synthon [37] used, amongst other things, to prepare the pheromone (−)-frontalin and the diol (5.261.e), a starting material for the synthesis of the chroman moiety of vitamin E [322].

5.6 Cyclization Reactions

There are numerous, scattered examples of cyclizations carried out with baker's yeast (Fig. 5.25).

a)

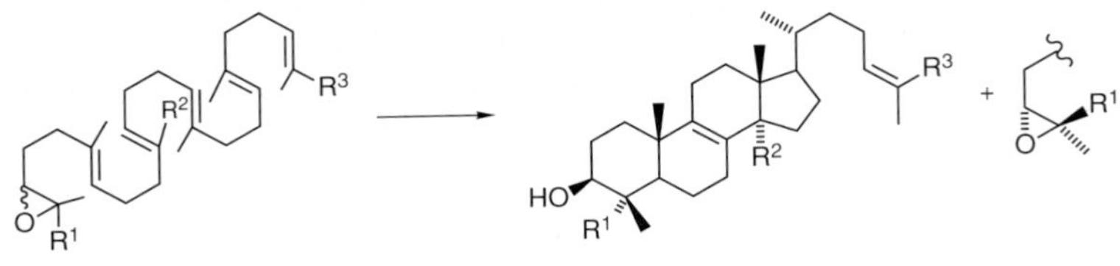

	R¹	R²	R³
5.262	Me	Me	Me
5.263	Me	Me	COOEt
5.264	Me	CH=CH₂	Me
5.265	CH₂OH	Me	Me

b)

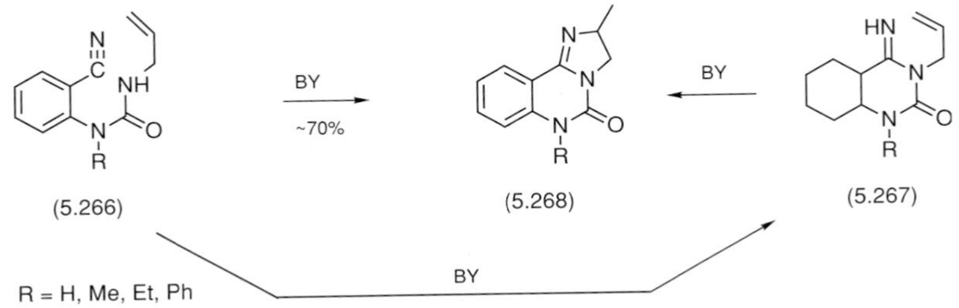

(5.266)   (5.268)   (5.267)

R = H, Me, Et, Ph

Fig. 5.25.   Cyclizations catalyzed by baker's yeast

Sonicated baker's yeast is a cheap source of lanosterol cyclase [323] and Fig. 5.25.a illustrates a number of cyclizations catalyzed by this enzyme. Racemic squalene-2,3-oxide and its analogs are transformed by this system in an enantiomer selective manner, i.e. only the (3S)-epoxides are affected. In addition, cyclization, which is in fact a cascade of reactions, is outstandingly regio- and diastereotope selective. Using the baker's yeast enzyme system a number of analogous epoxides other than lanosterol can be converted in good yield, e.g. to (Z)-ganoderic acid ethyl ester (5.263) [323], or to the potential HMG-CoA inhibiting steroid (5.264) [324]. It is of note, that whilst (2S,3S)-1-hydroxy-2,3-squalene oxide gives rise in an analogous process to the hydroxylanosterol (5.265), the 2R,3S diastereomer leads to a bicyclic product [325].

Using baker's yeast, tricyclic imidazoquinazolines (5.268) can be obtained in good yield from N-allylcarbamoyl-anthranonitriles (5.266) (Fig. 5.25.b) [326]. The intermediate is the bicyclic compound (5.267), a proposal confirmed by the fact that this compound, when used as substrate, yields the same product. Unfortunately however, the products are racemic. Nevertheless the procedure is useful owing to the mild conditions applied. Chemical reactions failed to provide any tricyclic products at all.

Recently, regioselective cycloaddition of nitrile oxides to cinnamic esters in the presence of baker's yeast was reported [327], but optical purities for the chiral products were low [355].

6 Other Enzymes and Microorganisms

In addition to the hydrolases and oxidoreductases already discussed in the preceding chapters many other enzymes have been used or have the potential to be used in synthesis.[1] As before, our discussion focuses primarily on methods that have greatest general synthetic utility. Enzymic transformations of a more restricted or specialised nature have only been touched on briefly with reference to pertinent reviews. With regard to microbial transformations, the situation is similar. Nowadays, certainly on a laboratory scale, baker's yeast can be regarded as having attained "reagent status" (cf. Chapter 5). With larger scale or industrial applications, however, the commercial availability of yeast is a less decisive factor, since finding the best microorganism and producing the necessary cell mass pays off. However a detailed discussion of this field would go beyond the scope of this book, so only a few selected examples have been presented.

6.1 Isolated Enzymes Other Than Hydrolases or Oxidoreductases

Amongst such enzymes, a number of liases and transferases can be used predominantly for synthetic purposes.

Despite their outstanding industrial importance though, isomerases are rarely used in laboratory practice since the processes they catalyze can usually be carried out as efficiently by purely chemical means. In this respect, it should be mentioned that isomerase catalyzed glucose-fructose conversion for the production of high fructose corn syrup is one of the enzymic processes that is realized on the largest scale. Enzyme catalyzed racemization has already been dealt with in connection with the preparation of amino acids in Sections 2.3.2.6 and 3.3.5.1. Additionally isomerases participate in several multienzymic processes used mainly for the preparation of carbohydrates and related compounds [1] (see also refs. [11,37,45] in Chapter 1).

---

[1] General references on biotransformations in Chapter 1: [4-6,9-11,14-16,50,110]. References concerned primarily with enzymic transformations in Chapter 1: [41,37,38,44,45,52,53,79]. References concerned primarily with microbial transformations in Chapter 1: [3,67,107-109,112,114-116,121,127,133].

6.1.1   Enzymes with Transferase Activity

Amongst the enzymes that possess transferase activity, three groups have some degree of practical potential in synthetic chemistry, namely the glycosyl transferases, i.e. enzymes that establish glycosidic linkages, aminotransferases, and kinases.

Examples can also be cited for applications of other transferase enzymes such as acetyl transferase and phosphotransacetylase, both of which are used for the regeneration of acetyl-CoA [2].

6.1.1.1   Enzymes Catalyzing Glycosyl Transfer [2]

Often, distinguishing between of transferases and hydrolases is not trivial since hydrolases are, in fact, special transferases which catalyze the transfer of a given group to water as the acceptor species [3]. In Section 3.2.2 it was mentioned that enzymes catalyzing the hydrolysis of esters can, under special conditions, function as transferases. The same can be said about glycosyl hydrolases (glycosidases) (EC 3.2), the in vivo function of which is to cleave glycosidic bonds (Fig. 6.1). Despite their differing functions in vivo, both glycosidases and glycosyl transferases (EC 2.4) can be used as catalysts in the formation of glycosidic bonds. In enzymic glycosidation the two types of enzymes are easily distinguished because glycosyl transferases require activated sugar nucleoside phosphates to function.

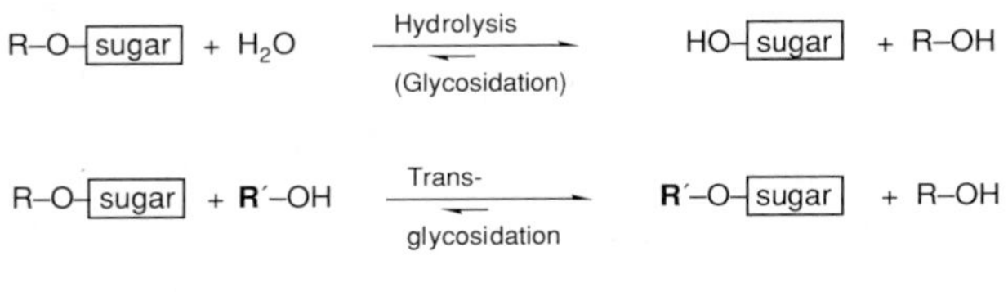

R–O–[sugar] + H_2O $\xrightarrow[\text{(Glycosidation)}]{\text{Hydrolysis}}$ HO–[sugar] + R–OH

R–O–[sugar] + **R′**–OH $\xrightarrow[\text{glycosidation}]{\text{Trans-}}$ **R′**–O–[sugar] + R–OH

Fig. 6.1.   Glycosidase catalyzed reactions with water and other nucleophiles

One example of an industrial application of enzymic glycosyl transfer is the production of cyclodextrins from starch with the assistance of cyclodextrin transglucosidase [4]. Cyclodextrins, i.e. cyclic oligosaccharides comprising 6-8 glucosyl units, are valuable hosts for inclusion complex formation. A number of analytical applications are detailed in Section 2.2.

Further applications of enzymic glycosyl transfer for the preparation of other oligosaccharides are covered by a recent review, and as such, we would rather use the space here to discuss examples of the other synthetic uses of these enzymes (Fig. 6.2).

---

[2]   For a very recent review of enzymic carbohydrate syntheses see ref. [1].

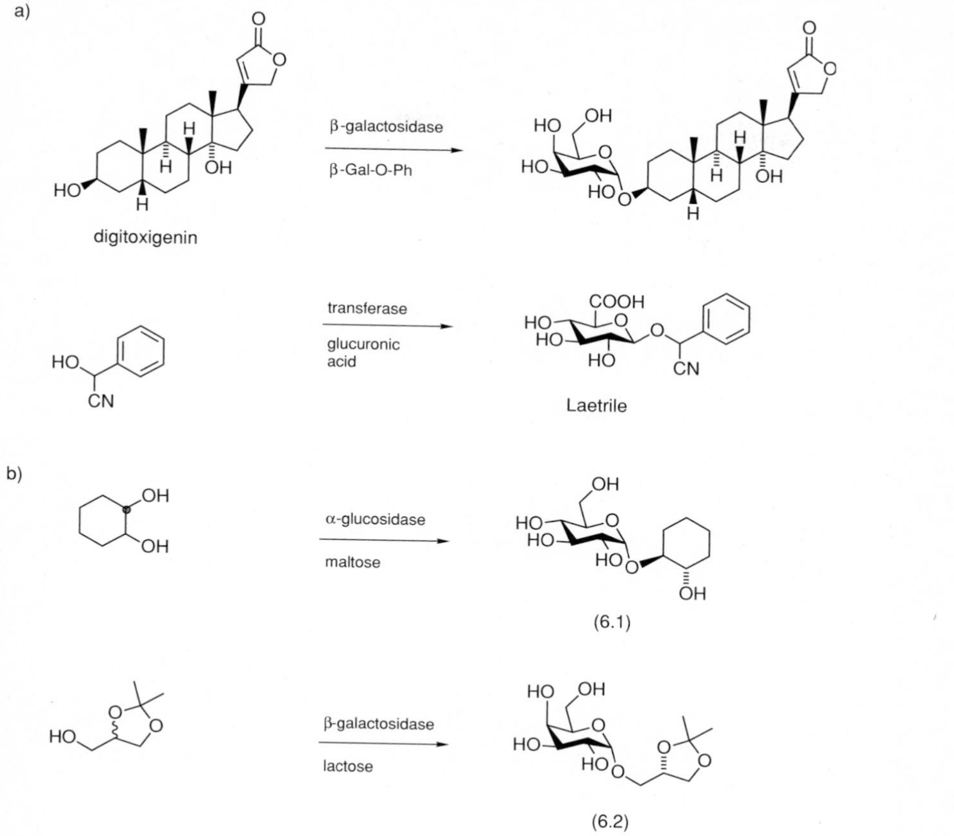

a)

β-galactosidase

β-Gal-O-Ph

digitoxigenin

transferase

glucuronic
acid

Laetrile

b)

α-glucosidase

maltose

(6.1)

β-galactosidase

lactose

(6.2)

Fig. 6.2.   Examples of glycosidations

In Fig. 6.2.a, transformation of digitoxigenine and other steroid aglycons to glycosides possessing a positive inotropic effect is shown using β-galactosidase catalyzed transglycosylation to drive the reaction [5].

Laetrile, an anti-cancer agent, was prepared by enzymic glycosidation from benzaldehyde cyanohydrin and glucuronic acid [6].

When glycosidases function as transferases, the reaction can be enantiomer selective with respect to the aglycon as illustrated in Fig. 6.2.b. Transglycosylation of both racemic *trans*-cyclohexane-1,2-diol with a crude α-glucosidase from *Aspergillus oryzae* [7] and of racemic glycerol acetonide with β-galactosidase [8] yielded glycosides of high stereochemical integrity (6.1 and 6.2). The selectivity of glycosidation depends greatly on the choice of enzyme. Other enzymes were reported to perform the same transformations but with lower diastereoselectivity [9,10]. Transglycosylation of *meso*-aglycons with relatively high enantiotope selectivity, the degree of which is again dependent on the source of the enzyme, has also been reported ([535] in Table 3.19 in Chapter 3).

The selectivities observed indicate the great potential for glycosidases in preparative work.

6.1.1.2 Aminotransferase

The preparative significance of aminotransferases is that they facilitate the preparation of unnatural and/or complex amino acids using simple amino acids as templates.

Fig. 6.3.a illustrates this approach by showing the preparation of D-selenomethionine by means of a complex of four enzymes [11]. The key enzyme is a D-aminoacid aminotransferase (D-AAT) which catalyzes transamination between 4-selenomethyl-2-oxobutyrate and D-alanine. In the presence of ammonium ions and NADH as coenzyme, alanine dehydrogenase (AlaDH) transforms the pyruvate formed to L-alanine, which in turn is then racemised by alanine racemase (AlaR). Finally, NADH is regenerated with formate dehydrogenase (FDH) utilizing the oxidation of formate to carbon dioxide.

The transformation of racemic 4-hydroxy-2-oxoglutaric acid to *anti-* and *syn*-4-hydroxyglutamic acids (6.3) is shown in Fig. 6.3.b [12]. Here the enzyme is glutamic oxalacetic aminotransferase converting the amine donor, cysteine sulphinic acid, to pyruvate and sulfur dioxide, whereby the process becomes irreversible. The result of this reaction, which can be realized even on a 100 mmol scale, is two separable diastereomers that are formed in equal amounts.

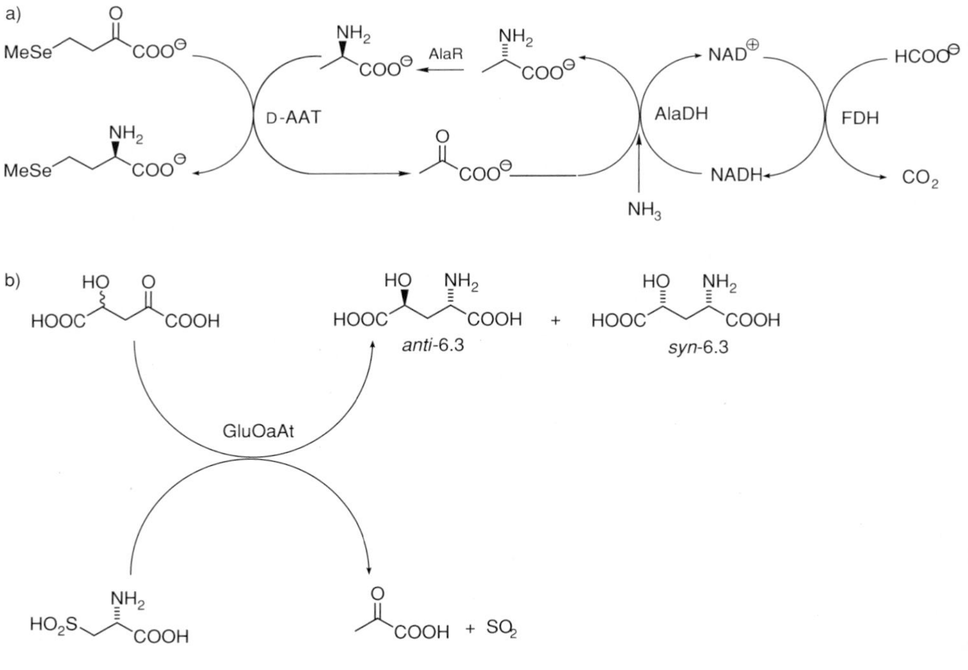

Fig. 6.3.   Examples of preparative use of aminotransferases

Irreversible decarboxylation is fully exploited in situations with other ketoacids formed as products in transaminations, the amino-donor most often being L-glutamate. A typical example is the preparation of L-phenylalanine from phenylpyruvic acid [13].

6.1.1.3 Enzymic Phosphorylations

Of the phosphokinases, glycerokinase (GK) qualifies as the primary, simple, "general reagent". This enzyme not only performs *pro-R* selective phosphorylation of glycerol, but can also be used for the enantiomer selective phosphorylation of a relatively wide range of substrates (Fig. 6.4) yielding phosphate esters of high enantiomeric purity. Purity of the unchanged substrate is, of course, dependent on conversion [14,15].

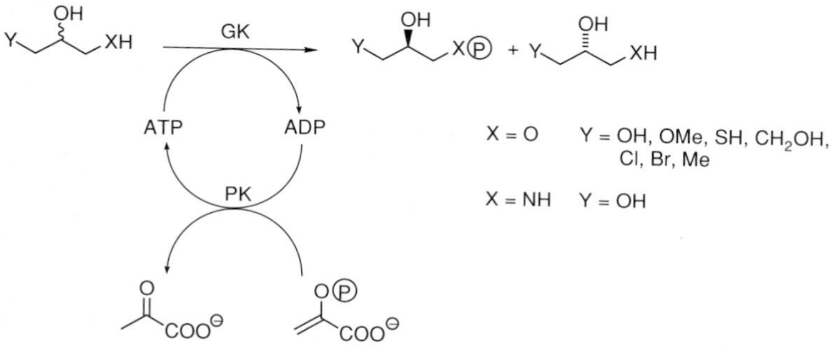

Fig. 6.4. Glycerokinase catalyzed phosphorylations

In this, as in many other phosphorylations, the phosphate donor is ATP. On the larger scale, regeneration of ATP must be provided and for both the initial preparation and regeneration of ATP, various kinase enzymes can be used [16,17]. The problems of regeneration can be solved, amongst others, with a pyruvate kinase/phosphoenol pyruvate system (Fig.6.4) [18] although it has been shown that phosphorylation of glycerol can also be realized using an acetate kinase/acetylphosphate ATP regenerating system [9].

Kinases can readily transform unnatural substrates too, as exemplified by the phosphorylation of sugars and sugar-like compounds or the regeneration of nucleoside triphosphates other than ATP (e.g. of CTP and UTP) [1,17].

A number of kinases and phosphorylases are used as members of multienzyme complexes for the preparation of important biomolecules or their analogs, such as in the synthesis of NAD(P) and its congeners [19].

Enzyme catalyzed base exchange of nucleosides is rapidly acquiring prime importance not so much for its role in the preparation of natural nucleosides, but rather for its participation in the synthesis of analogs that form the most potent class of antiviral agents [20-24]. Fig. 6.5 shows examples of three basic strategies of base exchange:

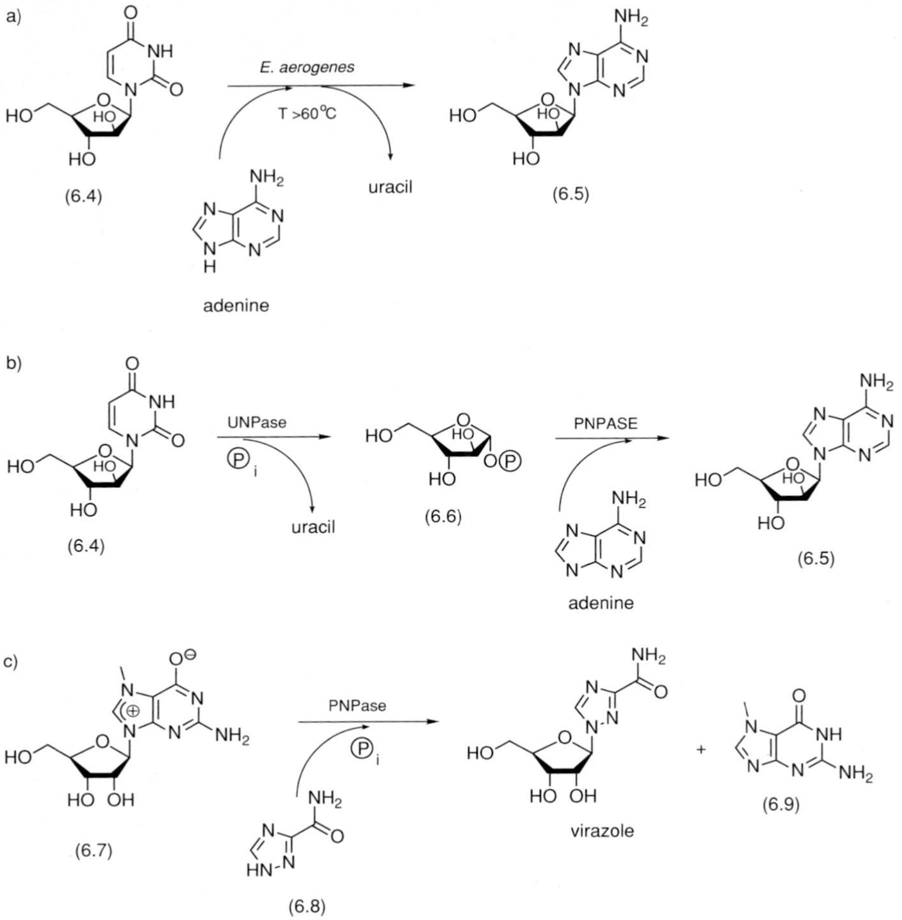

Fig. 6.5.   Enzymic base exchange of nucleosides

i) Pyrimidine nucleosides (6.4), *under the combined action of pyrimidine nucleoside phosphorylase and purine nucleoside phosphorylase (PNPase)* (both found in *Enterobacter aerogenes*), give in one operation the purine nucleoside (6.5) (Fig.6.5.a) [20].

ii) The same exchange can be accomplished *in two steps* using the purified enzymes *and by isolating the intermediate sugar phosphate* (Fig. 6.5.b) [22]. The product, arabinosyl adenine (6.5), is an antiviral agent.

iii) Base exchange between purine and purine analogs *can be affected by PNPase alone* (Fig. 6.5.c) [24]. In analogy with irreversible transesterification agents used in the field of hydrolases (cf. Section 3.2.2), here the replaced base (6.9) is also removed from the equilibrium by isomerization. The method has been used successfully by means of a base exchange with the triazole derivative (6.8) for the synthesis of the antiviral agent virazol.

It should be mentioned at this point that a number of these isolated enzymes (e.g. phosphodiesterases, restriction endonucleases, nucleotide and polynucleotide phosphorylases, DNA and RNA ligases etc.) are indispensable tools in gene technology [25] and in the sequencing of nucleic acids [27], as well as in their enzymic syntheses [28].

6.1.2   Lyases

Lyases catalyze additions onto or the actual formation of double bonds as well as the non-hydrolytic elimination of certain groups, although in terms of synthetic applications the former activity is the more important.

It is for this reason that decarboxylases for example, are rarely used in syntheses, although in the preparation of labelled compounds their stereoselectivity can be of use [29] (for general references cf. [7,11] in Chapter 1). It is very interesting in this respect that decarboxylation of racemic 2-alkyl-2-hydroxy-3-oxobutanoates led to chiral products [30]. This acetolactate decarboxylase (EC 4.1.5) catalyzed transformation involves inversion in the case of the *S* enantiomer, whilst with the *R* enantiomer decarboxylation is preceded by a 2,3-shift of the carboxyl group and only after this will take place.

Accordingly, in the case of methyl substitution, only one product is formed in contrast to the ethyl substituted enantiomers where two constitutional isomers arise, both in high enantiomeric purity.

The aspartase catalyzed addition of ammonia onto fumaric acid yielding L-aspartic acid (6.10) [31] or fumarase catalyzed addition of water onto the same substrate to yield L-malic acid (6.11) are good examples of lyase catalyzed transformations expanded to large scale industrial processes [32]:

(6.10) (X = NH$_2$)

(6.11) (X = OH)

Enzymes with both fumarase [33] and aspartase [34] activity also accept a number of 2-substituted fumarates as substrates.

Transformations relevant to the biosynthesis of terpenoids have been carried out on a small scale using farnesyl pyrophosphate synthetase (Fig. 6.6). Unambiguous stereoselective syntheses of the 4-methyl analogue of juvenile hormone (4-Me-JH-I) [35] and of the pheromone (+)-faranal [124] have been accomplished this way. Carbon-carbon bonds have been formed with high enantiotope selectivity with the aid of this enzyme

[36], for example the bond at C-4 in both the (Z)- and (E)-pyrophosphates (6.12a and b) was established from the *re* face to yield the (R)- and (S)-trienes (6.14a and b.)

In synthesis, lyase catalyzed reactions generally have a rather limited scope, with the exception of aldolases and oxynitrilases, which show a wider synthetic applicability.

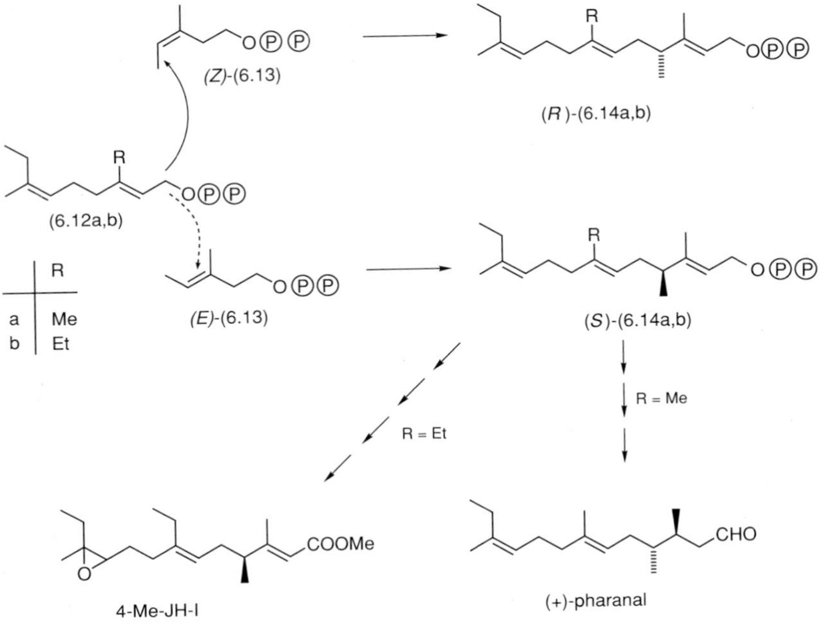

Fig. 6.6.   Farnesyl pyrophosphate synthase catalyzed carbon-carbon bond formations

6.1.2.1   Aldolases

Aldolases mediating carbohydrate metabolism are universally present in living organisms and catalyze a great variety of aldol reactions with excellent stereoselectivity [1]. When comparing the products of purely chemical asymmetric syntheses [37] and of preparative enzymic reactions, it is noticeable that they are more complementary to, than competitors of one another.

Several aldolases have been isolated and studied, primarly for the purpose of elucidating carbohydrate biosynthesis and metabolism although to date only a few have found preparative applications. Their role in the synthesis of monosaccharides and related compounds has recently been reviewed [1]. The most often, and at the same time most versatile enzyme used, is fructose-1,6-diphosphate aldolase from rabbit muscle (EC 4.1.2.13, abbreviated as FDPA or RAMA). It should be noted that the gene for the Zn^{2+}

containing FDPA has recently been cloned and the enzyme can now be expressed in *Escherichia coli* [38], and whilst this preparation has a substrate specificity similar to that of the mammalian enzyme, it is more stable. 2-Deoxyribose-5-phosphate aldolase (DERA, EC 4.1.2.4) [39] or a bacterial fuculose-1-phosphate aldolase (EC 4.1.2.17) [127] has also been prepared using molecular biology technology. The following discussion has been restricted to only include applications of FDPA directed towards non-carbohydrate products.

FDPA catalyzes the aldol reaction of aldehydes and dihydroxyacetone phosphate (DHAP) (cf. Section 2.3.2.5) and while the process is very specific to this ketone, the aldehyde component is almost freely variable [1,40,41] (Table 6.1) which renders FDPA very valuable for syntheses. DHAP can be prepared either by chemical [42] or glycerokinase catalyzed enzymic phosphorylation [14] of dihydroxyacetone, or enzymatically using fructose-1,6-diphosphate [46]. Similarly, the phosphate group can be removed from the products both by chemical and enzymic hydrolysis [41]. An interesting variation of this procedure is the addition of inorganic arsenate to the medium [52] (Fig. 6.7) which enables the use of free dihydroxyacetone as substrate which in turn facilitates the isolation of the product in a free hydroxy containing form. This method also works with other aldolases, but has the inherent disadvantage of the toxicity of the additive [38,52].

With FDPA enzyme, it can be concluded from the data in Table 6.1 that the products invariably have an (*S*)-*syn* configuration at the centre marked with an asterisk and that selectivity at this site is complete. If the aldehyde component is racemic (6.38-41, 6.46, 6.48-50), enantiomer selectivity becomes variable but never complete and a mixture of diastereomers in a ratio other than 1:1 is produced.

From a synthetic point of view, the useful aldol reactions are those which involve aldehydes with chemically transformable functionalities (e.g. 6.20-6.30, 6.33-39, 6.48-6.50).

As examples of synthetic exploitation of the products, synthesis of (+)-exo-brevicomin, a pheromone with a spiroketal structure, from aldols (6.36.b) or (6.37.b) [45,128] or the synthesis of biologically active polyhydroxypiperidines from 6.48.b [49,50] may be cited.

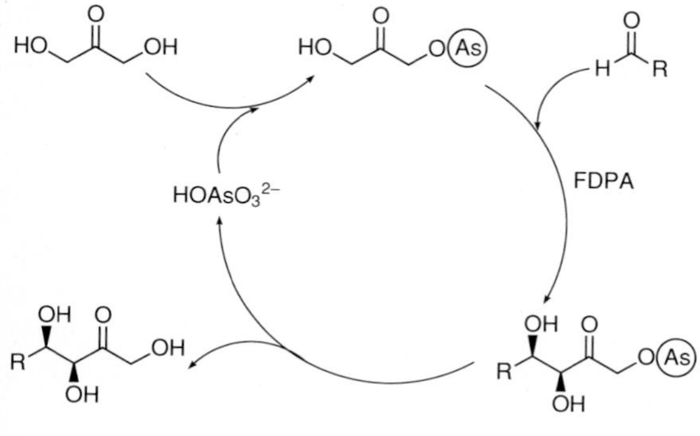

Fig. 6.7.   Aldolase catalysed reactions between aldehydes and labile arsenate of dihydroxyacetone

Table 6.1. Non-carbohydrate like chiral building blocks prepared by FDPA.

No.	R	Ref.	No.	R	Ref.
6.15- 6.19	H-(CH$_2$)$_n$ (n=0-4)	[41]	6.33 6.34 6.35 6.36	Cl(CH$_2$)$_3$ Cl(CH$_2$)$_3$ OHC(CH$_2$)$_3$ CH$_3$(O)C(CH$_2$)$_3$	[41] [41] [42] [45]
6.20	HO-CH$_2$	[41]			
6.21	BnO-CH$_2$	[41]	6.37		[45]
6.22	Cl-CH$_2$	[41]	6.38		[41]
6.23	Br-CH$_2$	[41]	6.39		[42]
6.24	(EtO)$_2$PO-CH$_2$	[42]	6.40		[44]
6.25	OHC	[42]	6.41		[44,46]
6.26	(EtO)$_2$CH	[43]	6.42		[41]
6.27	HOOC	[42]	6.43		[44,47]
6.28	HO-(CH$_2$)$_2$	[44]	6.44		[47]
6.29		[43]	6.45		[48]
6.30	MeS(CH$_2$)$_2$	[41]	6.46		[48]
6.31	Ph-CH$_2$	[41,42]	6.47		[41]
6.32	PhCO-	[42]	6.48		[49,50]

Table 6.1.

No.	R	Ref.	No.	R	Ref.
6.49	CF$_3$–C(=O)–NH–(chain)	[48]	6.50	MeOOC–(chain with NHAc)	[51]

The ever widening selection of optically pure synthons that can be prepared with the aid of FDPA and the increasing number of aldolases with special selectivity prepared by genetic manipulation methods [38,39,52] suggests that the synthetic importance of this approach is already increasing and will continue to increase dramatically.

6.1.2.2 Oxynitrilases

Cyanohydrins are versatile precursors, convertible to, among others, amino acids, α-hydroxy acids, vicinal diols, ethanolamines, and α-hydroxyketones. (For recent results on such transformations from enzymatically prepared cyanohydrins see ref. [130].) A simple access to optically active cyanohydrins provides a convenient route to these compounds in an optically active form too.

As mentioned in Section 3.3.5.3, both enantiomerically enriched cyanohydrins and acyl cyanohydrins can be prepared using purified hydrolases or various microorganisms [53-58]. Since these are enantiomer selective reactions, their yield is, of course, limited to 50%. Oxynitrilase assisted HCN addition to aldehydes is, in turn, an enantiotope selective reaction providing one of the enantiomers in high yield. The use of (R)-oxynitrilase, isolated from almonds ((R)-ON, EC 4.1.2.10, also called mandelonitrile lyase) has a relatively long tradition and proved its usefulness in the sixties in the transformation of more than 50 aldehydes [59,60]. Purification and immobilization of this enzyme is simple [61], but it is now possible to apply the crude extract [62], and using a column containing the immobilized enzyme the process can even be carried out on a kilogram scale [59].

(S)-Cyanohydrins are also accessible when (S)-oxynitrilase ((S)-ON, EC 4.1.2.11) is used, although the acceptable substrates are restricted to aromatic aldehydes [63,131].

Optimization experiments revealed that enantiomeric purity can be substantially enhanced by using organic solvents [63,64]. Another way to obtain products of higher purity is to work in an ethanol-water-acetic acid-cyanide system at high ethanol content [62,65]. Some of the results obtained under such optimized conditions are presented in Fig. 6.8.

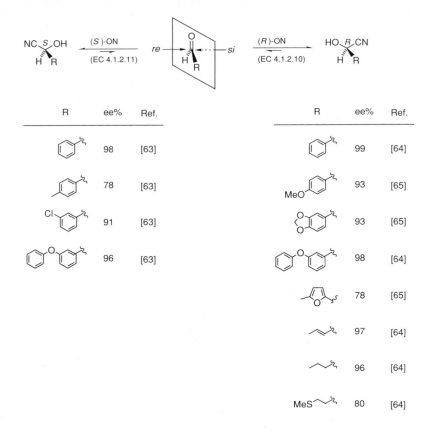

Fig. 6.8.   Preparation of optically active cyanohydrins by *(R)*- or *(S)*-oxynitrilases

6.2 Microbial Transformations: Selected Examples

In the past few decades a vast amount of information has accumulated about microbial transformations. While a book about the microbial transformations of non-steroids published in 1976 [66] quoted about 1700 publications, now the number of papers in this field counted in the ten thousands. Since several reviews are available on this subject (see Chapter 6), only a few typical examples of microbial hydrolyses and oxidoreductions have been presented here.

Often research on microbial transformations not only solves the targeted problem, but also yields more general results that permit the use of the particular organisms as a reagent or the product as a chiral building block for other syntheses. The improvement of selectivity in microbial transformations can be achieved, with the necessary circumspection of course, using all the methods described in previous chapters.

6.2.1   Hydrolytic Transformations

In principle, microorganisms are capable of accomplishing the same hydrolytic transformations as isolated enzymes (cf. Chapter 3). Enantiomer selective hydrolyses, (e.g. of esters of racemic alcohols [67-73], racemic acids [74], or further of racemic binaphthol diacetate [75]), as well as enantiotope selective hydrolyses of esters of prochiral dicarboxylic acids have been solved with the aid of microorganisms.

A typical example is the synthesis of a precursor of the polyether antibiotic monensin A (Fig. 6.9) [76,77]. This triene (6.51) containing nine chiral centers, was assembled from three subunits. Key intermediates for each subunit (6.52-54) were prepared by biocatalysis, and for two of these microorganisms were used. The ester (6.53) was the unchanged fraction in the partial hydrolysis of the racemic ester with *Bacillus subtilis*, whereas the hemiester (6.54) was obtained by hydrolysis of the corresponding *meso*-diester with *Gliocadolium roseum*.

Transformations of the antitumor agent anguidine (6.55) to other acetylated derivatives is of interest for several reasons [80]:

	R^1	R^2	R^3	
(6.55)	Ac	Ac	H	(anguidine)
(6.56)	Ac	Ac	Ac	
(6.57)	H	Ac	H	
(6.58)	H	H	Ac	

While regioselective hydrolysis with *Streptomyces griseus* of the acetate group at C-15 to the monoacetate (6.57) was not really unexpected, the fact that acylation with microbial assistance was possible in an aqueous medium was rather surprising. With the aid of *Mucor mucedo*, anguidine could be acetylated at C-3 to give a triacetate (6.56) whereas with *Fusarium oxysporum* sequential acetylation and desacetylation yielded the triacetate (6.58).

One important process is the hydrolysis of nitriles which can be carried out with various enzymes (Fig. 6.10.a)[3]. Nitrilase catalyzes hydrolysis to the acid stage, whereas nitrile hydratase takes it only as for as the amides, although amidases that are often present may carry hydrolysis further to the acid. One important advantage of this approach is that all these transformations proceed at almost neutral pH, i.e. under very mild conditions.

---

[3] Recently, further results mainly with isolated nitrilases revealed that structurally diverse nitriles [132], among them dinitriles [133,134] or racemic nitriles [135], can be efficiently transformed to carboxylic acids.

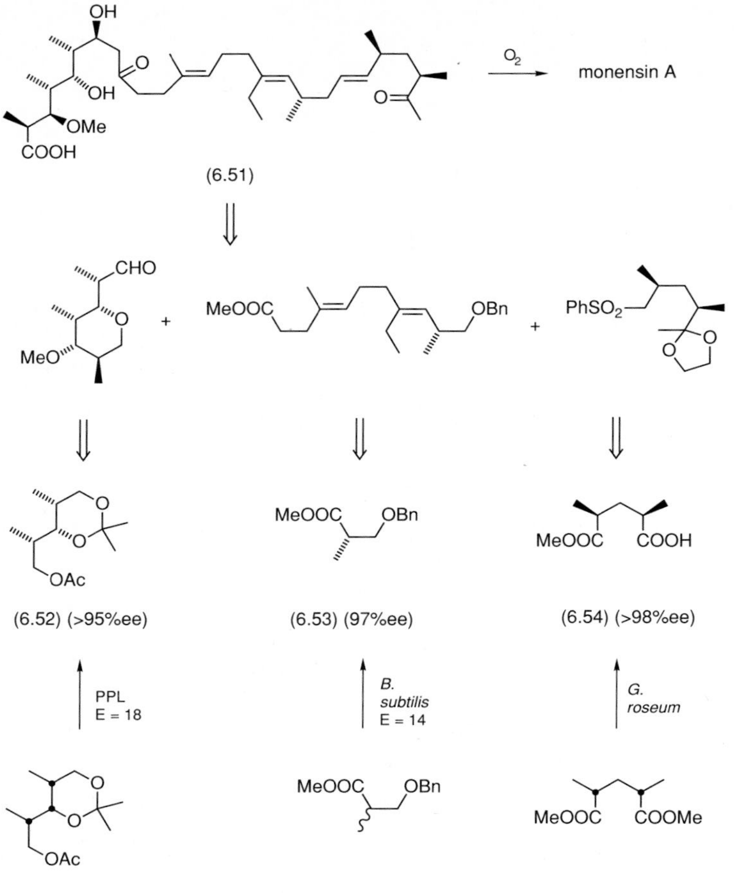

(6.51)

(6.52) (>95%ee)   (6.53) (97%ee)   (6.54) (>98%ee)

PPL
E = 18

B.
subtilis
E = 14

G.
roseum

Fig. 6.9.   Synthesis of the triene precursor of monensin A

A very important application of this process is the hydrolysis of acrylonitrile to the amide. Several microorganisms are capable of catalyzing this reaction, but *Pseudomonas chlororaphis* proved to be the best [81] and thus was utilized for the industrial scale application of this reaction [125]. The corresponding enzyme has also been isolated and adequately characterized [82] since with enzymes, transformation of nitriles can be taken under control (Fig. 6.10.b). When grown on a cobalt containing medium, *Rhodococcus rhodocrous* cells transform nicotinic nitrile to the amide, whereas when the additive is 2-methylbutyronitrile [83] hydrolysis proceeds to the acid stage. Cobalt containing media permit the controlled hydrolysis of other nitriles to amides too [84].

Fig. 6.10.c illustrates the selective hydrolysis of dinitriles. *R. rhodocrous* catalyzes the selective hydrolysis of 1,3-dicyanobenzene to 3-cyanobenzoic acid [85.86]. Similarly 3-cyanoacrylic acid can be prepared from fumaronitrile [86].

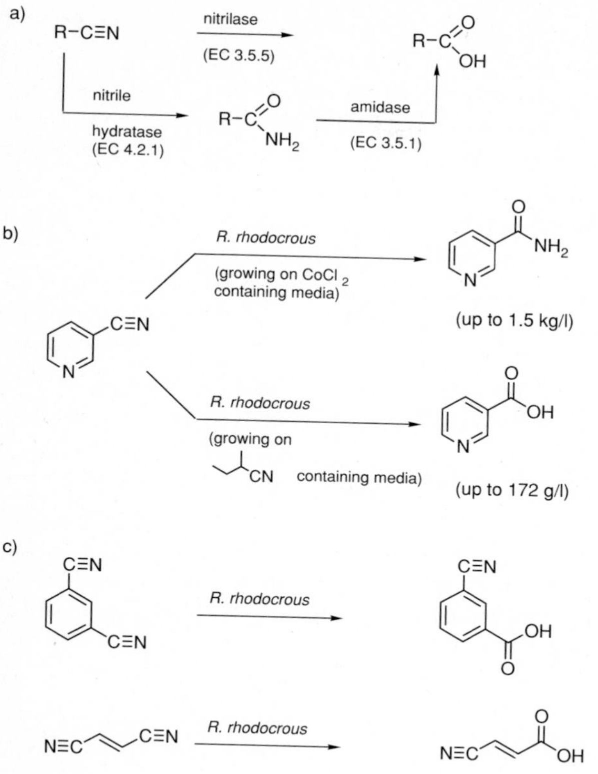

Fig. 6.10.   Transformations by microbial nitrile converting enzymes

The transformations described above are important for synthesis too, since chemical methods only permit the hydrolysis of dinitriles to either diamides or diacids, whereas biocatalysis renders any of the mixed derivatives accessible.

6.2.2   Oxidoreductions with Microorganisms

A general discussion of biocatalytic oxidoreductions was provided in earlier chapters, thus only selected examples of microbial carbonyl reductions, as well as mono- or dioxygenase catalyzed transformations have been described here.

i) Carbonyl reductions

Any transformation which can be realized with isolated NAD(P)H dependent dehy-drogenases (cf. Section 4.2) or with baker's yeast is also possible using a variety of microorganisms.

The range of feasible reactions includes enantiotope selective reduction of ketones [87,88] and α-ketoacids [89,90] to secondary alcohols, enantiomer selective reduction of girochiral ketones [91], as well as the enantiomer and enantiotope selective reduction of racemic and *meso*-metallocene aldehydes respectively [92]. Inverse transformations have also been reported such as the enantiomer selective oxidation of alcohols to the corresponding carbonyl compounds [93,94].

In Fig. 6.11, a few applications of the above reactions are shown for the synthesis of natural products.

(*S*)-3-(2-Hydroxyethyl)-pyridine was the product of microbial reduction of 3-acetylpyridine and serves as a chiral building block for the synthesis of allo-heteroyohimbine alkaloids (Fig. 6.11.a) [95].

Reduction of 2,2,5,5-tetramethylcyclohexane-1,4-dione yielded an intermediate used in the synthesis of (1*R*)-*cis*-chrysanthemic acid, the acid component of insecticidal pyrethroids (Fig. 6.11.b) [96,97].

As illustrated in Fig. 6.11.c, a diketone of C_2 symmetry was reduced to a diol that was both practically optically pure and of the same symmetry, ultimately be used in the synthesis of (+)-compactin, a potent anticholesteremic agent [98,99]. It is worth noting that, although reduction proceeds with poor enantiomer selectivity, diastereotope selectivity is different for the individual enantiomers and in this way a separable mixture of four stereoisomers is obtained containing about 33% of the desired diol.

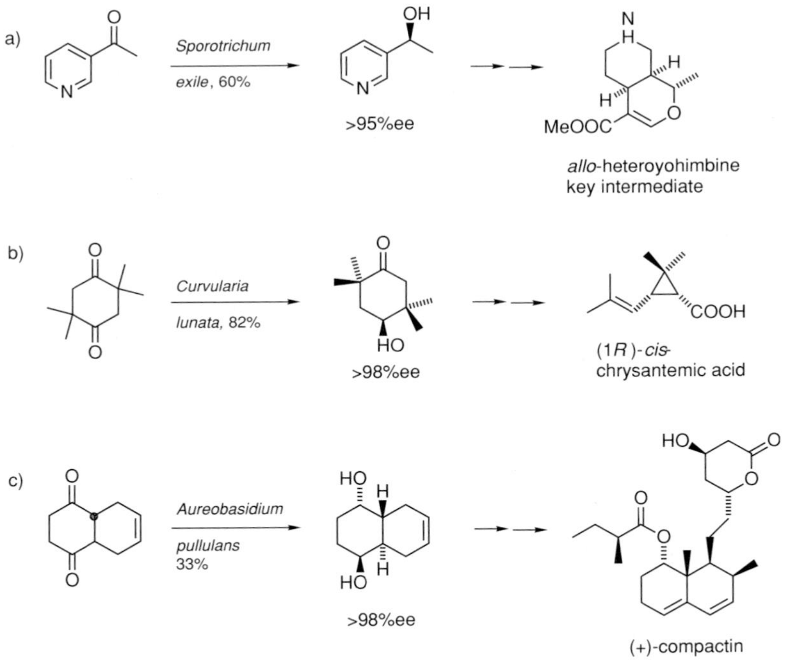

Fig. 6.11.   Microbial carbonyl reductions in syntheses of some complex natural products

Fig. 6.12.a illustrates a case in which microbial transformations were used in several steps of a natural product synthesis. In the synthesis of PGE$_2$, cyclopentanetrione and iodoketone intermediates were reduced in a microbial manner [100] and ultimately the methyl ester of the assembled molecule was converted by mild microbial hydrolysis to the end product.

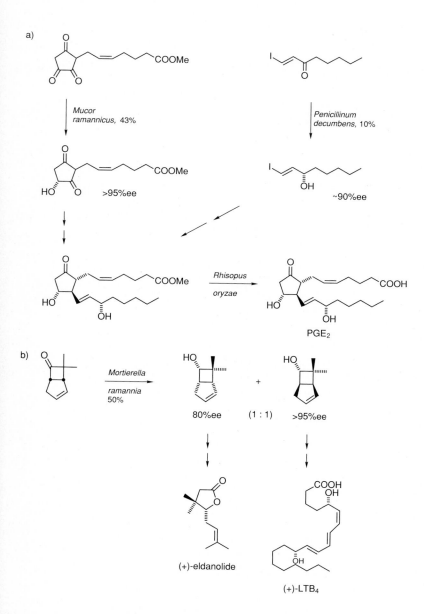

Fig. 6.12.   Use of microbial carbonyl reductions in organic syntheses

In a similar way, a version of the cyclopentanetrione derivative with a saturated side chain was converted to PGE [101]. It may be of interest to note that a recent synthesis of PGE, based on a quite different scheme, also involved a microbial reduction of a highly substituted cyclopentanone derivative as the key step [102].

In Fig. 6.12.b an enantiomer non-selective microbial reduction is depicted, in which two diastereomeric products of high enantiomeric purity were produced and each converted to different optically active natural products. Thus the *endo*-alcohol was an advanced intermediate of the pheromone (+)-eldanolide , the *exo*-alcohol in contrast was converted to a key intermediate of (+)-leukotriene B$_4$ (LTB$_4$) synthesis [103].

ii) Functionalization of organic molecules by microbial oxidations

A reaction which is rarely accomplished by chemical means is oxidative functionalization at non-activated positions of a carbon chain. In contrast, very many examples of such biotransformations are known, with a wide variety of acceptable substrates (steroids, terpenoids, alkaloids, antibiotics, etc.) (see refs. in Chapter 6).

Thus a number of reactions are described below which, apart from regioselective functionalization, involve the preparation of optically active products.

A typical example is biooxidation using *Aspergillus niger* at the terminal double bond of a hydroxy protected geraniol derivative to yield an (S)-diol of high optical purity [104]:

When the substrate was the corresponding Z isomer, i.e. the phenylcarbamoyl ester of nerol, the result was similar. The (S)-diols can be readily transformed to (R)-epoxides and thus a pair of enantiomeric chiral building blocks becomes available.

Other useful chiral synthons can be prepared via the oxidation of sulfides, e.g oxidation of (6.59) by *Rhodococcus equi* [105] (Fig. 6.13) yielding the corresponding sulfoxides in high optical purity. The latter can be transformed to (R)-phenyl-vinyl sulfoxide or (R)-phenyl-(2-hydroxyethyl)-sulfoxide.

Epoxides obtained with microorganisms from various alkenes have a special synthetic relevance ([106] and references therein) in that the preparation of such epoxides from non-functionalized alkenes is difficult by purely chemical means (Fig. 6.14). The epoxides (6.61) and (6.62) can be used in the synthesis of prostaglandin analogs [108].

The microbial oxidation of benzene derivatives to various substituted chiral *cis*-1,2-dihydroxycyclohexa-3,5-dienes (Fig. 6.15.a: 6.64-6.70) with high enantiomeric purity is of substantial synthetic interest. Such oxidations can be carried out with mutant *Pseudomonas* strains deficient in *cis*-benzeneglycol dehydrogenase. Similar oxidations can take place, not only with aromatic compounds but also in the non-aromatic ring of benzocycloalkenes [118] or even at norbornadiene [136]. Oxidation of 7-phenylnorbornadiene

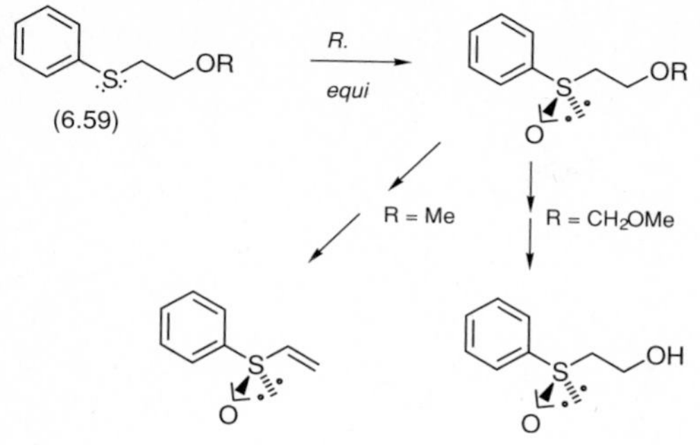

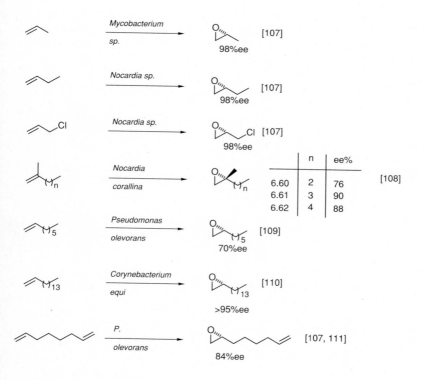

R	Yield (%)	ee. (%)
H	7	32
Me	80	>99
allyl	73	98
Bn	11	n.d.
CH$_2$OMe	78	>99

Fig. 6.13. Chiral sulfone building blocks obtained via microbial oxidations of sulfides

		n	ee%	
	6.60	2	76	[108]
	6.61	3	90	
	6.62	4	88	

Fig. 6.14. Microbial epoxidation of alkenes

[136] demonstrated that the phenyl ring was the more susceptible moiety for the enzyme system of a mutant *Pseudomonas* strain. Note that in the case of acetophenone, monooxygenation rather than the expected dioxygenation was experienced [137].

The synthetic utility of dihydrodiols prepared from benzene derivatives is illustrated in Fig. 6.15.b. The *cis*-diol (6.63) obtained from benzene was a starting material

a)

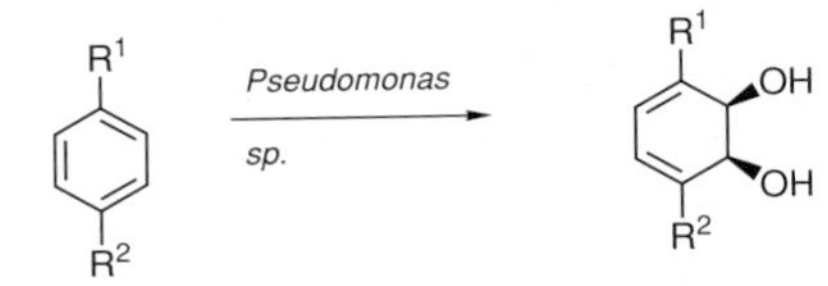

	R1	R2	Ref.
6.63	H	H	[112, 113]
6.64	Me	H	[114, 115]
6.65	Cl	H	[123]
6.66	$CH=CH_2$	H	[123]
6.67	CH_2OEt	H	[116]
6.68	COOH	Me	[117]
6.69	COOH	CF_3	[117]
6.70	COOH	Br	[117]

b)

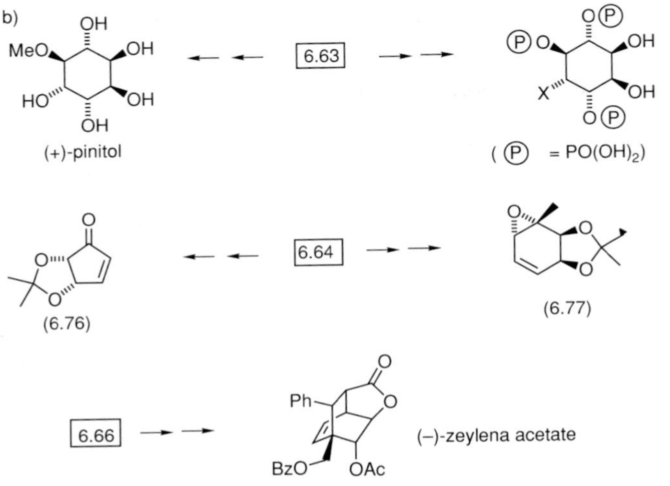

	X
6.71	OH
6.72	OMe
6.73	H
6.74	F
6.75	Me

Fig. 6.15. Microbial oxidation of benzene derivatives to *cis*-dihydrodiols

for the synthesis of both racemic [118] and (+)-pinitol [120], a pheromone. The same diol was also converted to *myo*-inositol-1,2,4-triphosphate (6.71) [121] and its analogs (6.72-6.75) [122]. Toluene yields a similar optically pure diol which was also a valuable chiral building block, used for example to prepare the acetonides (6.76) and (6.77) [123]. The diol (6.66) arose from the oxidation of styrene and was used for the synthesis of a tricyclic lactone, (−)-zeylena acetate, an antitumor agent [116]. Moreover, synthesis of (+)-conduritol C from fluorobenzene [138] and preparation of the stereoisomers of the pyrrolizidine alkaloid trihydroxyhelioridane from chlorobenzene [139] via microbial dioxygenation with *Pseudomonas putida* were accomplished.

Literature

Chapter 1

[1] Plimmer, R. H. A., *The Chemical Changes and Products Resulting from Fermentation*, London: Longmans, Green and Co., 1903.

[2] Neuberg, C., *Adv. Carbohydr. Chem. 4* (1949) 75.

[3] Gutcho, S. J., *Chemicals by Fermentation*, Park Ridge,N.J.: Noyes Data Corp, 1973.

[4] Jones, J. B., Sih, C. J., Perlmann, D. (Eds.), *Applications of Biochemical Systems in Organic Chemistry* (Techniques of Chemistry, Vol. 10), New York: Wiley, 1976.

[5] Tramper, J., Plas, H. C., Linko, P. (Eds.), *Biocatalysts in Organic Synthesis* (Stud. Org. Chem., Vol. 22), Amsterdam: Elsevier, 1985.

[6] Porter, R., Clark, S. (Eds.), *Enzymes in Organic Synthesis*, CIBA Foundation Symposium III, London: Pitman, 1984.

[7] Rétey, J., Robinson, J. A., *Stereospecificity in Organic Chemistry and Enzymology* (Monographs in Modern Chemistry, Vol. 13), Weinheim: Verlag Chemie, 1982.

[8] Jones, J. B., in : *Asymmetric Synthesis*, Vol. 5, Morrison, J. D. (Ed.), New York: Academic Press, 1986; pp 309-344.

[9] Schneider, M. P. (Ed.), *Enzymes as Catalysts in Organic Synthesis*, Dordrecht: D. Reidel Publishing, 1986.

[10] Butt, S., Roberts, S. M., *Nat. Prod. Reports* (1986) 481.

[11] Jones, J. B., *Tetrahedron 42* (1986) 3351.

[12] Butt, S., Roberts, S. M., *Chem. Br.* (1987) 127.

[13] Altenbach, H. J., *Nachr. Chem. Tech. Lab. 36* (1988) 1114.

[14] Yamada, H., Shimizu, S., *Angew. Chem., Int. Ed. Engl. 27* (1988) 622.

[15] Crout. H. G., Christen, M., "Biotransformations in Organic Synthesis" in : *Modern Synthetic Methods*, Vol. 5, Scheffold, R. (Ed.), New York: Wiley, 1989; p 1.

[16] Davies, H. G., Green, R. H., Kelly, D. R., Roberts, S. M. (Eds.), *Biotransformations in Preparative Organic Chemistry*, London: Academic Press, 1989.

[17] Bergmeyer, H. U., Bergmeyer, J., Grassl, M. (Eds.), *Methods of Enzymatic Analysis*, 3rd Ed., Weinheim: Verlag Chemie, 1986.

[18] Izumi, Y., Chibata, I., Itoh, T., *Angew. Chem., Int. Ed. Engl. 17* (1978) 176.

[19] Aida, K., Chibata, I., Nakayama, K., Takinami, K., Yamada, H. (Eds.), *Biotechnology of Amino Acid Production*, Tokyo/Amsterdam: Kodanska/Elsevier, 1986.

[20] Bucke, C., in : *Enzymes and Food Processing*, Birch, G. G., Blakebrough, N., Parker, K. J. (Eds.), London: Applied Science, 1981; p 51.

[21] Poulsen, P. B., *Enzyme Microb. Technol. 3* (1981) 271.

[22] Linko, Y. Y., Phjola, L., Linko, P., *Process Biochem. 12* (1977) 14.

[23] Vandamme, E. J., in : *Microbial Enzymes and Bioconversions*, Rose, A. H. (Ed.), London: Academic Press, 1980; p 467.

[24] Abbott, B. J., *Adv. Appl. Microbiol. 20* (1976) 203.

[25] Sato, T., Tosa, T., Chibata, I., *Eur. J. Appl. Microbiol. 2* (1976) 153.

[26] Neuberg, C., Hirsch, J., *Biochem. Z. 133* (1921) 282; Hildebrandt, C., Klavehn, W., Ger. Pat., 548'459 (1930).

[27] Thomas, J. M., *Angew. Chem., Int. Ed. Engl. Adv. Mater. 28* (1989) 1079.

[28] Naoshima, Y., Nishiyama, T., Munakata, Y, *Chem. Lett.* (1989) 1517.

[29] Klibanov, A. M., *CHEMTECH 16* (1986) 354.

[30] Tao, B. Y., *Enzyme Microb. Technol. 11* (1989) 317.

[31] Kaiser, E. T., *Angew. Chem., Int. Ed. Engl. 27* (1988) 913.

[32] Schultz, P. G., *Angew. Chem., Int. Ed. Engl. 28* (1989) 1283.

[33] Tabushi, I., *Tetrahedron 40* (1984) 269.

[34] Bender, M. L., Bergeron, R. J., Komiyama, M., *The Bioorganic Chemistry of Enzymic Catalysis*, New York: Wiley, 1984.

[35] Breslow, R., "Artificial Enzymes and Enzyme Models" in : *Advances in Enzymology and Related Areas of Molecular Biology*, Vol. 58, Meister, A. (Ed.), New York: Wiley, 1986; pp 1-60.

[36] D'Souza, V. T., Bender, M. L., *Acc. Chem. Res. 20* (1987) 146.

[37] Pike, V. W., "Synthetic Enzymes" in *Biotechnology, Vol. 7A: Enzyme Technology,* Rehm, H. J., Reed, G. (Eds.), Weinheim: VCH Verlagsgesellschaft, 1987; pp 465-487.

[38] Akiyama, A., Bednarski, M., Kim, M. J., Simon, E., Waldmann, H., Whitesides, G. M., *Chem. Br.* (1987) 647.

[39] Schneider, M., Reimerdes, E. H., *Forum Mikrob.* (1987) 302.

[40] Schneider, M., Reimerdes, E. H., *Forum Mikrob.* (1987) 65.

[41] Akiyama, A., Bednarski, M., Kim, M. J., Simon, E. H., Waldmann, H., Whitesides, G. M., *CHEMTECH 18* (1988) 627.

[42] Zaks, A., Empie, M., Gross, A., *Trends Biotechnol. 6* (1988) 272.

[43] Pratt, A. J., *Chem. Br.* (1989) 282.

[44] Wong, C. H., *Science 244* (1989) 1145.

[45] Rehm H. J., Reed, G. (Eds.), *Biotechnology, Vol. 7A: Enzyme Technology*, Weinheim: VCH Verlagsgesellschaft, 1987.

[46] International Union of Biochemistry, *Enzyme Nomenclature 1984*, New York: Academic Press, 1984.

[47] Dixon, M., Webb, E. C. (Eds.), *Enzymes*, 3rd Ed., Orlando, FL: Academic Press, 1979.

[48] Barman, T. E. (Ed.), *Enzyme Handbook*, Berlin: Springer Verlag, 1985.

[49] Sigma and Boehringer Mannheim catalogues, 1990.

[50] Laane, C., Tramper, J., Lilly, M. D. (Eds.), *Biocatalysts in Organic Media* (Stud. Org. Chem., Vol. 29), Amsterdam/New York: Elsevier, 1987.

[51] Dickinson, M., Flechert, P. D. I., *Enzyme Microb. Technol. 11* (1989) 55.

[52] Klibanov, A. M., *Trends Biochem. Sci. 14* (1989) 141.

[53] Dordick, J. S., *Enzyme Microb. Technol. 11* (1989) 194.

[54] Luisi, P. L., *Angew. Chem., Int. Ed. Engl. 24* (1985) 439.

[55] Martinek, K., Lerashov, A. V., Klyachko, N., Khmelnitsky, Y. L., Berezin, I. V., *Eur. J. Biochem. 155* (1986) 453.

[56] Shield, J. W., Ferguson, H. D., Bommarius, A. S., Hatton, T. A., *Ind. Eng. Chem., Fundam. 25* (1986) 603.

[57] Sheper, T., Likidis, Z., Makkyaleas, K., Nowottny, C., Schüberl, K., *Enzyme Microb. Technol. 9* (1987) 25.

[58] Miethe, P., Gruber, R., Voss, H., *Biotechnol. Lett. 11* (1989) 449.

[59] Pasta, P., Mazzola, G., Carrea, G., Riva, S., *Biotechnol. Lett. 11* (1989) 643.

[60] Barzana, E., Karel, M., Klibanov. A. M., *Biotechnol. Bioeng. 34* (1989) 1178.

[61] Ramos-Tombo, G. M., Schär, H. P., Fernandez, X., Busquets, I., Ghisalba, O., *Tetrahedron Lett. 27* (1986) 5707.

[62] Chang, T. M. S., *Artificial Cells*, Springfield, Ill.: Thomas, 1972.

[63] Zaborsky, O. R., *Immobilized Enzymes*, Cleveland: CRC Press, 1973.

[64] Messing, R. A., *Immobilized Enzymes for Industrial Reactors*, New York: Academic Press, 1975.

[65] Barker, S. A., Kennedy, J. F., in : *Handbook of Enzyme Biotechnology*, Part II, Wisemann, A. (Ed.), New York: Ellis Horwood Ltd., 1975; p 203.

[66] Mosbach, K., Ed., *Meth. Enzymol. 44* (1976).

[67] Chibata, I., *Immobilized Enzymes, Research and Development*, Tokyo/New York: Kodanska/Halsted, 1978.

[68] Wingard, L. B., Katchalski-Katzir, E., Goldstein, L. (Eds.), *Applied Biochemistry and Bioengineering*, Vol. 1: *Immobilized Enzyme Principles*, New York: Academic Press, 1976.

[69] Trevan, M. D., *Immobilized Enzymes: Introduction and Application in Biotechnology*, New York: Wiley, 1980.

[70] Sharma, B. P., Bailey, L. F., Messing, E. A., *Angew. Chem. Int. Ed. Engl. 21* (1982) 837.

[71] Klibanov, A. M., *Science 219* (1983) 722.

[72] Chang, T. M. S., *Microencapsulation and Artificial Cells*, Clifton, N.J.: Humana Press, 1984.

[73] Woodward, J. (Ed.), *Immobilized Cells and Enzymes: A Practical Approach*, Oxford: IRL Press, 1985.

[74] Gekas, V. C., *Enzyme Microb. Technol. 8* (1986) 450.

[75] Bednarski, M. D., Chenault, H. K., Simon, E. S., Whitesides, G. M., *J. Am. Chem. Soc. 109* (1987) 1283.

[76] Kennedy, J. F., Cabral, J. M. S., "Enzyme Immobilization" in *Biotechnology, Vol. 7A: Enzyme Technology*, Rehm, H. J., Reed, G. (Eds.), Weinheim: VCH Verlagsgesellschaft, 1987; pp 347-404.

[77] Hartmeier, W., *Immobilized Biocatalysts*, Berlin: Springer Verlag, 1988.

[78] Cornish-Bowdwen, A., Cardenas, M. L., "Chemistry of Enzymes" in *Biotechnology, Vol 7A: Enzyme Technology*, Rehm, H., Reed, G. (Eds.), Weinheim: VCH Verlagsgesellschaft, 1986; pp 3-33.

[79] Wong, C. H., Whitesides, G. M., *Aldrichim. Acta 16* (1984) 27.

[80] Simon, H., Bader, J., Günther, H., Neumann, S., Thanos, J., *Angew. Chem., Int. Ed. Engl. 24* (1985) 539.

[81] Whitesides, G. M., Wong, C. H., *Angew. Chem., Int. Ed. Engl. 24* (1985) 617.

[82] Nakamura, K., Aizawa, M., Miyawaki, O., *Electro-Enzymology Coenzyme Regeneration*, Berlin: Springer-Verlag, 1988.

[83] Willner, I., Mandler, D., *Enzyme Microb. Technol. 11* (1989) 467.

[84] Simon, E. S., Grabowski, S., Whitesides, G. M., *J. Am. Chem. Soc. 111* (1989) 8920.

[85] Liebmann, J. E. (Ed.), *Mechanistic Principles of Enzyme Activity*, Weinheim: VCH Verlagsgesellschaft, 1989.

[86] Segel, I. H., *Enzyme Kinetics*, New York: Wiley, 1975.

[87] Wong, J. T., *Kinetics of Enzyme Mechanisms*, London: Academic Press, 1975.

[88] Cornish-Bowden, A., *Fundamentals of Enzyme Kinetics*, London: Butterworths, 1979.

[89] Walsh, C., *Enzymatic Reaction Mechanisms*, New York: W. H. Freeman, 1979.

[90] Engel, P. C., *Enzyme Kinetics*, New York: Chapman and Hall, 1981.

[91] Fersht, A., *Enzyme Structure and Mechanism*, 2nd Ed., New York: W. H. Freeman, 1985.

[92] Hammes, G. G., *Enzyme Catalysis and Regulation*, New York: Academic Press, 1982.

[93] Michaelis, L., Menten, M. L., *Biochem. Z. 49* (1913) 333.

[94] Lineweaver, H., Burk, D., *J. Am. Chem. Soc. 56* (1934) 685.

[95] Lee, L. G., Whitesides, G. M., *J. Org. Chem. 51* (1986) 25.

[96] Ibrahim, C. O., Nishio, N., Nagai, S., *Agric. Biol. Chem. 52* (1988) 2923.

[97] Keinan, E., Hafeli, E. K., Seth, K. K., Lamed, R., *J. Am. Chem. Soc. 108* (1986) 162.

[98] Cho, H. Y., Tanizawa, K., Tanaka, H., Soda, K., *Agric. Biol. Chem. 51* (1987) 2793.

[99] Geutzen, I., Löffler, H. G., Schneider, F., *Z. Naturforsch., C. 35* (1980) 544.

[100] Chenault, H. K., Dahmer, J., Whitesides, G. M., *J. Am. Chem. Soc. 111* (1989) 6354.

[101] Nakamura, K., Kawai, Y., Oka, S., Ohno, A., *Tetrahedron Lett. 30* (1989) 2245.

[102] Kunugi, S., Tanabe, K., Yamashita, K., Morikawa, Y., Ito, T., Kondoh, T., Hirata, K., Nomura, A., *Bull. Chem. Soc., Japan 62* (1989) 514.

[103] Laane, C., Boeren, S., Vos, K., Veeger, C., *Biotechnol. Bioeng. 30* (1987) 81.

[104] Empie, M. W., Gross, A., *Annu. Rep Med. Chem. 23* (1988) 305.

[105] Tanaka, A., Sonomoto, K., *CHEMTECH* (1990) 112.

[106] Zaks, A., Klibanov, A. M., *Proc. Natl. Acad. Sci. USA 82* (1985) 3192.

[107] Rehm, H. J., Reed, G. (Eds.), *Biotechnology, Vol. 3: Biomass, Microorganisms for Special Applications, Microbial Products I, Energy from Renewable Resources*, Weinheim: VCH Verlagsgesellschaft, 1983.

[108] Rehm, H., Reed, G. (Eds.), *Biotechnology, Vol. 4: Microbial Products II*, Weinheim: VCH Verlagsgesellschaft, 1986.

[109] Rehm, H. J., Reed, G., *Biotechnology, Vol. 6A: Biotransformations*, Kieslich, K., Vol. Ed., Weinheim: VCH Verlagsgesellschaft, 1984.

[110] Moo-Young, M. (Ed.), *Comprehensive Biotechnology, Vol. 3: The Practice of Biotechnology: Current Commodity Chemicals*, Oxford: Pergamon Press, 1985.

[111] Moo-Young, M. (Ed.), *Comprehensive Biotechnology, Vol. 4: The Practice of Biotechnology: Speciality Products and Service Activities*, Oxford: Pergamon Press, 1985.

[112] Kieslich, J. K., *Microbial Transformations of Non-Steroid Cyclic Compounds*, Stuttgart: G. Thieme Verlag, 1976.

[113] Sih, C. J., Rosazza, J. P., "Microbial Transformations in Organic Synthesis" in : *Applications of Biochemical Systems in Organic Chemistry*, Jones, J. B., Sih, C. J., Perlman D. (Eds.), New York: Wiley, 1976; pp 69-106.

[114] Rose, A. H. (Ed.), *Economic Microbiology, Vol. 5: Microbial Enzymes and Bioconversions*, London: Academic Press, 1980.

[115] Rosazza, J. P. (Ed.), *Microbial Transformations of Bioactive Compounds*, Boca Raton, FL: CRC Press, 1982.

[116] Sariaslani, F. S., Rosazza, J. P. M., *Enzyme Microb. Technol. 6* (1984) 242.

[117] Sebek, O. K., "Historical Development of Microbial Transformations" in : *Microbial Transformations of Bioactive Compounds*, Rosazza, J. P. (Ed.), Boca Raton, FL: CRC Press, 1982; p 1.

[118] Pasteur, L., C. R. *Acad. Sci. 55* (1862) 28.

[119] Bourtroux, L., C. R. *Acad. Sci. 91* (1880) 236.

[120] Charney, W., Herzog, H. L., *Microbial Transformations of Steroids*, New York: Academic Press, 1967.

[121] Iizuka, H., Naito, A., *Microbial Transformations of Steroids and Alkaloids*, New York: Springer-Verlag, 1981.

[122] Holland, H. L., in : *The Alkaloids, Vol. 19*, Ch. 5, Rodrigo, R. G. A. (Ed.), New York: Academic Press, 1981; pp 324-400.

[123] Ogata, K., Kinoshita, S., Tsunoda, T., Aida, K. (Eds.), *Microbial Production of Nucleic Acid-Related Substances*, Tokyo: Kodanska, 1976.

[124] Kuninaka, A., in : *Biotechnology, Vol. 4: Microbial Products II*, Rehm, H. J., Reed, G. (Eds.), Weinheim: VCH Verlagsgesellschaft, 1986; p 71.

[125] Spencer, J. F. T., Gorin, P. A. J., "Microbiological Transformations of Sugars and Related Compounds" in : *Progr. Ind. Microbiol.*, Vol. 7, London: Heywood and Co., 1965; p 177.

[126] Collins, C. H., Lyne, P. M., Grange, G. M. (Eds.), *Microbiological Methods*, 6th Ed., London: Butterworth and Co., 1989.

[127] Rehm, H. J., Reed, G. (Eds.), *Biotechnology, Vol. 1: Microbial Fundamentals*, Weinheim: VCH Werlagsgesellschaft, 1981.

[128] Sih, C. J., Chen, C. S., *Angew. Chem., Int. Ed. Engl. 23* (1984) 570.

[129] Chen, C. S., Zhou, B. M., Girdaukas, G., Shieh, W. R., Van Middlesworth, F., Gopalan, A. S., Sih, C. J., *Bioorg. Chem. 12* (1984) 98.

[130] Shieh, W. R., Gopalan, A. S., Sih, C. J., *J. Am. Chem. Soc. 107* (1985) 2993.

[131] Boland, W., Niedermayer, U., *Synthesis* (1987) 28.

[132] Jack, T. R., Zajic, J. E., *Adv. Biochem. Eng. 5* (1977) 125.

[133] Chibata, I., Tosa, T., *Adv. Appl. Microbiol. 22* (1977) 1.

[134] Venkatsubramanian, K. (Ed.), *Immobilized Microbial Cells*, Washington D.C.: American Chemical Society, 1979.

[135] Mattiason, G., *Immobilized Cells and Organelles*, Cleveland: CRC Press, 1982.

[136] Nunez, H. J., Lema, J. M., *Enzyme Microb. Technol. 9* (1987) 642.

[137] Brodelius, P., Vandamme, E. J., "Immobilized Cell Systems" in: *Biotechnology, Vol. 7A: Enzyme Technology*, Rehm, H. J., Reed, G. (Eds.), Weinheim: VCH Verlagsgesellschaft, 1987; pp 405-464.

[138] Phillips, C. R., Poon, Y. C., *Immobilization of Cells* (Biotechnology Monographs, Vol. 5), Berlin: Springer Verlag, 1988.

[139] Vezina, C., Shegal, S. N., Singh, K., *Adv. Appl. Microbiol. 10* (1968) 211.

[140] Vezina, C., Singh, K., in : *The Filamentous Fungi, Vol. 1.*, Chap 9., Smith, J. E., Berry, D. R. (Eds.), London: Edward Arnold, 1975.

[141] Bull, A. T., "Mixed Culture and Mixed Substrate Systems" in : *Comprehensive Biotechnology, Vol. 1*, Chap 15, Moo-Young, M. (Ed.), London: Pergamon Press, 1985.

[142] Fukumura, T., *Agric. Biol. Chem. 40* (1976) 1687.

[143] Takamatsu, S., Umemura, I., Yamamoto, K., Sato, T., Tosa, T., Chibata, I., *Eur. J. Appl. Microbiol. Biotechnol 15* (1982) 147.

[144] Moo-Young, M. (Ed.), *Comprehensive Biotechnology, Vol. 1: The Principles of Biotechnology: Scientific Fundamentals*, Oxford: Pergamon Press, 1985.

[145] Moo-Young, M. (Ed.), *Comprehensive Biotechnology, Vol. 2: The Principles of Biotechnology: Engineering Considerations*, Oxford: Pergamon Press, 1985.

[146] Fukui, S., Tanaka, A., *Experientia 45* (1989) 1055.

[147] Guiseley, K. B., *Enzyme Microb. Technol. 11* (1989) 706.

[148] Webb, C., Black, G. M., Atkinson, B. (Eds.), *Process Engineering Aspects of Immobilized Cell Systems*, Elmsford, New York: Pergamon Press, 1986.

[149] Spier, R. E., Fowler, M. N., "Animal and Plant Cell Cultures" in : *Comprehensive Biotechnology, Vol.1*, Chap 16, Moo-Young, M. (Ed.), Oxford: Pergamon Press, 1985.

[150] Di Cosimo, F., Facchini, P. J., Kraml, M. M., *Chem. Br.* (1989) 1001.

[151] Hulst, A. C., Tramper, J., *Enzyme Microb. Technol. 11* (1989) 546.

[152] Neumann, K. H., Barz, W., Reinhard, E. (Eds.), *Primary and Secondary Metabolism of Plant Cell Cultures*, Berlin: Springer Verlag, 1985.

[153] Spieler, H., Alfermann, A. W., Reinhard, E., *Appl. Microbiol. Biotechnol. 23* (1985) 1.

[154] Furuya, T., Yoshikawa, T., Taira, T., *Phytochem. 23* (1984) 999.

[155] Naoshima, Y., Akakabe, Y., *J. Org. Chem. 54* (1989) 4237.

[156] Gu, K. F., Chang, T. M. S., *Biotechnol. Bioeng. 32* (1988) 363.

[157] Wong, C. H., Haynie, S. L., Whitesides, G. M., *J. Org. Chem. 47* (1982) 5416.

[158] De Luca, M., Cricka, L. J., *Arch. Biochem. Biophys.* 226 (1983) 285.

[159] Wong, C. H., Pollack, A., Mc Curray, S. D., Sue, J. M., Knowles, R. J., Whitesides, G. M., *Meth. Enzymol. 89* (1982) 108.

[160] Pantoliano, M. W., Ladner, R. C., Bryan, P. N., Rollence, M., Wood, J. F., Poulos, T. L., *Biochemistry 26* (1987) 2077.

[161] Wong, C. H., Chen, S. T., Hennen, W. J., Bibbs, J. A., Wang, Y. F., Liu, J. L. C., Pantoliano, M. W., Whitlow, M., Bryan, P. N., *J. Am. Chem. Soc. 112* (1990) 945.

[162] Oxender, D. L., Fox, C. F. (Eds.), *Protein Engineering*, New York: Alan R. Liss, Inc., 1987.

[163] Rehm, H. J., Reed, G. (Eds.), *Biotechnology, Vol. 7B: Gene Technology*, Weinheim: VCH Verlagsgesellschaft, 1989.

[164] Kaimal, T. N. B., Saroja, M., *Biotechnol. Lett. 11* (1989) 31.

[165] Nakatsuka, T., Sasaki, T., Kaiser, E. T., *J. Am. Chem. Soc. 109* (1987) 3809.

[166] West, J. B., Scholten, J., Stolowich, N. J., Hogg, J. L., Scott, A. I., Wong, C. H., *J. Am. Chem. Soc. 110* (1988) 3709.

[167] Wells, J. A., Estell, D. A., *Trends. Biochem. Sci. 13* (1988) 291.

[168] Luyten, M. A., Bur, D., Wynn, H., Parris, W., Gold, M., Friesen, J. D., Jones, J. B., *J. Am. Chem. Soc. 111* (1989) 6800.

[169] Hudlicky, T., Luna, H., Barbieri, G., Kwart, L. D., *J. Am. Chem. Soc. 110* (1988) 4735.

[170] Van den Tweel, W. J. J., de Bont, J. A. M., Vorage, M. J. A. W., Marsman, E. H., Tramper, J., Koppejan, J., *Enzyme Microb. Technol. 10* (1988) 134.

[171] Peberdy, J. F., "Genetic Engineering in Relation to Enzymes" in : *Biotechnology, Vol. 7A: Enzyme Technology*, Rehm, H. J., Reed, G. (Eds.), Weinheim: VCH Verlagsgesellschaft, 1987; pp 325-344.

[172] Anderson, S., Marks, C. B., Lazarus, R., Miller, J., Stafford, K., Seymour, J., Light, D., Rastetter, W., Estell, D., *Science 230* (1985) 144.

[173] Bigelis, R., Das, R. C., in : *Protein Transfer and Organelle Biogenesis*, Das, R. C., Robbins, P. W. (Eds.), New York: Academic Press, 1988; pp 771-808.

[174] Zemel, H., *J. Am. Chem. Soc. 109* (1987) 1875.

[175] Kokubo, T., Sassa, S., Kaiser, E. T., *J. Am. Chem. Soc. 109* (1987) 606.

[176] Sasaki, S., Takase, Y., Koga, K., *Tetrahedron Lett. 31* (1990) 6051.

[177] Schürmann, G., Diederich, F., *Tetrahedron Lett. 27* (1986) 4249.

[178] Ohno, A., Yasuma, T., Nakamura, K., Oka, S., *Bull. Chem. Soc., Japan 59* (1986) 2905.

[179] Breslow, R., Chmielewski, J., Foley, D., Johnson, B., Kumabe, N., Varney, M., Mehra, R., *Tetrahedron 44* (1988) 5515.

[180] Groves, J. T., Neumann, R., *J. Org. Chem. 53* (1988) 3893.

[181] Colonna, S., Gaggero, N., *Tetrahedron Lett. 30* (1989) 6233.

[182] Hanson, D. E., *Nature (London) 325* (1987) 304.

[183] Goding, J. W., *Monoclonal Antibodies: Principles and Practice*, New York: Academic Press, 1986.

[184] Shokat, K. M., Leumann, C. J., Sugasawara, R., Schultz, P. G., *Nature (London) 338* (1989) 269.

[185] Janjic, N., Tramontano, A., *J. Am. Chem. Soc. 111* (1989) 9109.

[186] Hilvert, D., Hill, K. W., Nared, K. D., Auditor, M. T. M., *J. Am. Chem. Soc. 111* (1989) 9261.

[187] Pollack, S. J., Hsiun, P., Schultz, P. G., *J. Am. Chem. Soc. 111* (1989) 5961.

[188] Janda, K. D., Weinhouse, M. I., Schloeder, D. M., Lerner, R. A., *J. Am. Chem. Soc. 112* (1990) 1274.

[189] Cech, T. R., *Nature (London) 339* (1989) 507.

[190] Plückthun, A., *Kontakte* (Darmstadt) (1991) 40.

[191] Rétey, J., *Angew. Chem., Int. Ed. Engl. 29* (1990) 355.

[192] Schuster, M., Aviksaar, A., Jakubke, H. D., *Tetrahedron 46* (1990) 8093.

[193] West, J. B., Hennen, W. J., Lalonde, J. L., Bibbs, J. A., Zhong, Z., Meyer, E. F., Wong, C. H., *J. Am. Chem. Soc. 112* (1990) 5313.

[194] Kamal, A., Ramalingam, T., Venugopal, N., *Tetrahedron: Asymmetry 2* (1991) 39.

[195] Cochran, A. G., Schultz, P. G., *J. Am. Chem. Soc. 112* (1990) 9414.

[196] Braisted, A. C., Schultz, P. G., *J. Am. Chem. Soc. 112* (1990) 7430

[197] Iverson, B. L., Cameron, K. E., Jahangiri, G. K., Pasternak, D. S., *J. Am. Chem. Soc. 112* (1990) 5320.

[198] Janda, K. D., Ashley, J. A., Jones, T. M., McLeod, D. A., Schloeder, D. M., Weinhouse, M. I., *J. Am. Chem. Soc. 112* (1990) 8886.

Chapter 2

[1] Eliel, C. L., *Stereochemistry of Carbon Compounds*, New York: Mc Graw-Hill, 1962; p434.

[2] Fuganti, C., Ghiringhelli, D., Graselli, P., *Chem. Commun.* (1975) 846.

[3] Rétey, J., Robinson, J. A., *Stereospecificity in Organic Chemistry and Enzymology* (Monographs in Modern Chemistry, Vol. 13), Weinheim: Verlag Chemie, 1982; Bartmann, W., Sharpless, K. B. (Eds.), *Stereochemistry of Organic and Bioorganic Transformations*, New York: Verlag Chemie, 1987.

[4] Jones, J. B., "Stereochemical Considerations and Terminologies of Biochemical Importance" in: *Applications of Biochemical Systems in Organic Chemistry*, Jones, J. B., Sih, C. J., Perlman, D. (Eds.), New York: Wiley, 1976; pp 479-490.

[5] Mislow, K., *Introduction to Stereochemistry*, New York: Benjamin, 1966.

[6] Eliel, E. L., *Elements of Stereochemistry*, New York: Wiley, 1969.

[7] Nógrádi, M., *Stereochemistry, Basic Concepts and Applications*, Weinheim/Budapest: Verlag Chemie/Akadémiai Kiadó, 1981.

[8] Alworth, W. A., *Stereochemistry and Its Application in Biochemistry*, New York: Wiley, 1972.

[9] Cahn, R. S., Ingold, C. K., Prelog, V., *Experientia 12* (1956) 81; Cahn, R. S., Ingold, C. K., Prelog, V., *Angew. Chem., Int. Ed. Engl. 5* (1966) 385.

[10] Prelog, V., Helmchen, G., *Angew. Chem., Int. Ed. Engl. 21* (1982) 567.

[11] IUPAC Tentative Nomenclature Rules, Section E. Fundamental Stereochemistry, *J. Org. Chem. 35* (1970) 2849.

[12] Heathcock, C. H., Base, C. T., Kleislich, W. A., Pirrung, M. C., Sohn. J. E., Lampe. J., *J. Org. Chem. 45* (1980) 1066.

[13] Hirschmann, H., Hanson, K. R., *Tetrahedron 30* (1974) 3649.

[14] Hanson, K. R., *J. Am. Chem. Soc. 88* (1966) 2731.

[15] Prelog, V., Helmchen, G., *Helv. Chim. Acta 55* (1972) 2581.

[16] Hirshmann, H., Hanson, K. R., *J. Org. Chem. 36* (1971) 3293.

[17] Mislow, K., Raban, M., *Top. Stereochem. 1* (1967) 1.

[18] Schurig, V., *Kontakte* (Darmstadt (1985) 1/54.

[19] Schurig, V., *Kontakte* (Darmstadt) (1985) 2/22.

[20] Lyle, G. G., Lyle, R. E, in : *Asymmetric Synthesis, Vol. 1*, Morrison, J. D. (Ed.), London: Academic Press, 1983; p 13.

[21] Mislow, K., *Einführung in die Stereochemie*, Weinheim: Verlag Chemie, 1967; p 155.

[22] Anderson, K. K., Gash, D. M., Robertson, J. D., in : *Asymmetric Synthesis*, Vol. I, Morrison, J. D. (Ed.), London: Academic Press, 1983; p 45.

[23] Jaques, J., Collet, A., Wilen, S. H., *Enantiomers, Racemates and Resolution*, New York: Wiley, 1981.

[24] Pirkle, W. H., Hoover, D. J., *Top. Stereochem. 13* (1982) 263.

[25] Weisman, G. R., in : *Asymmetric Synthesis, Vol. 1*, Morrison, J. D. (Ed.), London: Academic Press, 1983; p 153.

[26] Strekowski, L., Visnick, M., Battiste, M. A., *J. Org. Chem. 51* (1986) 4836.

[27] Pirkle, W. H., Pochapsky, T. C., *J. Am. Chem. Soc. 108* (1986) 5627.

[28] Rosini, C., Ucello-Barretta, G., Pini, D., Abete, C., Salvadori, P., *J. Org. Chem. 53* (1988) 4579.

[29] Benson, S., Cai, P., Colon, M., Haiza, M. A., Tokles, M., Snyder, J. K., *J. Org. Chem. 53* (1988) 5335.

[30] Sullivan, G. R., *Top. Stereochem. 10* (1976) 287.

[31] Fraser, R. R., in : *Asymmetric Synthesis, Vol. 1*, Morrison, J. D. (Ed.), London: Academic Press, 1983; p 173.

[32] Sweeting, L. M., Crans, D. C., Whitesides, G. M., *J. Org. Chem. 52* (1987) 2273.

[33] *Aldrichimica Acta 10* (1977) 54.

[34] Allenmark, S. G., *Chromatographic Enantioseparation: Methods and Applications*, New York: Wiley, 1988.

[35] Schurig, V., in : *Asymmetric Synthesis, Vol. 1*, Morrison, J. D. (Ed.), London: Academic Press, 1983; p 59.

[36] Schurig, V., *Kontakte* (Darmstadt (1986) 1/3.

[37] Pirkle, W. H., Finn, J., in : *Asymmetric Synthesis, Vol. 1*, Morrison, J. D. (Ed.), London: Academic Press, 1983; p 87.

[38] Pirkle, W. H., Pochapsky, T. C., *Adv. Chromatogr. 27* (1988) 73.

[39] Frank, H., Nicholson, G. J., Bayer, E., *J. Chromatogr. Sci. 15* (1977) 174.

[40] Bayer, E., *Z. Naturforsch. B38* (1983) 1281.

[41] Koppenhofer, B., Bayer, E., *J. Chromatogr. Library 32* (1985) 1.

[42] Koppenhofer, B., Bayer, E., *The Science in Chromatography* , Amsterdam: Elsevier, 1985.

[43] Koppenhofer, B., Allmendinger, H., *Chromatographia 21* (1986) 503.

[44] Schurig, V., *Angew. Chem., Int. Ed. Engl. 16* (1977) 110.

[45] Schurig, V., Bürkle, W., *Angew. Chem., Int. Ed. Engl. 17* (1978) 132.

[46] Schurig, V., *Chromatographia 13* (1980) 263.

[47] Schurig, V., Bürkle, W., *J. Am. Chem. Soc. 104* (1982) 7573.

[48] Davankov, V. A., Kurganov, A. A., Bochkov, A. S., *Adv. Chromatogr. 22* (1983) 71.

[49] Schurig, V., Nowotny, H. P., Schmalzing, D., *Angew. Chem., Int. Ed. Engl. 28* (1989) 736.

[50] König, W. A., Lutz, S., Wentz, G., Görgen, G., Neumann, C., Gäbler, A., Boland, W., *Angew. Chem., Int. Ed. Engl. 28* (1989) 178.

[51] Ward, T. J., Amstrong, D. W., *J. Liq. Chromatogr. 9* (1986) 407.

[52] Mannschreck, A., Koller, H., Wernicke, R., *Kontakte* (Darmstadt) (1985) 1/40.

[53] Shibata, T., Okamoto, I., Ishii, K., *J. Liq. Chromatogr. 9* (1986) 313.

[54] Werner, A., *Kontakte* (Darmstadt) (1989) 3/50.

[55] Yamaguchi, S., in : *Asymmetric Synthesis, Vol. I*, Morrison, J. D. (Ed.), London: Academic Press, 1983; p 125.

[56] Dale, J. A., Dull, D. L., Mosher, H. S., *J. Org. Chem. 34* (1969) 2543.

[57] Hull, W. E., Seeholzer, K., *Tetrahedron 42* (1986) 547.

[58] Nabeya, A., Endo, T., *J. Org. Chem. 53* (1988) 3358.

[59] Nishida, Y., Ohrni, H., Meguro, H., *Tetrahedron Lett. 30* (1989) 5277.

[60] Doolitl, R. E., Heath, R. R, *J. Org. Chem. 49* (1984) 5041.

[61] Johnson, C. R., Elliott, R. C., Penning, T. D., *J. Am. Chem. Soc. 106* (1984) 5019.

[62] Meyers, A. I., Williams, D. R., Erickson, G. W., White, S. K., Druelinger, M., *J. Am. Chem. Soc. 103* (1981) 3081.

[63] Cuvinot, D., Mangoney, P., Alexakis, A., Normant, J. F., *J. Org. Chem. 54* (1989) 2420.

[64] Gil-Av, E., Narok, D., *Adv. Chromatogr. 10* (1974) 99.

[65] Pirkern, W. H., Hoelstra, M. S., *J. Org. Chem. 39* (1974) 1904.

[66] Halpern, B., "Derivatives for Chromatographic Resolution of Optically Active Compounds" in : *Handb. Deriv. Chromatogr.*, Blau, K., King, G. S. (Eds.), London: Heyden, 1978; pp 201-233.

[67] Prelog, V., *Pure Appl. Chem. 9* (1964) 119.

[68] Nógrádi, M., *Stereoselective Synthesis*, Weinheim: VCH Verlagsgesellschaft, 1986.

[69] Trost, B. M., Salzmann, T. M., Hirdi, K., *J. Am. Chem. Soc. 98* (1976) 4887.

[70] Morrison, J. D., Mosher, H. S., *Asymmetric Organic Synthesis*, New York: Prentice-Hall, 1973.

[71] Iizumi, Y., Tai, A., *Stereodifferentiating Reactions*, New York: Academic Press, 1977.

[72] a) Chen, C. S., Fujimoto, Y., Girdaukas, G., Sih, C., J., *J. Am. Chem. Soc. 104* (1982) 7294; b) Sih, C. J., Chen, C. S., *Angew. Chem., Int. Ed. Engl. 23* (1984) 570.

[73] Eyring, H., *Chem. Rew. 17* (1935) 65.

[74] Chen, C. S., Wu, S. H., Girdaukas, G., Sih, C. J., *J. Am. Chem. Soc. 109* (1987) 2812;

[75] Chen, C. S., Sih, C. J., *Angew. Chem., Int. Ed. Engl. 28* (1989) 695.

[76] Jongejan, J. A., Duine, J. A., *Tetrahedron Lett. 28* (1987) 2767.

[77] Burger, U., Erne-Zellweger, D., Mayerl, C., *Helv. Chim. Acta 70* (1987) 587.

[78] Abbott, B. J., *Adv. Appl. Microbiol. 20* (1976) 203.

[79] Sakina, K., Kawazura, K., Morihara, K., Tajima, H., *Chem. Pharm. Bull. 36* (1988) 4345.

[80] Sih, C. J., Salomon, R. G., Price, P., Sood, R., Peruzzotti, G., *J. Am. Chem. Soc. 97* (1975) 875.

[81] Porter, N. A., Byers, J. D., Holden, K. M., Menzel, D. B., *J. Am. Chem. Soc. 101* (1979) 4319.

[82] Lin, C. H., Alexander, D. L., Chicester, C. G., Gorman, R. R., Johnson, R. A., *J. Am. Chem. Soc. 104* (1982) 1621.

[83] Michalski, T. J., Hunt, J. E., Bradshaw, C., Wagner, A. M., Norris, J. R., Katz, J. J., *J. Am. Chem. Soc. 110* (1988) 5888.

[84] a) Russell, S. T., Robinson, J. A., Williems, D. J., *Chem. Commun.* (1987) 351; b) Schaap, P. A., Handley, R. S., Giri, B. P., *Tetrahedron Lett. 28* (1987) 935.

[85] Rebolledo, F., Brieva, R., Gotor, V., *Tetrahedron Lett. 30* (1989) 5345.

[86] Lin, Y. X., Jones, J. B., *J. Org. Chem. 38* (1973) 3575.

[87] Taunton-Rigby, A., *J. Org. Chem. 38* (1973) 977.

[88] Chinsky, N., Margolin, A. L., Klibanov, A. M., *J. Am. Chem. Soc. 111* (1989) 386.

[89] Pelsy, G., Klibanov, A. M., *Biotechnol. Bioeng. 25* (1983) 919.

[90] Arai, Y., Yamada, K., *Agric. Biol. Chem. 33* (1969) 63.

[91] Davis, J. B., Raymond, R. L., *Appl. Microbiol. 9* (1961) 383.

[92] Gateau-Olesker, A., Cléophax, J., Géro, S. D., *Tetrahedron Lett. 27* (1986) 41.

[93] Rosazza, J. P., Reeg, S., Yang, L. M., *Enzyme Microb. Technol. 8* (1986) 161.

[94] Riva, S., Klibanov, A. M., *J. Am. Chem. Soc. 110* (1988) 3291.

[95] Uemura, A., Nozaki, K., Yamashita, J. I., Yasumoto, M., *Tetrahedron Lett. 30* (1989) 3819.

[96] Klibanov, A. M., Siegel, E. H., *Enzyme Microb. Technol. 4* (1982) 172.

[97] Schotten, T., Boland, W., Jaenicke, L., *Helv. Chim. Acta 68* (1985) 1186.

[98] Sicsic, S., Leroy, J., Wakselman, C., *Synthesis* (1987) 155.

[99] Charney, W., Herzog, H. L., *Microbial Transformations of Steroids*, New York: Academic Press, 1967; Iizuka, H., Naito, A., *Microbial Transformations of Steroids and Alkaloids*, New York: Springer-Verlag, 1981.

[100] Fonken, G. S., Johnson, R. A., *Chemical Oxidations with Microorganisms*, New York: Marcel Dekker, 1972.

[101] Sih, C. J., Rosazza, J. P., "Microbial Transformations in Organic Synthesis" in : *Applications of Biochemical Systems in Organic Chemistry*, Jones, J. B., Sih, C. J., Perlman D. (Eds.), New York: Wiley, 1976; pp 69-106.

[102] Blunt, J. W., Clarck, I. M., Evans, J. M., Jones, E. R., Meakins, G. D. Pinkey, J. T., *J. Chem. Soc.* C. (1971) 1136.

[103] Johnson, R. A., Murray, H. C., *Chem. Commun.* (1971) 989.

[104] Archelas, A., Fourneron, J. D., Furstoss, R., *J. Org. Chem. 53* (1988) 1797.

[105] Moran, J. R., Whitesides, G. M., *J. Org. Chem. 49* (1984) 704.

[106] Fuganti, C., Grasselli, P., Spreafico, F., Zirotti, C., *J. Org. Chem. 49* (1984) 543.

[107] Bednarski, M. D., Simon, E. S., Bischoffberger, N., Fessner, W. D., Kim, M. J., Lees, W., Saito, T., Waldmann, H., Whitesides, G. M., *J. Am. Chem. Soc. 111* (1989) 627.

[108] Borsyenko, C. W., Spaltenstein, A., Straub, J. A., Whitesides, G. M., *J. Am. Chem. Soc. 111* (1989) 9275.

[109] Wong, C. H., Whitesides, G. M., *J. Org. Chem. 48* (1983) 3199.

[110] Alphand, V., Archelas, A., Furstoss, R., *Tetrahedron Lett. 30* (1989) 3663.

[111] Chibata, I., Tosa, T., Sato, T., Mori, T., *Meth. Enzymol. 44* (1976) 746.

[112] Izumi, Y., Chibata, I., Itoh, T., *Angew. Chem., Int. Ed. Engl. 17* (1978) 176.

[113] Chibata, I., *Immobilized Enzymes, Research and Development*, Tokyo/New York: Kodanska/Halsted, 1978.

[114] Nakazaki, M., Chikamatsu, H., Sasaki, Y., *J. Org. Chem. 48* (1983) 2056.

[115] Fülling, A., Sih, C. J., *J. Am. Chem. Soc. 109* (1987) 2845.

[116] Fukumura, T., *Agric. Biol. Chem. 40* (1976) 1687.

[117] Fukumura, T., *Agric. Biol. Chem. 40* (1976) 1695.

[118] Kan, K., Miyama, A., Hamaguchi, S., Ohashi, T., Watanabe, K., *Agric. Biol. Chem. 49* (1985) 1669.

[119] Hamaguchi, S., Hasegawa, A., Kawaharada, H., Watanabe, K., *Agric. Biol. Chem. 48* (1984) 2055.

[120] Hamaguchi, S., Asada, M., Hasagawa, J., Watanabe, K., *Agric. Biol. Chem. 48* (1984) 2331.

[121] Hamaguchi, S., Yamamura, H., Hasegawa, J., Watanabe, K., *Agric. Biol. Chem. 49* (1985) 1509.

[122] Kan, K., Miyama, A., Hamaguchi, S., Ohashi, T., Watanabe, K., *Agric. Biol. Chem. 49* (1985) 207.

[123] Hamaguchi, S., Asada, M., Hasegawa, J., Watanabe, K., *Agric. Biol. Chem. 49* (1985) 1661.

[124] Wang, Y. F., Chen, C. S., Girdaukas, G., Sih. C. J., *J. Am. Chem. Soc. 106* (1984) 3695.

[125] Wang, Y. F., Lalonde, J. J., Momongau, M., Bergbreiter, D. E., Wong, C. H., *J. Am. Chem. Soc. 110* (1988) 7200.

[126] Ramos-Tombo, G. M., Schär, H. P., Fernandez, X., Busquets, I., Ghisalba, O., *Tetrahedron Lett. 27* (1986) 5707.

[127] Akhtar, M., Botting, M. P., Cohen, M. A., Gani, T., *Tetrahedron 43* (1987) 5899.

[128] Lee, L. G., Whitesides, G. M., *J. Org. Chem. 51* (1986) 25.

[129] Chen, C. S., Zhou, B. M., Girdaukas, G., Shieh, W. R., Van Middlesworth, F., Gopalan, A. S., Sih, C. J., *Bioorg. Chem. 12* (1984) 98.

[130] Long, A., Ward, O. P., *Biotechnol. Bioeng. 34* (1989) 933.

[131] Bally, C., Leuthardt, F., *Helv. Chim. Acta 53* (1970) 732.

[132] Tshamber, T., Waespe-Saracevic, N., Tamm, C., *Helv. Chim. Acta 69* (1986) 621.

[133] Ohno, M., *Ferment. Ind. (Japan) 37* (1979) 836.

[134] Yamazaki, Y., Hosono, K., *Biotechnol. Lett. 11* (1989) 627.

[135] Bucciarelli, M., Forni, A., Moretti, I., Torre, G., *Chem. Commun.* (1978) 456.

[136] Brousse, J., Roos, E. C., Van der Gren, A., *Tetrahedron Lett. 29* (1988) 4485.

[137] Utaka, M., Konishi, S., Okubo, T., Tsuboi, S., Takeda, A., *Tetrahedron Lett. 28* (1987) 1447.

[138] Roy, R., Rey, A. W., *Tetrahedron Lett. 28* (1987) 4935.

[139] Irwin, A. J., Jones, J. B., *J. Am. Chem. Soc. 99* (1977) 1625.

[140] Lowe, G., Swain, S., *J. Chem. Soc., Perkin I* (1985) 391.

[141] Naemura, K., Fujii, T., Chikamatsu, H., *Chem. Lett.* (1986) 923.

[142] Dai, W. M., Zhou, W. S., *Tetrahedron 41* (1985) 4475.

[143] Guo, Z. W., Wu, S. H., Chen, C. S., Girdaukas, G., Sih, C. J., *J. Am. Chem. Soc. 112* (1990) 4942.

[144] Miller, D. B., Raychaudhuri, S. R., Avashti, K., Lal, K., Levison , B., Salomon, R. G., *J. Org. Chem. 55* (1990) 3164.

[145] Marchand, A. P., Reddy, G. M., *Tetrahedron Lett. 31* (1990) 1811.

Chapter 3

[1] Poppe, L., Novák, L., *Magy. Kém. Lapja 43* (1988) 237.

[2] Whitesides, G. M., Wong, C. H., *Angew. Chem., Int. Ed. Engl. 24* (1985) 617.

[3] Butt, S., Roberts, S. M., *Nat. Prod. Reports* (1986) 481.

[4] Jones, J. B., *Tetrahedron 42* (1986) 3351.

[5] Schneider, M., Reimerdes, E. H., *Forum Mikrob.* (1987) 302.

[6] Zaks, A., Empie, M., Gross, A., *Trends Biotechnol. 6* (1988) 272.

[7] Altenbach, H. J., *Nachr. Chem. Tech. Lab. 36* (1988) 1114.

[8] Yamada, H., Shimizu, S., *Angew. Chem. Int. Ed. Engl. 27* (1988) 622.

[9] Tramper, J., Plas, H. C., Linko, P. (Eds.), *Biocatalysts in Organic Synthesies* (Stud. Org. Chem., Vol. 22), Amsterdam: Elsevier, 1985.

[10] Porter, R., Clark, S. (Eds.), *Enzymes in Organic Synthesis*, CIBA Foundation Symposium III, London: Pitman, 1984.

[11] Schneider, M. P. (Ed.), *Enzymes as Catalysts in Organic Synthesis*, Dordrecht: D. Reidel Publishing, 1986.

[12] Davies, H. G., Green, R. H., Kelly, D. R., Roberts, S. M. (Eds.), *Biotransformations in Preparative Organic Chemistry*, London: Academic Press, 1989.

[13] Ohno, M., Otsuka, M., in : *Organic Reactions, Vol. 37*, New York: Wiley, 1989; pp 1-55.

[14] Dixon, M., Webb, E. C. (Eds.), *Enzymes*, 3rd Ed., Orlando, FL: Academic Press, 1979.

[15] Barman, T. E. (Ed.), *Enzyme Handbook*, Berlin: Springer Verlag, 1985.

[16] Frost, G. M., Moss, D. A., in : *Biotechnology, Vol. 7A: Enzyme Technology*, Rehm, H. J., Reed, G. (Eds.), Weinheim: VCH Verlagsgesellschaft, 1987; pp 65-211.

[17] Fukumura, T., *Agric. Biol. Chem. 40* (1976) 1687; Fukumura, T., *Agric. Biol. Chem. 41* (1977) 1327.

[18] Fukumura, T., *Agric. Biol. Chem. 40* (1976) 1695.

[19] Glänzer, B. I., Faber, K., Griengl, H., *Tetrahedron Lett. 27* (1986) 4293.

[20] Fadnavis, N. W., Reddy, N. P., Bhalerao, U. T., *J. Org. Chem. 54* (1989) 3281.

[21] Adachi, K., Kobayashi, S., Ohno, M., *Chimia 40* (1986) 311.

[22] Poppe, L., Novák, L., Kolonits, P., Bata, Á., Szántay, Cs., *Tetrahedron 44* (1988) 1477.

[23] Foque, E., Rousseau, G., *Synthesis* (1989) 661.

[24] Ramos-Tombo, G. M., Schär, H. P., Fernandez, X., Busquets, I., Ghisalba, O., *Tetrahedron Lett. 27* (1986) 5707.

[25] Brünner, W., Gunzer, G., "Laboratory Techniques of Enzyme Recovery" in : *Biotechnology, Vol. 7A: Enzyme Technology*, Rehm, H. J., Reed, G. (Eds.), Weinheim: VCH Verlagsgesellschaft, 1987; pp 213-323.

[26] Krish, K., *The Enzymes*, 3rd Ed., Vol. 5, Chap 3., Boyer, P. D. (Ed.), New York: Academic Press, 1971.

[27] Gais, H. J., Bülow, G., Zatorski, A., Jentsch, M., Maidonis, P., Hemmerle, H., *J. Org. Chem. 54* (1989) 5115.

[28] Grrenzaid, P., Jencks, W. P., *Biochemistry 10* (1971) 1210.

[29] Stoops, J. K., Horgan, D. J., Runnegar, M. T. C., Jersey, J., Webb, E. C., Zerner, B., *Biochemistry 8* (1969) 2026.

[30] Duddmann, N. P. B., Zerner, B., *Meth. Enzymol. 35* (1975) 191.

[31] Farb, D., Jenks, W. P., *Arch. Biochem. Biophys. 203* (1980) 214.

[32] Lam, L. K. P., Brown, C. M., De Jeso, B., Lym, L., Toone, E., Jones, J. B., *J. Am. Chem. Soc. 110* (1988) 4409.

[33] Ohno, M., Kobayashi, S., Adachi, K., in : *Enzymes as Catalysts in Organic Synthesis*, Schneider, M. (Ed.), Dordrecht: D. Reidel Publishing, 1986; pp 123-142.

[34] Mohr, P., Waespe-Saracevic, N., Tamm, C., Gawronska, K., Gawronski, J. K., *Helv. Chim. Acta 66* (1983) 2501.

[35] Zemlicka, J., Craine, L. E., Heeg, M. J., Oliver, J. P., *J. Org. Chem. 53* (1988) 937.

[36] Horgan, D. J., Stoops, J. K., Webb, E. C., Zerner, B., *Biochemistry 8* (1969) 2000.

[37] Nagao, Y., Kume, M., Wakabayashi, R. C., Nakamura, T., Ochiai, M., Symposium of Progress in Organic Reactions and Synthesis, 13th, symposium paper, Tokushima/Japan, 1986; pp 34-38.

[38] Laumen, K., Reimerdes, E. H., Schneider, M., *Tetrahedron Lett. 26* (1985) 407.

[39] Desnuelle, P., *The Enzymes*, 3rd Ed., Vol. 7, Boyer, P. D. (Ed.), New York: Academic Press, 1972; p 575.

[40] Borgström, B., Brockmann, H. L. (Eds.), *Lipases*, Amsterdam: Elsevier, 1984.

[41] Verger, R., Mieras, M. C. F., De Haas, G. H., *J. Biol. Chem. 248* (1973) 4023.

[42] Benzonana, G., Esposito, S., *Biochim. Biophys. Acta 231* (1971) 15.

[43] Kokusho Y., Machida, H., Iwasaki, S., *Agric. Biol. Chem. 46* (1982) 1159.

[44] Arun, K., *Proc. Biochem. 4* (1985) 35.

[45] Sztajer, H., Maliszewska, I., Wieczorek, J., *Enzyme Microb. Technol. 10* (1988) 492.

[46] Sigiura, M., Isobe, M., *Biochim. Biophys. Acta 341* (1974) 195.

[47] Khor, H. T., Tau, N. H., Chura, C. L., *J. Am. Oil. Chem. Soc. 4* (1986) 538.

[48] Fox, P. F., Stepniak, L., *J. Dairy Res. 1* (1983) 172.

[49] Meito Sangyo Co. Ltd., Brit. Patent, 976'415, 1964.

[50] Dherbromez, M., Lacrampe, J. L., Larouquere, J., *Rev. Fr. Corps Gras 22* (1975) 147.

[51] Pal, N., Das, S., Kundu, A. K., *J. Ferment. Technol. 56* (1978) 593.

[52] Wouters, J. T. M., *Antonie van Leeuwenhoek 33* (1967) 365.

[53] Liu, W. H., Beppu, T., Arima, K., *Agric. Biol. Chem. 36* (1972) 1919.

[54] Ibrahim, C. O., Nishio, N., Nagai, S., *Agric. Biol. Chem. 52* (1988) 2923.

[55] Takeda Chemical Industries, Ltd., Brit. Patent 1'442'667, 1972.

[56] Laboureour, P., Labrousse, M., *Bull. Soc. Chim. Biol. 48* (1966) 747.

[57] Semeriva, M., Benzonana, G., Desnuelle, P., *Biochem. Biophys. Acta 191* (1969) 598.

[58] Iwai, M., Tsujisaka, Y., *Agric. Biol. Chem. 38* (1974) 1241.

[59] Singer, T. P., *J. Biol. Chem. 174* (1948) 11.

[60] Verger, R., De Haas, G. H., Sarda, L., Desnuelle, P., *Biochem. Biophys. Acta 188* (1969) 272.

[61] Shaw, J. F., Chang, C. H., *Biotechnol. Lett. 11* (1989) 779.

[62] Veeraragavan, K., Gibbs, B. F., *Biotechnol. Lett. 11* (1989) 345.

[63] Schmidt, W. E., Conlon, J. M., Mutt, V., Carlquist, M., Gallwitz, B., Creutzfeld, W., *Eur. J. Biochem. 162* (1987) 467.

[64] Schwert, G. W., Takeneka, Y., *Biochim. Biophys. Acta 16* (1955) 570.

[66] Ottesen, M., Svendsen, I., *Meth. Enzymol. 19* (1970) 199.

[67] Feder, J., et. al., *Dev. Ind. Microbiol. 18* (1977) 267.

[68] International Union of Biochemistry, *Enzyme Nomenclature 1984*, New York: Academic Press, 1984.

[69] Savidge, T. A., in : *Biotechnology of Industrial Antibiotics*, Vandamme, E. J. (Ed.), New York: Marcel Dekker, 1984; pp 373-377.

[70] Chibata, I., *Immobilized Enzymes, Research and Development*, Tokyo/New York: Kodanska/Halsted, 1978.

[71] Chibata, I., Tosa, T., Sato, T., Mori, T., *Meth. Enzymol. 44* (1976) 746.

[72] Izumi, Y., Chibata, I., Itoh, T., *Angew. Chem., Int. Ed. Engl. 17* (1978) 176.

[73] Amano Pharmaceutical Co. Ltd., Japan Kokai Tokkyo Koho, 58'209'985 (1983); *Chem. Abstr. 100* (1984) 155'216g.

[74] Geutzen, I., Löffler, H. G., Schneider, F., *Z. Naturforsch. C. 35* (1980) 544.

[75] Chenault, H. K., Dahmer, J., Whitesides, G. M., *J. Am. Chem. Soc. 111* (1989) 6354.

[76] Cho, H. Y., Tanizawa, K., Tanaka, H., Soda, K., *Agric. Biol. Chem. 51* (1987) 2793.

[77] Kasche, V, *Enzyme Microb. Technol. 8* (1986) 4.

[78] Jakubke, H. D., Kuhl, P., Könnecke, A., *Angew. Chem., Int. Ed. Engl. 24* (1985) 85.

[79] Jones, J. B, "Stereochemical Considerations and Terminologies of Biochemical Importance" in : *Applications of Biochemical Systems in Organic Chemistry* (Techniques of Chemistry, Vol. 10), Jones, J. B., Sih, C. J., Perlman, D. (Eds.), New York: Wiley, 1976; pp 479-490.

[80] Walsh, C., *Enzymatic Reaction Mechanisms*, New York: W. H. Freeman, 1979.

[81] Fersht, A., *Enzyme Structure and Mechanism*, 2nd Ed., New York: W. H. Freeman, 1985.

[82] Hammes, G. G., *Enzyme Catalysis and Regulation*, New York: Academic Press, 1982.

[83] Liebmann, J. E. (Ed.), *Mechanistic Principles of Enzyme Activity*, Weinheim: VCH Verlagsgesellschaft, 1989.

[84] Jones, J. B., Beck, B. F, "Asymmetric Syntheses and Resolutions Using Enzymes" in : *Applications of Biochemical Systems in Organic Chemistry*, Jones, J. B., Sih, C. J., Perlman, D. (Eds.), New York: Wiley, 1976; pp 107-401.

[85] Kitazume, T., Ikeya, T., Murata, K., *Chem. Commun.* (1986) 1331.

[86] Akita, H., Enoki, Y., Yamada, H., Oishi, T., *Chem. Pharm. Bull. 37* (1989) 2876.

[87] Empie, M. W., Gross, A., *Annu. Rep. Med. Chem. 23* (1988) 305.

[88] Lam, L. K. P., Hui, R. A. H. F., Jones, J. B., *J. Org. Chem. 51* (1986) 2047.

[89] Björkling, F., Boutelje, J., Gatenbeck, S., Hult, K., Norin, T., Szmulik, P., *Bioorg. Chem. 14* (1986) 176.

[90] Guanti, G., Banfi, L., Narisano, E., Riva, R., Thea, S., *Tetrahedron Lett. 27* (1986) 4639.

[91] Eggers, D. K., Blach, H. W., Prauschnitz, J. M., *Enzyme Microb. Technol. 11* (1989) 84.

[92] Halling, P. J., *Enzyme Microb. Technol. 6* (1984) 513.

[93] Kise, H., Shirato, H., *Tetrahedron Lett. 26* (1985) 6081.

[94] Kise, H., Shirato,H., *Enzyme Microb. Technol. 10* (1988) 582.

[95] Tai, D. F., Fu, S. L., Chuang, S. F., Tsai, H., *Biotechnol. Lett. 11* (1989) 173.

[96] Ibrahim, C. O., Saeki, H., Nishio, N., Nagai, S., *Biotechnol. Lett. 11* (1989) 161.

[97] Semenov, A. N., Cerovsky, V., Titov, M. I., Martinek, K., *Collect. Czech. Chem. Commun. 53* (1988) 2963.

[98] Kullmann, W., *Enzymatic Peptide Synthesis*, Boca Raton, FL: CRC Press, 1987.

[99] Vidaluc, J. L., Baboulene, M., Speciale, V., Lattes, A., Monsan, P., *Tetrahedron 39* (1983) 269.

[100] Klibanov, A. M., Samokhin, G. P., Martinek, K., Berezin, I. V., *Biotechnol. Bioeng. 19* (1977) 1351.

[101] Cantacuzene, D., Pascal, F., Guerreiro, C., *Tetrahedron 43* (1987) 1823.

[102] Semenov, A. N., Gachok, A. P., Titov, M. I., Martinek, K., *Biotechnol. Lett. 11* (1989) 875.

[103] Shield, J. W., Ferguson, H. D., Bommarius, A. S., Hatton, T. A., *Ind. Eng. Chem., Fundam. 25* (1986) 603.

[104] Luisi, P. L., *Angew. Chem., Int. Ed. Engl. 24* (1985) 439.

[105] Martinek, K., Lerashov, A. V., Klyachko, N., Khmelnitsky, Y. L., Berezin, I. V., *Eur. J. Biochem. 155* (1986) 453.

[106] Okahata, Y., Fujimoto, Y., Ijiro, K., *Tetrahedron Lett. 29* (1988) 5133.
[107] Okahata, Y., Ijiro, K., *Chem. Commun.* (1988) 1392.
[108] Ekberg, B., Lindbladh, C., Kempe, M., Mosbach, K., *Tetrahedron Lett. 30* (1989) 583.
[109] Tanaka, T., Ohno, E., Ishihira, M., Yamanaka, S., Takinami, K., *Agric. Biol. Chem. 45* (1981) 2387.
[110] Seo, C. W., Yamada, Y., Okada, H., *Agric. Biol. Chem. 46* (1982) 405.
[111] Zaks, A., Klibanov, A. M., *Science 224* (1984) 1249.
[112] Marlot, C., Langrand, G., Triantaphylides, C., Baratti, J., *Biotechnol. Lett. 7* (1985) 647.
[113] Langrand, G., Baratti, J., Bouno, G., Triantaphylides, C., *Tetrahedron Lett. 27* (1986) 29.
[114] Zaks, A., Klibanov, A. M., *Proc. Natl. Acad. Sci. USA 82* (1985) 3192.
[115] Klibanov, A. M., *CHEMTECH 16* (1986) 354.
[116] Laane, C., Tramper, J., Lilly, M. D. (Eds.), *Biocatalysts in Organic Media* (Stud. Org. Chem., Vol. 29), Amsterdam/New York: Elsevier, 1987.
[117] Klibanov, A. M., *Trends Biochem. Sci. 14* (1989) 141.
[118] Dickinson, M., Flechert, P. D. I., *Enzyme Microb. Technol. 11* (1989) 55.
[119] Dordick, J. S., *Enzyme Microb. Technol. 11* (1989) 194.
[120] Kodera, Y., Takahashi, K., Nishimura, H., Matsushima, A., Saito, Y., Inada, Y., *Biotechnol. Lett. 8* (1986) 881.
[121] Nishio, T., Takahashi, K., Yoshimoto, T., Kodera, Y., Saito, Y., Inada, Y., *Biotechnol. Lett. 9* (1987) 187.
[122] Ferjancic, A., Puigserver, A., Gaertner, H., *Biotechnol. Lett. 10* (1988) 101.
[123] Kawase, M., Tanaka, A., *Biotechnol. Lett. 10* (1988) 393.
[124] Yamane, T., *Biocatalysis 2* (1988) 1.
[125] Yamane, T., Kojima, Y., Ichiryu, T., Nagata, M., Shimizu, S., *Biotechnol. Bioeng. 34* (1989) 383.
[126] Zaks, A., Klibanov, A. M., *J. Biol. Chem. 263* (1988) 8017.
[127] Zaks, A., Klibanov, A. M., *J. Am. Chem. Soc. 108* (1986) 2767.
[128] Kitaguchi, H., Klibanov, A. M., *J. Am. Chem. Soc. 111* (1989) 9272.
[129] Burke, P. A., Smith, S.O., Bachovchin, W. W., Klibanov, A. M., *J. Am. Chem. Soc. 111* (1989) 8290.
[130] Kanerva, L. T., Klibanov, A. M., *J. Am. Chem. Soc. 111* (1989) 6864.
[131] Chen, C. S., Sih, C. J., *Angew. Chem., Int. Ed. Engl. 28* (1989) 695.
[132] Herschlag, D., *Bioorg. Chem. 10* (1987) 62.
[133] Wu, S. H., Guo, Z. W., Sih, C. J., *J. Am. Chem. Soc. 112* (1990) 1990.
[134] Stahl, M., Mansson, M. O., Mosbach, K., *Biotechnol. Lett. 12* (1990) 161.
[135] Sakurai, T., Margolin, A. L., Russell, A. J., Klibanov, A. M., *J. Am. Chem. Soc. 110* (1988) 7236.
[136] Kirchner, G., Scollar, M. P., Klibanov, A. M., *J. Am. Chem. Soc. 107* (1985) 7072.
[137] Bevinakatti, H. S., Banerji, A. A., *Biotechnol. Lett. 10* (1988) 397.
[138] Degueil-Castaing, M., De Jeso, B., Drouillard, S., Maillard, B., *Tetrahedron Lett. 28* (1987) 953.
[139] Wang, Y. F., Lalonde, J. J., Momongau, M., Bergbreiter, D. E., Wong, C. H., *J. Am. Chem. Soc. 110* (1988) 7200.
[140] Ghogare, A., Kumar, G. S., *Chem. Commun.* (1989) 1533.
[141] Terao, Y., Tsuji, K., Murata, M., Achiwa, K., Nishio, T., Watanabe, N., Seto, K., *Chem. Pharm. Bull. 37* (1989) 1653.
[142] Norin, M., Boutelje, J., Holmberg, E., Hult, K., *Appl. Microbiol. Biotechnol. 28* (1988) 527.
[143] Tanaka, A., Sonomoto, K., *CHEMTECH* (1990) 112.
[144] Mozhaev, V. V., Sergeyeva, M. V., Belova, A. B., Khemelnitsky, Y. L., Biotechnol. Bioeng. 35 (1990) 653.
[145] Therisod, M., Klibanov, A. M., *J. Am. Chem. Soc. 108* (1986) 5638.
[146] Shaw, J. F., Klibanov, A. M., *Biotechnol. Bioeng. 29* (1987) 648.
[147] Carrera, G., Riva, S., Secundo, F., Danieli, B., *J. Chem. Soc., Perkin I* (1989) 1057.
[148] Reinhoudt, D. N., Eendeback, A. M., Nijenhuis, W. F., Verboom, W., Koolsterman, M., Schoemacker, H. E., *Chem. Commun.* (1989) 399.
[149] Macrae, A. R., *J. Am. Oil. Chem. Soc. 60* (1983) 291.

[150] Macrae, A. R., in : *Biocatalysts in Organic Synthesis*, Tramper, J., v. d. Plaas, H. C., Linko, P. (Eds.), Amsterdam: Elsevier, 1985; pp 195-208.

[151] Haraldsson, G. G., Höskuldsson, P. A., *Tetrahedron Lett. 30* (1989) 1671.

[152] Ergan, F., Trani, M., André, G., *Biotechnol. Bioeng. 35* (1990) 195.

[153] Iwai, M., Okumura, S., Tsuisaka, Y., *Agric. Biol. Chem. 44* (1980) 2731.

[154] Langrand, G., Triantaphylides, C., Baratti, J., *Biotechnol. Lett. 10* (1988) 549.

[155] Ibrahim, C. O., Nishio, N., Nagai, S., *Biotechnol. Lett. 10* (1988) 799.

[156] Morihara, K., *Trends Biotechnol. 5* (1987) 164.

[157] Barbas, C. F., Matos, J. R., West, J. B., Wong, C. H., *J. Am. Chem. Soc. 110* (1988) 5162.

[158] Petkov, D. D., Stoineva, I. B., *Tetrahedron Lett. 25* (1984) 3751.

[159] Khmelniski, Y. L., Dien, F. K., Semenov, A. N., Martinek, K., Vernovic, B., Kubanek, Y., *Tetrahedron 40* (1984) 4425.

[160] Homandberg, G. A., Mattis, J. A., Lakowski, M., *Biochemistry 17* (1978) 5220.

[161] Ricca, J. M., Crout, D. H. G., *J. Chem. Soc., Perkin I* (1989) 2126.

[162] Lüthi, P., Luisi, P. L., *J. Am. Chem. Soc. 106* (1984) 7285.

[163] Ferjancic, A., Pigserver, A., Gaertner, H., *Appl. Microb. Biotechnol. 32* (1990) 651.

[164] Fadnavis, N. W., Luisi, P. L., *Biotechnol. Bioeng. 33* (1989) 1277.

[165] Barbas, C. F., Wong, C. H., *Chem. Commun.* (1987) 533.

[166] Shin, C. S., Takahashi, N., *Chem. Lett.* (1989) 747.

[167] Jost, R., Brambilla, E., Monti, J. C., Luisi, P. L., *Helv. Chim. Acta 63* (1980) 375.

[168] Chen, S. T., Wang, K. T., *J. Org. Chem. 53* (1988) 4589.

[169] Barbas, C. F., Wong, C. H., *Tetrahedron Lett. 29* (1988) 2907.

[170] Stehle, P., Baksitta, H. P., Monter, B., Fürst, P., *Enzyme Microb. Technol. 12* (1990) 56.

[171] Nakatsuka, T., Sasaki, T., Kaiser, E. T., *J. Am. Chem. Soc. 109* (1987) 3808.

[172] Oyama, K., Nishimura, S., Nonaka, Y., Kihara, K., Hashimoto, T., *J. Org. Chem. 46* (1981) 5242.

[173] Isowa, Y., Ohmori, M., Ichikawa, T., Mori, K., *Tetrahedron Lett.* (1979) 2611.

[174] Sakina, K., Kawazura, K., Morihara, K., Tajima, H., *Chem. Pharm. Bull. 36* (1988) 4345.

[175] Matos, J. R., West, J. B., Wong, C. H., *Biotechnol. Lett. 9* (1987) 233.

[176] West, J. B., Wong, C. H., *Tetrahedron Lett. 28* (1987) 1629.

[177] Margolin, A. L., Klibanov, A. M., *J. Am. Chem. Soc. 109* (1987) 3802.

[178] Jensen, S., Wolfe, S., Westlake, D. W. S., *Appl. Microbiol. Biotechnol. 30* (1989) 111.

[179] Margolin, A. L., Tai, D. F., Klibanov, A. M., *J. Am. Chem. Soc. 109* (1987) 7885.

[180] Asamo, Y., Nakazawa, A., Kato, Y., Kondo, K., *Angew. Chem. Int. Ed. Engl. 28* (1989) 450.

[181] Kato, Y., Asano, A., Nakazawa, A., Kondo, K., *Tetrahedron 45* (1989) 5743.

[182] Kullmann, W., *J. Org. Chem. 47* (1982) 5300.

[183] Sakina, K., Kawazura, K., Morihara, K., Yajima, H., *Chem Pharm. Bull. 36* (1988) 3915.

[184] Könnecke, A., Schönfels, C., Hänsler, M., Jakubke, H. D., *Tetrahedron Lett. 31* (1990) 989.

[185] Stoinova, I. B., Petkov, D. D., *FEBS Lett. 183* (1985) 103.

[186] Kleinkauf, H., Dören, H. (Eds.), *Peptide Antibiotics, Biosynthesis and Functions*, Berlin: de Gruyter, 1982.

[187] Clingmann, K. A., Hajdu, J., *J. Chem. Education 64* (1987) 358.

[188] Regen, S. L., Singh, A., Oehme, G., Singh, M., *J. Am. Chem. Soc. 104* (1982) 791.

[189] Shuto, S., Itoh, H., Ueda, S., Imamura, S., Fukukawa, K., Tsujino, M., Watsuda, A., Ueda, T., *Chem. Pharm. Bull. 36* (1988) 209.

[190] Schär, H. P., Gygax, D., Ramos-Tombo, G. M., Ghisalba, O., *Appl. Microbiol. Biotechnol. 27* (1988) 451.

[191] Sime, J. T., Pool, C. R., Tyler, J. W., *Tetrahedron Lett. 28* (1987) 5169.

[192] Kuroda, H., Miyadera, A., Imura, A., Suzuki, A., *Chem. Pharm. Bull. 37* (1989) 2929.

[193] Waldmann, H., *Liebigs Ann. Chem.* (1988) 1175.

[194] Holla, W., *J. Carbohydr. Chem. 9* (1990) 113.

[195] Altamura, M., Cesti, P., Francalanci, F., Marchi, M., Cambianghi, S., *J. Chem. Soc., Perkin I* (1989) 1225.

[196] De Jeso, B., Druillard, S., Lafarge, C., Maillard, B., *Tetrahedron Lett. 26* (1985) 6063.
[197] De Jeso, B., Drouillard, S., Degueil-Castaing, M., Saux, A., Maillard, B., *Synth. Commun. 18* (1988) 1699.
[198] De Jeso, B., Drouillard, S., Degueil-Castaing, M., Maillard, B., *Synth. Commun. 18* (1988) 1691.
[199] Fuganti, C., Grasselli, P., *Tetrahedron Lett. 27* (1986) 3191.
[200] Rossi, D., Romeo, A., Lucente, G., *J. Org. Chem. 43* (1978) 2576.
[201] Waldmann, H., *Tetrahedron Lett. 29* (1988) 1131.
[202] Pessina, A., Lüthi, P., Luisi, P. L., Prenosil, J., Zhang, Y. S., *Helv. Chim. Acta 71* (1988) 631.
[203] Milne, H. B., Carpenter, F. H., *J. Org. Chem. 33* (1968) 4476.
[204] Xaus, N., Clapés, P., Bardaá, E., Torres, J. L. Jorba, X., Mata, J., Valencia, G., *Biotechnol. Lett. 11* (1989) 393.
[205] Xaus, N., Clapés, P., Bardaji, E., Torres, J. L., Jorba, X., Mata, J., Valencia, G., *Tetrahedron 45* (1989) 7421.
[206] Chen, S. T., Wang, K. T., *Synthesis* (1987) 581.
[207] Chen, S. T., Wang, K. T., *Chem. Commun.* (1988) 327.
[208] Shin, C., Takahashi, N., Yonezawa, Y., *Chem. Lett.* (1988) 2001.
[209] Wu, S. H., Lo, L. C., Chen, S. T., Wang, K. T., *J. Org. Chem. 54* (1989) 4220.
[210] Kitaguchi, H., Tai, D. F., Klibanov, A. : *M., Tetrahedron Lett. 29* (1988) 5489.
[211] Riva, S., Klibanov, A. M., *J. Am. Chem. Soc. 110* (1988) 3291.
[212] Riva, S., Bovara, R., Ottolina, G., Secundo, F., Carrea, G., *J. Org. Chem. 54* (1989) 3161.
[213] Njar, V. C. O., Caspi, E., *Tetrahedron Lett. 28* (1987) 6549.
[214] Björkling, F., Godtfredsen, S. E., Kirk, O., *Chem. Commun.* (1989) 934; Adelhorst, K. Björkling, F., Godtfredsen, S. E. Kirk, O., *Synthesis* (1990) 112.
[215] Therisod, M., Klibanov, A. M., *J. Am. Chem. Soc. 109* (1987) 3977.
[216] Ciuffreda, P., Ronchetti, F., *J. Carbohydr. Chem. 9* (1990) 125.
[217] Gateau-Olesker, A., Cléophax, J., Géro, S. D., *Tetrahedron Lett. 27* (1986) 41.
[218] Holla, W., *Angew. Chem., Int. Ed. Engl. 28* (1989) 220.
[219] Hennen, W. J., Sweers, H. M., Wang, Y. F., Wong, C. H., *J. Org. Chem. 53* (1988) 4939.
[220] Nicotra, E., Riva, S., Secundo, F., Zucchelli, L., *Tetrahedron Lett. 30* (1989) 1703.
[221] Uemura, A., Nozaki, K., Yamashita, J. I., Yasumoto, M., *Tetrahedron Lett. 30* (1989) 3817.
[222] Riva, S., Chopineau, J., Kieboom, A. P. G., Klibanov, A. M., *J. Am. Chem. Soc. 110* (1988) 584.
[223] Sweers, H. M., Wong, C. H., *J. Am. Chem. Soc. 108* (1986) 6421.
[224] Csuk, R., Glänzer, B. I., *Z. Naturforsch. B. 43* (1988) 1355.
[225] Kloosterman, M., de Nijs, M. P., Weijnen, J. G. J., Schoemaker, H. E., Meijer, E. M., *J. Carbohydr. Chem. 8* (1989) 333.
[226] Margolin, A. L., Delinck, D. L., Whalon, M. R., *J. Am. Chem. Soc. 112* (1990) 2849.
[227] Ottolina, G., Carrea, G., Riva, S., *J. Org. Chem. 55* (1990) 2366.
[228] Papageorgiu, C., Benezra, C., *J. Org. Chem. 50* (1985) 1144.
[229] Schneider, M., Engel, M., Boensmann, H., *Angew. Chem. Int. Ed. Engl. 23* (1984) 64.
[230] Poppe, L., C. Sc. Thesis (in Hungarian), Budapest, 1986.
[231] Yamazaki, T., Ichikawa, S., Kitazume, T., *Chem. Commun.* (1989) 253.
[232] Nori, K., Takeuchi, T., *Tetrahedron 44* (1988) 333.
[233] Bucciarelli, M., Forni, A., Morretti, I., Prati, F., *Tetrahedron:Asymmetry 1* (1990) 5.
[234] Bucciarelli, M., Forni, A., Moretti, I., Prati, F., *Chem. Commun.* (1988) 1614.
[235] Yamada, H., Shimizu, S., in : *Biocatalysts in Organic Synthesis*, Tramper, H. C., Plas, P., Linko, P. (Eds.), Amsterdam: Elsevier, 1985; pp 19-37.
[236] Turk, J., Pause, G. T., Marschall, R., *J. Org. Chem. 40* (1975) 953.
[237] Chenault, H. K., Kim, M. J., Akiyama, A., Miyazawa, T., Simon, E., Whitesides, G. M., *J. Org. Chem. 52* (1987) 2608.
[238] Bosch, R., Brückner, H., Jung, G., Winter, W., *Tetrahedron 38* (1982) 3579.
[239] Masaoka, Y., Sakakibara, M., Mori, K., *Agric. Biol. Chem. 46* (1982) 2319.
[240] Sugai, T., Mori, K., *Agric. Biol. Chem. 48* (1984) 2497.

[241] Mori, K., Otsuka, T., *Tetrahedron 41* (1985) 547.

[242] Tsushima, T., Kawada, K., Ishihara, S., Uchida, N., Shiratori, O., Higaki, J., Hirata, M., *Tetrahedron 44* (1988) 5375.

[243] Keller, J. W., Jo Nell-Hamilton, B., *Tetrahedron Lett. 27* (1986) 1249.

[244] Mori, K., Iwasawa, H., *Tetrahedron 36* (1980) 2209.

[245] Vidal-Cros, A., Gaudry, M., Marquet, A., *J. Org. Chem. 50* (1985) 3163.

[246] Baldwin, J. E., Christie, M. A., Haber, S. B., Cruse, L I., *J. Am. Chem. Soc. 98* (1976) 3045.

[247] Miyazawa, T., Iwanaga, H., Ueji, S., Yamada, T., Kuwata, S., *Chem. Lett.* (1989) 2219.

[248] Cohen, S. G., Spriznak, Y., Khedouri, E., *J. Am. Chem. Soc. 83* (1961) 4225.

[249] Cohen, S., Crossley, J., *J. Am. Chem. Soc. 86* (1964) 4999.

[250] Chénevert, R., Léturneau, M., *Chem. Lett.* (1986) 1151.

[251] Chénevert, R., Thiboutot, S., *Synthesis* (1989) 444.

[252] Lankiewicz, L., Kasprzykowski, F., Grzonka, Z., Kettmann, U., Hermann, P., *Bioorg. Chem. 17* (1989) 275.

[253] Drueckhammer, D. G., Barbas, C. F., Nozak, K., Wong, C. H., Wood, C. Y., Ciufolini, M. A., *J. Org. Chem. 53* (1988) 1607.

[254] Ciufolini, M. A., Hermann, C. W., Whitmire, K. H., Byrne, N. E., *J. Am. Chem. Soc. 111* (1989) 3473.

[255] Miyazawa, T., Takitani, T., Ueji, S., Yamada, T., Kuwata, S., *Chem. Commun.* (1988) 1214.

[256] Ananthramaiah, G. M., Roeske, R. W., *Tetrahedron Lett. 23* (1982) 3335.

[257] Guibe-Jampel, E., Rousseau, G., Salaün, J., *Chem. Commun.* (1987) 1080.

[258] Cohen, S. G., *Trans. N.Y. Acad. Sci. 31* (1969) 705.

[259] Natchev, I., *Bull. Chem. Soc., Japan 61* (1988) 3699.

[260] Natchev, I., *J. Chem. Soc., Perkin I* (1989) 125.

[261] Natchev, I., *Bull. Chem. Soc., Japan 61* (1988) 3711.

[262] Natchev, I., *Synthesis* (1987) 1079.

[263] Natchev, I., *Bull. Chem. Soc., Japan 61* (1988) 4447.

[264] Natchev, I., *Bull. Chem. Soc., Japan 61* (1988) 3705.

[265] Natchev, I., *Bull. Chem. Soc., Japan 61* (1988) 4488.

[266] Natchev, I., *Bull. Chem. Soc., Japan 61* (1988) 4491.

[267] Solodenko, V. A., Kukhar, V. P., *Tetrahedron Lett. 30* (1989) 6917.

[268] Möller, A., Syldatk, C., Schulze, M., Wagner, F., *Enzyme Microb. Technol. 10* (1988) 618.

[269] Runser, S., Ohleyer, E., *Biotechnol. Lett. 12* (1990) 259.

[270] Wilson, W. K., Baca, S. B., Barber, Y. J., Scallen, T. J., Morrow, C. J., *J. Org. Chem. 48* (1983) 3960.

[271] Watanabe, S., Fujita, T., Sakamoto, M., Arai, T., Kitazume, T., *J. Am. Oil Chem. Soc. 66* (1989) 1312.

[272] Bianchi, D., Cabri, W., Cesti, P., Francalanci, F., Ricci, M., *J. Org. Chem. 53* (1988) 104.

[273] Mohr, P., Rösslein, L., Tamm, C., *Helv. Chim. Acta 30* (1989) 2513.

[274] Schneider, M., Engel, N., Hönicke, P., Heinemann, G., Görisch, H., *Angew. Chem., Int. Ed. Engl. 23* (1984) 67.

[275] Burgess, K., Henderson, I., *Tetrahedron Lett. 30* (1989) 3633.

[276] Cohen, S. G., Weinstein, S. Y., *J. Am. Chem. Soc. 86* (1964) 725.

[277] Salaun, J., Karkour, B., *Tetrahedron Lett. 28* (1987) 4669.

[278] Barnier, J. P., Blanco, L., Guibé-Jumpel, E., Rousseau, G., *Tetrahedron 45* (1989) 5051.

[279] Cohen, S. G., Milovanovic, A., *J. Am. Chem. Soc. 90* (1968) 3495.

[280] Cohen, S., Crosley, M., Khedouri, E., Zand, R., Klee, L. H., *J. Am. Chem. Soc. 85* (1963) 1685.

[281] Cohen, S. G., Weinstein, S. Y., *J. Am. Chem. Soc. 86* (1964) 5326.

[282] Chen, C. S., Fujimoto, Y., Girdaukas, G., Sih, C. J., *J. Am. Chem. Soc. 104* (1982) 7294.

[283] Kalaritis, P., Regenye, R. W., Partridge, J. J., Coffen, D. L., *J. Org. Chem. 55* (1990) 812.

[284] Ahmar, M., Girard, C., Bloch, R., *Tetrahedron Lett. 30* (1989) 7053.

[285] Yamazaki, T. Ohnogi, T., Kitazume, T., *Tetrahedron:Asymmetry 1* (1990) 215.

[286] Gu, Q. M., Chen, C. S., Sih, C. J., *Tetrahedron Lett. 27* (1986) 1763.

[287] Gu, Q. M., Reddy, D. R., Sih, C. J., *Tetrahedron Lett. 27* (1986) 5203.

[288] Dernoncour, R., Azerad, R., *Tetrahedron Lett. 28* (1987) 4661.

[289] Cambou, B., Klibanov, A. M., *Biotechnol. Bioeng. 26* (1984) 1449.

[290] Chénevert, R., D'Astous, L., *Can. J. Chem. 66* (1988) 1219.

[291] Guo, Z. W., Sih, C. J., *J. Am. Chem. Soc. 111* (1989) 6836.

[292] Cohen, S. G., Lo, L. W., *J. Biol. Chem. 245* (1970) 5718.

[293] Lalonde, J J, Bergbeiter, D. E., Wong, C. H., *J. Org. Chem. 53* (1988) 2323.

[294] Ramaswamy, S., Hui, R. A. F., Jones, J. B., *Chem. Commun.* (1986) 1545.

[295] Alcock, N. W., Crout, D. H. G., Henderson, C. M., Thomas, S. E., *Chem. Commun.* (1988) 746.

[296] Sicsic, S., Igbal, M., Le Goffic, F., *Tetrahedron Lett. 28* (1987) 1887.

[297] Morimoto, Y., Terao, Y., Achiwa, K., *Chem. Pharm. Bull. 35* (1987) 2266.

[298] Suemune, H., Tanaka, M., Obaishi, H., Sakai, K., *Chem. Pharm. Bull. 36* (1988) 15.

[299] Pottie, M., Van der Eycken, J., Vanderwalle, M., Dewanckele, J. M., Röper, H., *Tetrahedron Lett. 30* (1989) 5319.

[300] Jones, J. B., Marr, P. W., *Tetrahedron Lett.* (1973) 3165.

[301] Björkling, F., Boutelje, J., Gatenbeck, S., Hult, K., Norin, T., *Appl. Microbiol. Biotechnol. 21* (1985) 16.

[302] Klunder, A. J. H., Van Gastel, F. J. L., Zwanenburg, B., *Tetrahedron Lett. 29* (1988) 2697.

[303] Van der Eycken, J., Vandewalle, M., Heinemann, G., Laumen, K., Schneider, M. P., Kredel, J., Sauer, J., *Chem. Commun.* (1989) 306.

[304] Ramos-Tombo, G. M., Schär, H. P., Ghisalba, O., *Agric. Biol. Chem. 51* (1987) 1833.

[305] Lawson, W. B., *J. Biol. Chem. 242* (1967) 3397.

[306] Fülling, A., Sih, C. J., *J. Am. Chem. Soc. 109* (1987) 2845.

[307] Schwartz, H. M., Wu, W. S., Marr, P. W., Jones, J. B., *J. Am. Chem. Soc. 100* (1978) 5199.

[308] Hayashi, Y., Lawson, W. B., *J. Biol. Chem. 244* (1969) 4158.

[309] Pattabiraman, T. N., Lawson, W. B., *J. Biol. Chem. 247* (1972) 3029.

[310] Cohen, S. G., Milovanovic, A., Schultz, R. M., Weinstein, S. Y., *J. Biol. Chem. 244* (1969) 2664.

[311] Hein, G. E., Niemann, C., *J. Am. Chem. Soc. 84* (1962) 4487.

[312] Klunder, A. J. H., Huizinga, W. B., Hulschoff, A. J. M., Zwanenburg, B., *Tetrahedron Lett. 27* (1986) 2543.

[313] Guibe-Jampel, E., Rousseau, G., *Tetrahedron Lett. 28* (1987) 3563.

[314] Patel, D. V., Van Middlesworth, F., Donaubauer, J., Gaunnett, P., Sih, C. J., *J. Am. Chem. Soc. 108* (1986) 4603.

[315] Chen, C. S., Wu, S. H., Girdaukas, G., Sih, C. J., *J. Am. Chem. Soc. 109* (1987) 2812.

[316] Klunder, A. J. H., Huizinga, W. B., Sessink, P. J. M., Zwanenburg, B., *Tetrahedron Lett. 28* (1987) 357.

[317] Lavayre, J., Verrier, J., Baratti, J., *Biotechnol. Bioeng. 24* (1982) 2175.

[318] Iriuchijima, S., Keiyu, A., Kojima, N., *Agric. Biol. Chem. 46* (1982) 1593.

[319] Francalanci, F., Cesti, P., Cabri, W., Bianchi, D., Martinengo, T., Foa, M., *J. Org. Chem. 52* (1987) 5079.

[320] Ladner, W. E., Whitesides, G. M., *J. Am. Chem. Soc. 106* (1984) 7250.

[321] Classen, A., Wershofen, S., Yusufoglu, A., Scharf, H. D., *Liebigs Ann. Chem.* (1987) 629.

[322] Bianchi, D., Cabri, W., Cesti, P., Francalanci, F., Rama, F., *Tetrahedron Lett. 29* (1988) 2455.

[323] Fuganti, C., Graselli, P., Lazzarini, A., Casati, P., *Chem. Commun.* (1987) 538.

[324] Fuganti, C., Grasselli, P., Servi, S., Lazzarini, A., Casati, P., *Tetrahedron 44* (1988) 2575.

[325] Pawlak, J. L., Berchtold, G. A., *J. Org. Chem. 52* (1987) 1765.

[326] Van Middlesworth, F., Patel, D. V., Donaubauer, J., Gaunnett, P., Sih, C. J., *J. Am. Chem. Soc. 107* (1985) 2996.

[327] Hamaguchi, S., Yamamura, H., Hasegawa, J., Watanabe, K., *Agric. Biol. Chem. 49* (1985) 1509.

[328] Hamaguchi, S., Asada, M., Hasegawa, J., Watanabe, K., *Agric. Biol. Chem. 49* (1985) 1661.

[329] Hamaguchi, S., Asada, M., Hasagawa, J., Watanabe, K., *Agric. Biol. Chem. 48* (1984) 2331.

[330] Suemune, H., Hizuka, M., Kamashita, T., Sakai, K., *Chem. Pharm. Bull. 37* (1989) 1379.

[331] Pfeuniger, A., *Synthesis* (1986) 89.

[332] Sonnet, P., Antoniani, E., *J. Agric. Food. Chem. 36* (1988) 856.

[333] Kloosterman, M., Elfering, V. H. M., et. al., *Trends Biotechnol. 6* (1988) 251.

[334] Cambou, B., Klibanov, A. M., *J. Am. Chem. Soc. 106* (1984) 2687.

[335] Bianchi, D., Cesti, P., Battistel, E., *J. Org. Chem. 53* (1988) 5531.

[336] Laumen, K., Schneider, M. P., *Chem. Commun.* (1988) 598.

[337] Rasor, P., Rüchardt, C., *Chem. Ber. 122* (1989) 1375.

[338] Bianchi, D., Cesti, P., Golini, P., *Tetrahedron 45* (1989) 869.

[339] Scilimati, A., Ngooi, T. K., Sih, C. J., *Tetrahedron Lett. 29* (1988) 4927.

[340] Lin, J. T., Yamazaki, T., Kitazume, T., *J. Org. Chem. 52* (1987) 3211.

[341] Kutsuki, A., Sawa, I., Hasegawa, J., Watanabe, K., *Agric. Biol. Chem. 50* (1986) 2369.

[342] Mori, K., Bernotas, R., *Tetrahedron:Asymmetry 1* (1990) 87.

[343] Hamaguchi, S., Ohashi, T., Watanabe, K., *Agric. Biol. Chem. 50* (1986) 1629.

[344] Hamaguchi, S., Ohashi, T., Watanabe, K., *Agric. Biol. Chem. 50* (1986) 375.

[345] Chen, C. S., Liu, Y. C., *Tetrahedron Lett. 30* (1989) 7165.

[346] Chenevert, R., Desjardins, M., Gagnon, R., *Chem. Lett.* (1990) 33.

[347] Almsick, A., Buddrus, J., Hönicke-Schmidt, P., Laumen, K., Schneider, M. P., *Chem. Commun.* (1989) 1391.

[348] Hirohara, H., Mitsuda, S., Ando, E., Komaki, R., in : *Biocatalysts in Organic Synthesis*, Tramper, J., v. d. Plaas, H. C., Linko, P. (Eds.), Amsterdam: Elsevier, 1985; pp 119-134.

[349] Matsou, N., Ohno, N., *Tetrahedron Lett. 26* (1985) 5533.

[350] Saf, R., Faber, K., Penn, G., Griengl, H., *Tetrahedron 44* (1988) 389.

[351] Itoh, T., Tagaki, Y., *Chem. Lett.* (1989) 1505.

[352] Tsuboi, S., Sakamoto, J., Sakai, T., Utaka, M., *Chem. Lett.* (1989) 1427.

[353] Burgess, K., Henderson, I., *Tetrahedron:Asymmetry 1* (1990) 57.

[354] Waldmann, H., *Tetrahedron Lett. 30* (1989) 3057.

[355] Suemune, H., Mizuhara, Y., Akita, H., Oishi, T., Sakai, K, *Chem. Pharm. Bull. 35* (1987) 3112.

[356] Bevinakatti, H. S., Banerji, A. A., Newadkar, R. V., *J. Org. Chem. 54* (1989) 2453.

[357] Feichter, C., Faber, K., Griengl, H., *Tetrahedron Lett. 30* (1989) 551.

[358] Akita, H., Matsukara, H., Oishi, T., *Tetrahedron Lett. 27* (1986) 5241.

[359] Foelsche, E., Hickel, A., Hönig, H., Seufel-Wasserthal, P., *J. Org. Chem. 55* (1990) 1749.

[360] Marples, B. A., Rogers-Evans, M., *Tetrahedron Lett. 30* (1989) 261.

[361] Ambramowicz, D. A., Keese, C. R., *Biotechnol. Bioeng. 33* (1989) 149.

[362] Laumen, K., Breitgoff, D., Schneider, M. P., *Chem. Commun.* (1988) 1459.

[363] Wang, Y. F., Chen, S. T., Lin, K. K. C., Wong, C. H., *Tetrahedron Lett. 30* (1989) 1917.

[364] Sonnet, T., *J. Org. Chem. 52* (1987) 3477.

[365] Sonnet, P. E., Baillargeon, M. W., *J. Chem. Ecol. 13* (1987) 1279.

[366] Stokes, T., Oehlschlager, A. C., *Tetrahedron Lett. 28* (1987) 2091.

[367] Belan, A., Bolte, J., Fauve, A., Gourcy, J. G., Veschambre, H., *J. Org. Chem. 52* (1987) 256.

[368] Theisen, P. D., Heathcock, C. H., *J. Org. Chem. 53* (1988) 2374.

[369] Boaz, N. W., *Tetrahedron Lett. 30* (1989) 2061.

[370] Burgess, K., Jennings, L. D., *J. Org. Chem. 55* (1990) 1138.

[371] De Amici, M., De Micheli, C., Carrea, G., Spezia, S., *J. Org. Chem. 54* (1989) 2646.

[372] Gil, G., Ferre, E., Meou, A., Le Petit, J., Triantaphylides, G., *Tetrahedron Lett. 28* (1987) 1647.

[373] Baba, N., Mimura, M., Hiratake, J., Uchida, K., Oda, J., *Agric. Biol. Chem. 52* (1988) 2685.

[374] Xie, Z. F., Suemune, H., Sakai, K., *Chem. Commun.* (1987) 838.

[375] Hönig, H., Seufer-Wasserthal, P., Fülöp, F., *J. Chem. Soc., Perkin I* (1989) 2341.

[376] Xie, Z. F., Nakamura, I., Suemune, H., Sakai, K., *Chem. Commun.* (1988) 966.

[377] Laumen, K., Breitgoff, D., Seemayer, R., Schneider, M. P., *Chem. Commun.* (1989) 148.

[378] Whitesell, J. K., Chen, H. H., Lawrence, R. M., *J. Org. Chem. 50* (1985) 4663.

[379] Whitesell, J. K., Lawrence, R. M., *Chimia 40* (1986) 318.

[380] Mitsuda, S., Nabeshima, S, Hirohara, H., *Appl. Microb. Biotechnol. 31* (1989) 334.

[381] Mori, K., Ogoche, J. I. J., *Liebigs Ann. Chem.* (1988) 903.

[382] Ganey, M. V., Padykula, R. E., Berchtold, G. A., Braun, A. G., *J. Org. Chem. 54* (1989) 2787.

[383] Fritsche, K., Syldatk, C., Wagner, F., Hengelsberg, H., Tacke, R., *Appl. Microbiol. Biotechnol. 31* (1989) 107.

[384] Klempier, N., Faber, K., Griengl, H., *Biotechnol. Lett. 11* (1989) 685.

[385] Klempier, N., Geymayer, P., Stadler, P., P., Faber, K., Griengl, H., *Tetrahedron:Asymmetry 1* (1990) 111.

[386] Cotteril, I. C., Finch, H., Reynolds, D. R., Roberts, S. M., Rzepa, H. S., Short, K. M., Slawin, A. M. Z., Wallis, C. J., Williams, D. J., *Chem. Commun.* (1988) 470.

[387] Washausen, P., Grebe, H., Kieslich, K., Winterfeldt, E., *Tetrahedron Lett. 30* (1989) 3777.

[388] Dumortier, L., Van der Eycken, J., Vanderwelle, M., *Tetrahedron Lett. 30* (1989) 3201.

[389] Eichberger, G., Penn, G., Faber, K., Griengl, H., *Tetrahedron Lett. 27* (1986) 2843.

[390] Oberhauser, T., Bodenteich, M., Faber, K., Penn, G., Griengl, H., *Tetrahedron 43* (1987) 3931.

[391] Hoshino, O., Itoh, K., Umezawa, B., Akita, H., Oishi, T., *Tetrahedron Lett. 29* (1988) 567.

[392] Hirose, Y., Anzai, M., Saitoh, M., Naemura, K., Chikamatsu, H., *Chem. Lett.* (1989) 1939.

[393] Tamai, Y., Nakano, T., Koike, S., Kawahara, K., Miyano, S., *Chem. Lett.* (1989) 1135.

[394] Inagaki, M., Hiratake, J., Nishioka, T., Oda, J., *Agric. Biol. Chem. 53* (1989) 1879.

[395] Naemura, K., Matsumura, T., Komatsu, M., Hirose, Y., Chikamatsu, H., *Bull. Chem. Soc., Japan 62* (1989) 3523.

[396] Takano, S., Inomata, K., Ogasawara, K., *Chem. Commun.* (1989) 271.

[397] Oberhauser, T., Faber, K., Griengl, H., *Tetrahedron 45* (1989) 1679.

[398] Faber, K., Hönig, H., Seufer-Wasserthal, P., *Tetrahedron Lett. 29* (1988) 1903.

[399] Baumgartner, H., Bodenteich, M., Griengl, H., *Tetrahedron Lett. 29* (1988) 5745.

[400] Langrand, D., Secchi, M., Buono, G., Baratti, J., Triantaphylides, C., *Tetrahedron Lett. 26* (1985) 1857.

[401] Lokotsch, W., Fritsche, K., Syldatk, C., *Appl. Microb. Biotechnol. 31* (1989) 467.

[402] Iriuchijima, S., Kojima, N., *Agric. Biol. Chem. 46* (1982) 1153.

[403] Yamamoto, K., Ando, H., Chikamatsu, H., *Chem. Commun.* (1987) 334.

[404] Crout, D. H. G., Gaundet, V. S. B., Laumen, K., Schneider, M., *Chem. Commun.* (1986) 808.

[405] Lin, Y. C., Chen, C. S., *Tetrahedron Lett. 30* (1989) 1617.

[406] Djadchenko, M. A., Pivnitsky, K. K., Theil, F., Shick, H., *J. Chem. Soc., Perkin I* (1989) 2001.

[407] Blanco, L., Guibe-Jampel, E., Rousseau, G., *Tetrahedron Lett. 29* (1988) 1915.

[408] Gutman, A. L., Zuobi, K., Guibe-Jampel, E., *Tetrahedron Lett. 31* (1990) 2037.

[409] Guibe-Jampel, E., Rousseau, G., Blanco, L., *Tetrahedron Lett. 30* (1989) 67.

[410] Yamada, H., Ohsawa, S., Sugai, T., Ohta, H., Yoshikawa, S., *Chem. Lett.* (1989) 1775.

[411] Makita, A., Nihira, T., Yamada, Y., *Tetrahedron Lett. 28* (1987) 805.

[412] Gutman, A. L., Zuobi, K., Boltansky, A., *Tetrahedron Lett. 28* (1987) 3861.

[413] Ngooi, T. K., Scilimati, A., Guo, Z., Sih, C. J., *J. Org. Chem. 54* (1989) 911.

[414] Gutman, A. L., Oren, D., Boltansky, A., Bravdo, T., *Tetrahedron Lett. 28* (1987) 5367.

[415] Gutman, A. L., Bravdo, T., *J. Org. Chem. 54* (1989) 4263.

[416] Margolin, A. L., Crenne, J. R., Klibanov, A. M., *Tetrahedron Lett. 28* (1987) 1607.

[417] Gutman, A. L., Bravdo, T., *J. Org. Chem. 54* (1989) 5645.

[418] Wei, G. Z., Sih, C. J., *J. Am. Chem. Soc. 110* (1988) 1999.

[419] Romeo, A., Lucente, G., Rossi, D., Zanotti, G., *Tetrahedron Lett.* (1971) 1799.

[420] Fuganti, C., Graselli, P., Seneci, P. F., Servi, S., *Tetrahedron Lett. 27* (1986) 2061.

[421] Gotor, V., Brieva, R., Rebolledo, F., *Chem. Commun.* (1988) 957.

[422] Chinsky, N., Margolin, A. L., Klibanov, A. M., *J. Am. Chem. Soc. 111* (1989) 386.

[423] Kitaguchi, H., Fitzpatrick, P. A., Huber, J. E., Klibanov, A. M., *J. Am. Chem. Soc. 111* (1989) 3094.

[424] Barili, P., Berti, G., Catelani, G., Colonna, F., Mastrorilli, E., *Chem. Commun.* (1986) 7.

[425] Witsuba, D., Schurig, V., *Angew. Chem. 98* (1986) 1008.

[426] Jerina, D. M., Ziffer, H., Daly, J. W., *J. Am. Chem. Soc. 92* (1970) 1056.

[427] Bellucci, G., Berti, G., Catelani, G., Mastrorilli, E., *J. Org. Chem. 46* (1981) 5148.

[428] Prestwich, G. D., Graham, S., König, W. A., *Chem. Commun.* (1989) 575.

[429] Thakker, D. R., Yagi, H., Levin, W., Lu, A. Y. H., Conney, A. H., Jerina, D. M., *J. Biol. Chem. 252* (1977) 6328.

[430] Sayer, J. M., Yagi, H., Van Blanderen, P. J., Levin, W., Jerina, D. M., *J. Biol. Chem. 260* (1985) 1630.

[431] Bellucci, G., Chiappe, C., Conti, L., Marioni, F., Pierini, G., *J. Org. Chem. 54* (1989) 5978.

[432] Bellucci, G., Berti, G., Bianchini, R., Cetera, P., Mastrorilli, E., *J. Org. Chem. 47* (1982) 3105.

[433] Bellucci, G., Berti, G., Ferretti, M., Mastrorilli, E., Silvestri, L., *J. Org. Chem. 50* (1985) 1471.

[434] Bellucci, G., Berti, G., Ingrosso, G., Mastrorilli, E., *J. Org. Chem. 45* (1980) 299.

[435] Bellucci, G., Ferretti, M., Lippi, A., Marioni, F., *J. Chem. Soc., Perkin I* (1988) 2715.

[436] Bellucci, G., Chiappe, C., Marioni, F., *J. Chem. Soc., Perkin I* (1989) 2369.

[437] Cateleni, G., Mastrorilli, E., *J. Chem. Soc., Perkin I* (1983) 2117.

[438] Barili, P. L., Berti, G., Catelani, G., Colonna, F., Mastrorilli, E., Paoli, M., *Tetrahedron 45* (1989) 1553.

[439] Barili, P. L., Berti, G., Catelani, G., Colonna, F., Mastrorilli, E., *J. Org. Chem. 52* (1987) 2887.

[440] Borthwick, A. D., Butt, S., Biggadigke, K., Exall, A., Roberts, S., Youds, P., Kirk, B., Booth, B., Cameron, J., Cox, S., Marr, C., Shill, M., *Chem. Commun.* (1988) 656.

[441] Secrist, J. A., Montgomery, J. A., Shealy, Y. F., O'dell, C. A., Clayton, S. J., *J. Med. Chem. 30* (1987) 746.

[442] Björkling, F., Godtfredsen, S. E., *Tetrahedron 44* (1988) 2957.

[443] Björkling, F., Boutelje, J., Gatenbeck, S., Hult, K., Norin, T., Szmulik, P., *Tetrahedron 41* (1985) 1347.

[444] Schneider, M., Engel, N., Boensmann, H., *Angew. Chem., Int. Ed. Engl. 23* (1984) 66.

[445] Björkling, F., Boutelje, J., Hult, K., Norin, T., *Tetrahedron Lett. 26* (1985) 4957.

[446] Luyten, M., Müller, S., Herzog, B., Keese, R., *Helv. Chim. Acta 70* (1987) 1250.

[447] De Jeso, B., Belair, N., Deluze, H., Rascle, M. C., Maillard, B., *Tetrahedron Lett. 31* (1990) 653.

[448] Kitazume, T., Sato, T., Ishikawa, N., *Chem. Lett.* (1984) 1811.

[449] Kitazume, T., Sato, T., Kobayashi, T ., Lin, J. T., *J. Org. Chem. 51* (1986) 1003.

[450] Iriuchijima, S., Hasegawa, K., Tsuchihashi, G., *Agric. Biol. Chem. 46* (1982) 1907.

[451] Morimoto, Y., Achiwa, K., *Chem. Pharm. Bull. 35* (1987) 3845.

[452] Cohen, S. G., Klee, L. H., *J. Am. Chem. Soc. 82* (1960) 6038.

[453] Van Middlesworth, F., Wang, Y. F., Zhou, B. N., Di Tullio, D., Sih, C. J., *Tetrahedron Lett. 26* (1985) 961.

[454] Ackermann, A., Waespe-Saracevic, N., Tamm, C., *Helv. Chim. Acta 67* (1984) 254.

[455] Herold, P., Mohr, P., Tamm, C., *Helv. Chim. Acta 66* (1983) 744.

[456] Nakada, M., Kobayashi, S., Ohno, M., *Tetrahedron Lett. 29* (1988) 3951.

[457] Roy, R., Rey, A. W., *Tetrahedron Lett. 28* (1987) 4935.

[458] Baader, E., Bartmann, W., Beck, G., Bergmann, A., Fehlhaber, H., Jendralla, H., Kessler, K., Saric, R., Schüssler, H., Teetz, W., Weber, M., Wess, G., *Tetrahedron Lett. 29* (1988) 2563.

[459] Cohen, S. G., Khedouri, E., *J. Am. Chem. Soc. 83* (1961) 4228.

[460] Wang, Y. F., Izawa, T., Kobayashi, S., Ohno, M., *J. Am. Chem. Soc. 104* (1982) 6465.

[461] Rosen, T., Watanabe, M., Heathcock, C. H., *J. Org. Chem. 49* (1984) 3657.

[462] Santaniello, E., Chiari, M., Ferraboschi, P., Trave, S., *J. Org. Chem. 53* (1988) 1567.

[463] Lam, L. K. P., Jones, J. B., *Can. J. Chem. 66* (1988) 1422.

[464] Huang, F. C., Lee, L. F. H., Mittal, R. S., Ravikumar, P. R., Chan, J. A., Sih, C. J., *J. Am. Chem. Soc. 97* (1975) 4144.

[465] Shuto, A., Kuwano, E., Eto, M., *Agric. Biol. Chem. 52* (1988) 915.

[466] Ohno, M., Kobayashi, S., Imori, T., Wang, Y. F., Izawa, T., *J. Am. Chem. Soc. 103* (1981) 2405.

[467] Cohen, S. G., Crossley, J., *J. Am. Chem. Soc. 86* (1963) 1217.

[468] Cohen, S. G., Khedouri, E., *J. Am. Chem. Soc. 83* (1961) 1093.

[469] Yamamoto, Y., Yamamoto, K., Nishioka, T., Oda, J., *Agric. Biol. Chem. 52* (1988) 3087.

[470] Mohr, P., Tori, M., Grossen, P., Herold, P., Tamm, C., *Helv. Chim. Acta 65* (1982) 1412.

[471] Francis, C. J., Jones, J. B., *Chem. Commun.* (1984) 579.

[472] Poppe, L., Novák, L., Kolonits, P., Bata, Á., Szántay, Cs., *Tetrahedron Lett. 27* (1986) 5769.

[473] Brooks, D. W., Palmer, J. T., *Tetrahedron Lett. 24* (1983) 3059.

[474] Iimori, T., Kobayashi, S., Okano, K., Izawa, T., Ohno, K., *J. Am. Chem. Soc. 105* (1983) 1659.

[475] Okano, K., Izawa, T., Ohno, M., *Tetrahedron Lett. 24* (1983) 217.

[476] Yamashita, H., Minami, M., Sakakibara, K., Kobayashi, S., Ohno, M., *Chem. Pharm. Bull. 36* (1988) 469.

[477] Yamamoto, K., Nishioka, T., Oda, J., Yamamoto, Y., *Tetrahedron Lett. 29* (1988) 1717.

[478] Xie, Z. F., Suemuene, H., Sakai, K., *Chem. Commun.* (1988) 1638.

[479] Tsuji, K., Terao, Y., Achiwa, K., *Tetrahedron Lett. 30* (1989) 6189.

[480] Wang, Y. F., Sih, C. J., *Tetrahedron Lett. 25* (1984) 4999.

[481] Mori, K., Chiba, N., *Liebigs Ann. Chem.* (1989) 957.

[482] Guanti, G., Banfi, L., Narisano, E., *Tetrahedron Lett. 30* (1989) 2697.

[483] Murata, M., Terao, Y., Achiwa, K., Nishio, T., Seto, K., *Chem. Pharm. Bull. 37* (1989) 2670.

[484] Breitgoff, D., Laumen, K., Schneider, M., *Chem. Commun.* (1986) 1523.

[485] Kerscher, V., Kreiser, W., *Tetrahedron Lett. 28* (1987) 531.

[486] Wang, Y. F., Wong, C. H., *J. Org. Chem. 53* (1988) 3127.

[487] Terao, Y., Murata, M., Achiwa, K., *Tetrahedron Lett. 29* (1988) 5173.

[488] Ehler, J., Seebach, D., *Liebigs Ann. Chem.* (1990) 397.

[489] Suemune, H., Harabe, T., Xie, Z. F., Sakai, K., *Chem. Pharm. Bull. 36* (1988) 4337.

[490] Suemune, H., Mizuhara, Y., Akita, H., Sakai, K., *Chem. Pharm. Bull. 34* (1986) 3440.

[491] Bellucci, G., Berti, G., Chiappe, C., Fabri, F., Marioni, F., *J. Org. Chem. 54* (1989) 969.

[492] Bellucci, G., Capitani, I., Chiappe, C., Marioni, F., *Chem. Commun.* (1989) 1170.

[493] Armstrong, R. N., Kedzierski, B., Levin, W., Jerina, D. M., *J. Biol. Chem. 256* (1981) 4726.

[494] Bellucci, G., Bert, G., Chiappe, C., Lippi, A., Marioni, F., *J. Med. Chem. 30* (1987) 786.

[495] Chen, C. S., Fujimoto, Y., Sih, C. J., *J. Am. Chem. Soc. 103* (1981) 3580.

[496] Sabbioni, G., Jones, J. B., *J. Org. Chem. 52* (1987) 4565.

[497] Sabbioni, G., Shea, M. L., Jones, J. B., *Chem. Commun.* (1984) 236.

[498] Kobayashi, S., Kamijama, K., Iimori, T., Ohno, M., *Tetrahedron Lett. 25* (1984) 2557.

[499] Yamazaki, Y., Hosono, K., *Tetrahedron Lett. 29* (1988) 5769.

[500] Mohr, P., Rösslein, L., Tamm, C., *Helv. Chim. Acta 70* (1987) 142.

[501] Nagao, Y., Kume, M., Wakabayashi, R. C., Nakamura, T., Ochiai, M., *Chem. Lett.* (1989) 239.

[502] Jones, J. B., Hinks, R. S., Hultin, P. G., *Can. J. Chem. 63* (1985) 452.

[503] Jones, J. B., *F.E.C.S. Int. Conf. Chem. Biotechnol. Biol. Act. Nat. Prod. [Proc.]*,3rd 1985 (pub. 1987), Vol 1., Weinheim: VCH Verlagsgesellschaft, 1987; p 18.

[504] Kurihara, M., Kamiyama, K., Kobayashi, S., Ohno, M., *Tetrahedron Lett. 26* (1985) 5831.

[505] Björkling, F., Boutelje, J., Hjalmarsson, H., Hult, K., Norin, T., *Chem. Commun.* (1987) 1041.

[506] Arita, M., Adachi, K., Ito, Y., Sawai, H., Ohno, M., *J. Am. Chem. Soc. 105* (1983) 4049.

[507] Kuhn, T., Tamm, C., *Tetrahedron Lett. 30* (1989) 693.

[508] Bloch, R., Guibe-Jampel, E., Girard, C., *Tetrahedron Lett. 26* (1985) 4087.

[509] Ito, Y., Shibata, T., Arita, M., Sawai, H., Ohno, M., *J. Am. Chem. Soc. 103* (1981) 6739.

[510] Ohno, M., *Nucleosides and Nucleotides 4* (1895) 21.

[511] Metz, P., *Tetrahedron 45* (1989) 7311.

[512] Gais, H. J., Lied, T., *Angew. Chem. Int. Ed. Engl. 23* (1984) 511.

[513] Ohno, M., Ito, Y., Arita, M., Shibata, T., Adachi, K., Sawai, H., *Tetrahedron 40* (1984) 145.

[514] Cotteril, I. C., Roberts, S. M., William, S. J. O., *Chem. Commun.* (1988) 1628.

[515] Lam, L. K. P., Jones, J. B., *J. Org. Chem. 53* (1988) 2637.

[516] Holmes, D. S., Sherringham, J. A., Dyer, U. C., Russel, S. T., Robinson, J. A., *Helv. Chim. Acta 73* (1990) 239.

[517] Tshamber, T., Waespe-Saracevic, N., Tamm, C., *Helv. Chim. Acta 69* (1986) 621.

[518] Born, M., Tamm, C., *Tetrahedron Lett. 30* (1989) 2083.

[519] Kaga, H., Kobayashi, S., Ohno, M., *Tetrahedron Lett. 30* (1989) 113.

[520] Kaga, H., Kobayashi, S., Ohno, M., *Tetrahedron Lett. 29* (1988) 1057.

[521] Kamiyama, K., Kobayashi, S., Ohno, M., *Chem. Lett.* (1987) 29.

[522] Kobayashi, S., Kamiyama, K., Ohno, M., *J. Org. Chem. 55* (1990) 1169.

[523] Shimada, M., Kobayashi, S., Ohno, M., *Tetrahedron Lett. 29* (1988) 6961.

[524] Gais, H. J., Lucas, K. L., Ball, W. A., Braun, S., Lindner, H. J., *Liebigs Ann. Chem.* (1986) 687.

[525] Gais, H. J., Lindner, H. L., Lied, T., Lukas, K. L., Ball, W. A., Rosenstock, B., Sliwa, H., *Liebigs Ann. Chem.* (1986) 1179.

[526] Gais, H. J., Ball, A. W., Bund, J., *Tetrahedron Lett. 29* (1988) 781.

[527] Gais, H. J., Schmiedl, G., Ball, A. W. Bund, J., Hellmann, G., Erdelmeier, I., *Tetrahedron Lett. 29* (1988) 1773.

[528] Erdelmeier, I., Gais, H. J., *J. Am. Chem. Soc. 111* (1989) 1125.

[529] Hemmerle, H., Gais, H. J., *Angew. Chem. Int. Ed. Engl. 28* (1989) 349.

[530] Gais, H. J., Lukas, K. L., *Angew. Chem. Int. Ed. Engl. 23* (1984) 142.

[531] Gais, H. J., Lied, T., *Angew. Chem. Int. Ed. Engl. 23* (1984) 145.

[532] Gais, H. J., Lukas, K. L., Ball, W.A., Braun, S., Lindner, H. J., *Liebigs Ann. Chem.* (1986) 687.

[533] Wang, Y. F., Chen, C. S., Girdaukas, G., Sih. C. J., *J. Am. Chem. Soc. 106* (1984) 3695.

[534] Seebach, D., Eberle, M., *Chimia 40* (1986) 315.

[535] Gais, H. J., Zeissler, A., Maidonis, P., *Tetrahedron Lett. 29* (1988) 5743.

[536] Hemmerle, H., Gais, H. J., *Tetrahedron Lett. 28* (1987) 3471.

[537] Laumen, K., Schneider, M., *Tetrahedron Lett. 26* (1985) 2073.

[538] Kasel, W., Hultin, P. G., Jones, J. B., *Chem. Commun.* (1985) 1563.

[539] Adler, U, Breitgoff, D., Klein, P., Laumen, E. K., Schneider, M., *Tetrahedron Lett. 30* (1989) 1793.

[540] Riva, R., Banfi, L., Danieli, B., Guanti, G., Lesma, G., Palmisano, G., *Chem. Commun.* (1987) 299.

[541] Laumen, K., Schneider, M., *Tetrahedron Lett. 25* (1984) 5875.

[542] Laumen, K., Schneider, M., *Chem. Commun.* (1986) 1298.

[543] Sugai, T., Mori, K., *Synthesis* (1988) 19.

[544] Deardoff, D. R., Matthews, A. J., Mc Heekin, D. S., Carney, C. L., *Tetrahedron Lett. 27* (1986) 1255.

[545] Theil, F., Ballschuh, S., Schick, H., Haupt, M., Häfner, B., Schwarz, S., *Synthesis* (1988) 540.

[546] Johnson, C. R., Penning, T. D., *J. Am. Chem. Soc. 108* (1986) 5655.

[547] Johnson, C. R., Penning, T. D., *J. Am. Chem. Soc. 110* (1988) 4726.

[548] Carda, M., Van der Eycken, J., Vanderwalle, M., *Tetrahedron: Asymmetry 1* (1990) 17.

[549] Hönig, H., Seufer-Wassertahl, P., Stütz, A. E., Zenz, E., *Tetrahedron Lett. 30* (1989) 811.

[550] Pearson, A. M., Bansal, H. S., Lay, Y. S., *Chem. Commun.* (1987) 519.

[551] Johnson, C. R., Senanayake, C. H., *J. Org. Chem. 54* (1989) 735.

[552] Suemune, H., Okano, K., Akita, H., Sakai, K., *Chem. Pharm. Bull. 35* (1987) 1741.

[553] Naemura, K., Takahashi, N., Chikamatsu, H., *Chem. Lett.* (1988) 1717.

[554] Esterman, H., Prasad, K., Shapiro, M. J., Repiz, O., Hardtmann, G. E., Bolsterli, J. J., Walkinshaw, M. D., *Tetrahedron Lett. 31* (1990) 445.

[555] Jones, J. B., Hinks, R. S., *Can. J. Chem. 65* (1987) 704.

[556] Xie, Z. F., Sakai, K., *Chem. Pharm. Bull. 37* (1989) 1650.

[557] Cornforth, J. W., Cornforth, R. H., Popjak, G., Yengoyan, L. S., *J. Biol. Chem. 241* (1966) 3970.

[558] Paquette, L. A., Nelson, N. A., *J. Org. Chem. 27* (1962) 2272.

[559] Yoon, N. M., Pak, C. S., Brown, H. C., Krishnamurthy, S., Stocky, T. P., *J. Org. Chem. 38* (1973) 2786.

[560] Fujisawa, T., Mori, T., Sato, T., *Chem. Lett.* (1983) 835.

[561] Sugai, T., Kakeya, H., Ohta, H., Morooka, M., Ohta, S., *Tetrahedron 45* (1989) 6135.

[562] Matsumoto, K., Ohta, H., *Chem. Lett.* (1989) 1109.

[563] Ohta, H., Matsumoto, K., Tsutsumi, S., Ihori, T., *Chem. Commun.* (1989) 485.

[564] Matsumoto, K., Ohta, H., *Chem. Lett.* (1989) 1589.

[565] Toone, E. J., Werth, M. J., Jones, J. B., *J. Am. Chem. Soc. 112* (1990) 4946.

[566] Zhu, L. M., Tedford, M. C., *Tetrahedron 46* (1990) 6587.

[567] Fulcrand, V., Jacquier, R., Lazaro, Viallefont, P., *Tetrahedron 46* (1990) 3909.

[568] Kuhl, P., Halling, P. J., Jakubke, H. D., *Tetrahedron Lett. 31* (1990) 5213.

[569] Kitaguchi, H., Itoh, I., Ono, M., *Chem.Lett.* (1990) 1203.

[570] Berger, B., Rabiller, C. G., Königsberger, K., Faber, K., Griengl, H., *Tetrahedron: Asymmetry 1* (1990) 541.

[571] Hoshino, T., Yamane, T., Shimizu, S., *Agric. Biol. Chem. 54* (1990) 1459.

[572] Braun, P., Waldmann, H., Vogt, W., Kunz, H., *Liebigs Ann. Chem.* (1991) 165.

[573] Ishii, H., Funabashi, K., Mimura, Y., Inoue, Y., *Bull. Chem. Soc., Japan. 63* (1990) 3042.

[574] Shin, C., Seki, M., Takahashi, N., *Chem. Lett.* (1990) 2089.

[575] Delinck, D. L., Margolin, A. L., *Tetrahedron Lett. 21* (1990) 3093.

[576] Natoli, M., Nicolosi, G., Piattelli, M., Nozaki, K., Uemura, A., Yamashita, J., Yasumoto, M., *Tetrahedron Lett. 31* (1990) 7371.

[577] Fuganti, C., Pedrocchi-Fantoni, G., Servi, S., *Chem. Lett.* (1990) 1137.

[578] Ramaswamy, S., Morgan, B., Oehlschlager, A. C., *Tetrahedron Lett. 21* (1990) 3405.

[579] Colombo, D., Ronchetti, F., Toma, L., *Tetrahedron 47* (1991) 103.

[580] Ciuffreda, P., Colombo, D., Ronchett, F., Toma, L., *J. Org. Chem. 55* (1990) 4187.

[581] Danieli, B., De Bellis, P., Carrea, G., Riva, S., *Helv. Chim. Acta. 73* (1990) 1837.

[582] Murakami, N., Imamura, H., Morimoto, T., Ueda, T., Nagai, S., Sakakiraba, J., Yamada, N., *Tetrahedron Lett. 32* (1991) 1331.

[583] Fourneron, J. D., Chiche, M., Pieroni, G., *Tetrahedron Lett. 31* (1990) 4875.

[584] Rabiller, C. G., Königsberger, K., Faber, K., Griengl, H., *Tetrahedron 46* (1990) 4231.

[585] Wallace, J. S., Reda, K. B., Williams, M. E., Morrow, C. J., *J. Org. Chem. 55* (1990) 3544.

[586] Sugai, T., Kakeya, H., Ohta, H., *J. Org. Chem. 55* (1990) 4643.

[587] Moorlag, H., Kellog, R. M., Kloosterman, M., Kaptein, B., Kamphuis, J., Schoemaker, H. E., *J. Org. Chem. 55* (1990) 5878.

[588] Sugai, T., Kakeya, H., Ohta, H., *Tetrahedron 46* (1990) 3463.

[589] Gu, R. L., Sih, C. J., *Tetrahedron Lett. 21* (1990) 3283.

[590] Gu, R. L., Sih, C. J., *Tetrahedron Lett. 21* (1990) 3287.

[591] Ferraboschi, P., Grisenti, P., Manzocchi, A., Santaniello, E., *J. Org. Chem. 55* (1990) 6214.

[592] Jannsen, A. J. M., Klunder, A. J. H., Zwanenburg, B., *Tetrahedron Lett. 31* (1990) 7219.

[593] Nakamura, K., Ishihara, K., Ohno, A., Uemura, M., Nishimura, H., Hayashi, Y., *Tetrahedron Lett. 21* (1990) 3603.

[594] Effenberger, F., Gutterer, B., Ziegler, T., Eckhardt, E., Aicholz, R., *Liebigs Ann. Chem.* (1991) 47.

[595] Mitsuda, S., Yamamoto, H., Umeura, T., Hirohara, H., Nabeshima, S., *Agricol. Biol. Chem. 54* (1990) 2907.

[596] Lu, Y., Miet, C., Kunesch, N., Poisson, J., *Tetrahedron: Asymmetry 1* (1990) 707.

[597] Itoh, T., Ohta, T., *Tetrahedron Lett. 31* (1990) 6407; Itoh, T., Ohta, T., Sano, M., *Tetrahedron Lett. 31* (1990) 6387;

[598] Tsukamato, T., Yamazaki, T., Kitazume, T., *Synth. Commun. 20* (1990) 3181.

[599] Kitazume, T., Yamazaki, T., Ito, K., *Synth. Commun. 20* (1990) 1469.

[600] Yamazaki, T., Okamura, N., Kitazume, T., *Tetrahedron: Asymmetry 1* (1990) 521.

[601] Basavaiah, D., Dharma Rao, P., *Synth. Commun. 20* (1990) 2945.

[602] Hönig, H., Seufer-Wassertahl, P., Weber, H., *Tetrahedron 46* (1990) 3841.

[603] Nideuzak, T. R., Carr, A. A., *Tetrahedron: Asymmetry 1* (1990) 535.

[604] Burgess, K., Jennings, L. D., *J. Am. Chem. Soc. 112* (1990) 7435.

[605] Jeromin, G. E., Weise, S., *Liebigs Ann. Chem.* (1990) 1045.

[606] Ramaswamy, S., Oehlschlager, A. C., *Tetrahedron 47* (1991) 1157.

[607] Kanerva, L. T., Vihanto, J., Pajunen, E., Euranto, E. K., *Acta. Chem. Scand. 44* (1990) 489.

[608] Lutz, D., Güldner, A., Thums, R., Schreier, P., *Tetrahedron: Asymmetry 1* (1990) 783.

[609] Hsu, S. H., Wu, S. S., Wang, Y. F., Wong, C. H., *Tetrahedron Lett. 31* (1990) 6403.

[610] Peterson, R. L., Liu, K. K. C., Rutan, J. F., Chen, L., Wong, C. H., *J. Org. Chem. 55* (1990) 4897.

[611] Chinchilla, R., Najera, C., Pardo, J., Yus, M., *Tetrahedron: Asymmetry 1* (1990) 575.

[612] Nakamura, K., Inoue, Y., Kitayama, T., Ohno, A., *Agric. Biol. Chem. 54* (1990) 1569.

[613] Xie, Z. H., Suemune, H., Sakai, K., *Tetrahedron: Asymmetry 1* (1990) 395.

[614] Bhide, R., Mortezaei, R. Scilimati, A., Sih, C. J., *Tetrahedron Lett. 31* (1990) 4827.

[615] Basahviah, D., Rama Krishna, P., Bharati, T. K., *Tetrahedron Lett. 31* (1990) 4347.

[616] Cregge, R. J., Wagner, E. R., Freedman, J., Margolin, A. L., *J. Org. Chem. 55* (1990) 4237.

[617] Babiak, K. A., Ng, J. S., Dygos, J. H., Weyker, C. L., Wang, Y. F., Wong, C. H., *J. Org. Chem. 55* (1990) 3377.

[618] Laumen, K., Seemayer, R., Schneider, M. P., *Chem. Commun.* (1990) 49.
[619] Macfarlane, E. L. A., Roberts, S. M., Turner, N. J., *Chem. Commun.* (1990) 569.
[620] Naemura, K., Takahashi, N., Tanaka, S., Ueno, M., Chikamatsu, H., *Bull. Chem. Soc., Japan. 63* (1990) 1010.
[621] Seemayer, R., Schneider, M. P., *Chem. Commun.* (1991) 49.
[622] Yamada, H., Sugai. T., Ohta, H., Yoshikawa, S., *Agric. Biol. Chem. 54* (1990) 1579.
[623] Gutman, A. L., Zuobi, K., Bravdo, T., *J. Org. Chem. 55* (1990) 3546.
[624] Sugai, T., Ohsawa, S., Yamada, H., Ohta, H., *Synthesis* (1990) 1112.
[625] Brieva, R., Rebolledo, F., Gotor, V., *Chem. Commun.* (1990) 1386; Gotor, V., Garcia, M. J. Rebolledo, F., *Tetrahedron: Asymmetry 1* (1990) 277.
[626] Onda, M., Motosugi, K., Nakajima, H., *Agric. Biol. Chem. 54* (1990) 3031.
[627] Bianchi, D., Cesti, P., *J. Org. Chem. 55* (1990) 5657.
[628] Monteiro, J., Braun, J., Le Goffic, F., *Synth. Commun. 20* (1990) 315.
[629] Guanti, G., Banfi, L., Ghiron, C., Narisano, E., *Tetrahedron Lett. 32* (1991) 267.
[630] Guanti, G., Banfi, L., Narisano, E., *Tetrahedron: Asymmetry 1* (1990) 721.
[631] Prasad, K., Estermann, H., Chen, C. P., Repic, O., Hardtmann, G. E., *Tetrahedron: Asymmetry 1* (1990) 421.
[632] Baba, N., Yoneda, K., Tahara, S., Iwasa, J., Kaneko, T., Matsuo, M., *Chem. Commun.* (1990) 1281.
[633] Atsuumi, S., Nakano, M., Koike, Y., Tanaka, S., Ohkubo, M., *Tetrahadron Lett, 31* (1990) 1601.
[634] Santaniello, E., Ferraboschi, P., Grisenti, P., *Tetrahedron Lett. 31* (1990) 5657.
[635] Petzold, K., Dahl, H., Skuballa, W., Gottwald, M., *Liebigs Ann. Chem.* (1990) 1087.
[636] Born, M., Tamm, C., *Helv. Chim. Acta 73* (1990) 2242.
[637] Kobayashi, S., Shibata, Y., Shimada, M., Ohno, M., *Tetrahedron Lett. 31* (1990) 1577.
[638] Kobayashi, S., Eguchi, Y., Shimada, M., Ohno, M., *Agric. Biol. Chem. 54* (1990) 1479.
[639] Gourcy, J. G., Dauphin, G., Jeminet, G., *Tetrahedron: Asymmetry 2* (1991) 31.
[640] Grandjean, D., Pale, P., Chuche, J., *Tetrahedron 47* (1991) 1215.
[641] Fuji, K., Kawataba, T., Kiryu, Y., Sugiura, Y., Taga, T., Miwa, Y., *Tetrahedron Lett. 31* (1990) 6663.
[642] Suemune, H., Takahashi, M., Maeda, S., Xie, Z. F., Sakai, K., *Tetrahedron: Asymmetry 1* (1990) 425.
[643] Laumen, K., Schneider, M. P., *Tetrahedron Lett. 25* (1984) 5875.
[644] Murata, M., Ikoma, S., Achiwa, K., *Chem. Pharm. Bull. 38* (1990) 2329.
[645] Matsumoto, K., Tsutsumi, S., Ihori, T., Ohta, H., *J. Am. Chem. Soc. 112* (1990) 9614.
[646] Matsumoto, K., Suzuki, N., Ohta, H., *Tetrahedron Lett. 31* (1990) 7159.
[647] Matsumoto, K., Suzuki, N., Ohta, H., *Tetrahedron Lett. 31* (1990) 7163.

Chapter 4

[1] International Union of Biochemistry, *Enzyme Nomenclature 1984*, New York: Academic Press, 1984.
[2] Whitesides, G. M., Wong, C. H., *Angew. Chem. Int. Ed. Engl. 24* (1985) 617.
[3] Walsh, C., *Enzymatic Reaction Mechanisms*, New York: W. H. Freeman, 1979.
[4] Godfrey, T., Reichelt, J. (Eds.), *Industrial Enzymology*, New York: Nature Press, 1983.
[5] Bergmeyer, H. U., Bergmeyer, J., Grassl, M. (Eds.), *Methods of Enzymatic Analysis*, 3rd Ed., Weinheim: Verlag Chemie, 1986.
[6] Alberti, B. N., Klibanov, A. M., *Enzyme Microb. Technol. 4* (1982) 47.
[7] Kosmann, D. J., "Galactose Oxidases" in : *Copper Proteins Copper Enzymes*, Vol. 2, Londrie, R. (Ed.), Boca Raton, FL: CRC Press, 1984; pp 1-26.
[8] Klibanov, A. M., Alberti, B. N., Marletta, M. A., *Biochem. Biophys. Res. Commun. 108* (1982) 804.
[9] Root, L. R., Durrwachter, J. R., Wong, C. H., *J. Am. Chem. Soc. 107* (1985) 2997.
[10] Tramper, J., Nagel, A., Van der Plaas, H. C., Muller, F., *J. R. Neth. Chem. Soc. 98* (1979) 224.

[11] Pelsy, G., Klibanov. A. M., *Biochim. Biophys. Acta 749* (1983) 352.

[12] Hecht, S. M., Rupprecht, K. M., Jacobs, P. M., *J. Am. Chem. Soc. 101* (1979) 3982.

[13] Frost, G. M., Moss, D. A., in : *Biotechnology, Vol. 7A: Enzyme Technology*, Rehm, H. J., Reed, G. (Eds.), Weinheim: VCH Verlagsgesellschaft, 1987; pp 65-211.

[14] Keilin, D., Hartree, E. F., *Biochem. J. 39* (1945) 293.

[15] Kamal, A., Sattur, P. B., *Tetrahedron Lett. 30* (1989) 1133.

[16] Kamal, A., Rao, A. B., Sattur, P. B., *Tetrahedron Lett. 28* (1987) 2425.

[17] Kamal, A., Rao, A. B., Sattur, P. B., *J. Org. Chem. 53* (1988) 4112.

[18] Schwartz, R. D., Hutchinson, D. B., *Enzyme Microb. Technol. 3* (1981) 361.

[19] Klibanov, A. M., Berman, Z., Alberti, B. N., *J. Am. Chem. Soc. 103* (1981) 6263.

[20] Dordick, J. S., Marletta, M. A., Klibanov, A. M., *Biotechnol. Bioeng. 30* (1987) 31.

[21] Klibanov, A. M., *CHEMTECH 16* (1986) 354.

[22] Takahashi, K., Nishimura, H., Yoshimoto, T., Saito, Y., Inada, Y., *Biochem. Biophys. Res. Commun. 121* (1984) 261.

[23] Neidlemann, S. L., Geigert, J., *Trends Biotechnol. 1* (1983) 21.

[24] Dawson, J. H., Sono, M., *Chem. Rev. 87* (1987) 1255.

[25] Ashley, P. L., Griffin, B. W., *Arch. Biochem. Biophys. 210* (1981) 167.

[26] Neidlemann, S. L, Levine, S. D., *Tetrahedron Lett.* (1968) 4057.

[27] Holland, H. L., *Chem. Rev. 88* (1988) 473.

[28] Colonna, S., Gaggero, N., Manfredi, A. Casella, L., Gullotti, M., *Chem. Commun.* (1988) 1451.

[29] Vick, B. A., Zimmermann, D. C., in: *Biochemistry of Plants*, Vol. 9, New York: Academic Press, 1987; p 53.

[30] Corey, E. J., Albright, J. D., Barton, A. E., Hashomoto, S., *J. Am. Chem. Soc. 102* (1980) 1435.

[31] Corey, E. J., Lansbury, P. Y., *J. Am. Chem. Soc. 105* (1983) 4093.

[32] Laakso, S., *Lipids 17* (1982) 667.

[33] Shak, S., Perez, H. D., Goldstein, I. M., *J. Biol. Chem.*, 14, (1983) 14948.

[34] Corey, E. J., Su, W. G., Cleaver, M. B., *Tetrahedron Lett. 30* (1989) 4181.

[35] Iacazio, G., Langrand, G., Baratti, J., Buono, G., Triantaphylides, C., *J. Org. Chem. 55* (1990) 1690.

[36] Pandey, G., Muralikrishna, C., Bhalerao, U. T., *Tetrahedron 45* (1989) 6867.

[37] Zhang, P., Kyler, K. S., *J. Am. Chem. Soc. 111* (1989) 9241.

[38] May, S. W., Padgette, S. R., *Biotechnology 1* (1983) 677.

[39] May, S. W., Phillips, R. S., *J. Am. Chem. Soc. 102* (1980) 5981.

[40] Sirimanne, S. R., May, S. W., *J. Am. Chem. Soc. 110* (1988) 7650.

[41] Baldwin, J. E., Abraham, E. P., *Nat. Prod. Reports 5* (1988) 129.

[42] Baldwin, J. E., Adlington, R. M., Mess, N., *Tetrahedron 45* (1989) 2841.

[43] Baldwin, J. E., Adlington, R. M., King, L. G., Parisi, M. F., Sobey, W. J., Sutherland, J. D., Ting, H. H., *Chem. Commun.* (1988) 1635.

[44] Katopodis, A. G., Smith, H. A., May, S. W., *J. Am. Chem. Soc. 110* (1988) 897.

[45] Fujimoto, Y., Chen, C. S., Szeleczky, Z., Di Tullio, D., Sih, C. J., *J. Am. Chem. Soc. 104* (1982) 4718.

[46] Battersby, A. R., *Chem. Br.* (1984) 611.

[47] Jones, J. B., *Tetrahedron 42* (1986) 3351.

[48] Rétey, J., Robinson, J. A., *Stereospecificity in Organic Chemistry and Enzymology* (Monographs in Modern Chemistry, Vol. 13), Weinheim: Verlag Chemie, 1982.

[49] Boyer, P. D. (Ed.), *The Enzymes*, 3rd Ed., Vol. 11, Part A, New York: Academic Press, 1975.

[50] Everse, J., Anderson, B. ,You, K. S. (Eds.), *The Pyridine Nucleotide Coenzymes*, New York: Academic Press, 1982.

[51] Dolphin, D., Avramovic, D. (Eds.), *Coenzymes and Cofactors*, New York: Wiley, 1986.

[52] Jones, J. B., Beck, B. F, "Asymmetric Syntheses and Resolutions Using Enzymes" in: *Applications of Biochemical Systems in Organic Chemistry*, Jones, J. B., Sih, C. J., Perlman, D. (Eds.), New York: Wiley, 1976; pp 107-401.

[53] Cornish-Bowdwen, A., Cardenas, M. L., "Chemistry of Enzymes" in: *Biotechnology, Vol 7A: Enzyme Technology*, Rehm, H., Reed, G. (Eds.), Weinheim: VCH Verlagsgesellschaft, 1986; pp 3-33.

[54] Sih, C. J., Chen, C. S., *Angew. Chem., Int. Ed. Engl. 23* (1984) 570.

[55] Lee, L. G., Whitesides, G. M., *J. Org. Chem. 51* (1986) 25.

[56] Cornforth, J. W., Ryback, G., Popjak, G., Donniger, C., Schroepfer, G., *Biochem. Biophys. Res. Commun. 9* (1962) 371.

[57] You, K., Arnold, L. J., Allison, W. S., Kaplan, N. O., *Trends Biochem. Sci. 3* (1978) 265.

[58] Prelog, V., *Pure Appl. Chem. 9* (1964) 179.

[59] Bowen, R., Pugh, S., *Chem. Ind.* (1985) 323.

[60] Wong, C. H., Whitesides, G. M., *Aldrichim. Acta 16* (1984) 27.

[61] Simon, H., Bader, J., Günther, H., Neumann, S., Thanos, J., *Angew. Chem., Int. Ed. Engl. 24* (1985) 539.

[62] Gorton, L., *J. C. S. Faraday Trans I 82* (1986) 1245.

[63] Nakamura, K., Aizawa, M., Miyawaki, O., *Electro-Enzymology Coenzyme Regeneration*, Berlin: Springer Verlag, 1988.

[64] Willner, I., Mandler, D., *Enzyme Microb. Technol. 11* (1989) 467.

[65] Wandrey, C., Buckmann, A. F., Kula, M. R., *Biotechnol. Bioeng. 23* (1981) 1789.

[66] Shaked, Z., Whitesides, G. M., *J. Am. Chem. Soc. 102* (1980) 7104.

[67] Tischer, W., Tiemeyer, W., Simon, H., *Biochemie 62* (1980) 331.

[68] Wong, C. H., Drueckhammer, D. G., Sweers, H. M., *J. Am. Chem. Soc. 107* (1985) 4028.

[69] Wong, C. H., Whitesides, G. M., *J. Am. Chem. Soc. 103* (1981) 4890.

[70] Wong, C. H., Gordon, J., Cooney, C. L., Whitesides, G. M., *J. Org. Chem. 46* (1981) 4676.

[72] Wang, S. S., King, C. K., *Adv. Biochem. Eng. 12* (1979) 119.

[73] Wong, C. H., Whitesides, G. M., *J. Org. Chem. 47* (1982) 2816.

[74] Wong, C. H., Daniels, W. H., Orme-Johnson, W. H., Whitesides, G. M., *J. Am. Chem. Soc. 103* (1981) 6227.

[75] Klibanov, A. M., Pugilski, A. V., *Biotechnol. Lett. 2* (1980) 445.

[76] Ergerer, P., Simon, H., Tanaka, A., Fukui, S., *Biotechnol. Lett. 4* (1982) 489.

[77] Payen, B., Segui, M., Monsan, P., Schneider, K., Friedrich, C. G., Sclegel, H. G., *Biotechnol. Lett. 5* (1983) 463.

[78] Shaked, Z., Barber, J. J., Whitesides, G. M., *J. Org. Chem. 46* (1981) 4100.

[79] Mandler, D., Willner, I. J., *J. Am. Chem. Soc. 106* (1984) 5352; Mandler, D., Willner, I. J., *J. Chem. Soc., Perkin II* (1986) 805.

[80] Franke, M., Steckham, E., *Angew. Chem., Int. Ed. Engl. 27* (1988) 265.

[81] Steckham, E., *Top. Curr. Chem. 142* (1987) 1.

[82] Jones, J. B., Sueddon, D. W., Higgins, W., Lewis, A. J., *Chem. Commun.* (1972) 856.

[83] Lehninger, A. L., *Meth. Enzymol. 3* (1957) 885.

[84] Keinan, E., Hafeli, E. K., Seth, K. K., Lamed, R., *J. Am. Chem. Soc. 108* (1986) 162.

[85] Lee, L. G., Whitesides, G. M., *J. Am. Chem. Soc. 107* (1985) 6999.

[86] Chenault, H. K., Whitesides, G. M., *Bioorg. Chem. 17* (1989) 400.

[87] Lemiere, G. L., Lepoivre, J. J., Aldelweireldt, F. C., *Tetrahedron Lett. 26* (1985) 4527.

[88] Chambers, R. P., Walle, E. M., Baricos, W. H., Cohen, W., *Enz. Eng. 3* (1978) 363.

[89] Juillard, M., Le Petit, J., Ritz, P., *Biotechnol. Bioeng. 28* (1986) 1774.

[90] Handman, J., Harriman, A., Porter, G., *Nature (London) 307* (1984) 534.

[91] Komoshinski, J., Steckhan, E., *Tetrahedron Lett. 29* (1988) 3299.

[92] Itoh, S., Kinugawa, M., Mita, N., Oshiro, Y., *Chem. Commun.* (1989) 694.

[93] Jones, J. B., Taylor, K. E., *Can. J. Chem. 54* (1976) 2969.

[94] Jones, J. B., Taylor, K. E., *Chem. Commun.* (1973) 205.

[95] Walsh, C. T., Chen, Y. C. J., *Angew. Chem. Int. Ed. Engl. 27* (1988) 333.

[96] Abril, O., Ryerson, C. C., Walsh, C., Whitesides, G. M., *Bioorg. Chem. 17* (1989) 41.

[97] Tjerneld, F., Johanson, G., Joelsson, M., *Biotechnol. Bioeng. 30* (1987) 809.

[98] Kim, M. J., Whitesides, G. M., *J. Am. Chem. Soc. 110* (1988) 2959.

[99] Luyten, M. A., Bur, D., Wynn, H., Parris, W., Gold, M., Friesen, J. D., Jones, J. B., *J. Am. Chem. Soc. 111* (1989) 6800.

[100] Bally, C., Leuthardt, F., *Helv. Chim. Acta 53* (1970) 732.

[101] Jones, J. B., *F.E.C.S. Int. Conf. Chem. Biotechnol. Biol. Act. Nat. Prod.* [Proc.], 3rd, 1985, Vol 1, Weinheim: VCH Verlagsgesellschaft, 1987; p 18.

[102] Jones, J. B., Jakovac, I. J., *Org. Synth. 63* (1984) 10.

[103] Keinan, E., Seth, K. K., Lamed, R., *J. Am. Chem. Soc. 108* (1986) 3474.

[104] Bryant, F. O., Wiegel, J., Ljungdahl, L. G., *Appl. Environ. Microbiol.* (1988) 460.

[105] Rella, R., Raia, C. A., Pensa, M., Pisani, F. M., Gambacorta, A., Rosa, M. De, Rossi, M., *Eur. J. Biochem. 167* (1987) 475.

[106] Yamada, H., Shimizu, S., *Angew. Chem. Int. Ed. Engl. 27* (1988) 622.

[107] Jones, J. B., Jakovac, I. J., *Can. J. Chem. 60* (1982) 19.

[108] Willaert, J. J., Lemiere, G. L., Joris, L. A., Lepoivre, J. A., Alderweireldt, F. C., *Bioorg. Chem. 16* (1988) 223.

[109] Alexandre, S., Butelet, I., Vincent, J. C., *Enzyme Microb. Technol. 10* (1988) 479.

[110] Pham, V. T., Phillips, R. S., Ljungdahl, L. G., *J. Am. Chem. Soc. 111* (1989) 1935.

[111] Skerker, P. S., Clark, D. S., *Biotechnol. Bioeng. 32* (1988) 148.

[112] Yamada, H., Ruyno, K., Nagasawa, T., Ehomoto, K., Watanabe, Y., *Agric. Biol. Chem. 50* (1986) 2859; Ruyno, K., Nagasawa, T., Yamada, H., *Agric. Biol. Chem. 52* (1988) 1813.

[113] Semenov, A. N., Cerovsky, V., Titov, M. I., Martinek, K., *Collect. Czech. Chem. Commun. 53* (1988) 2963.

[114] Matos, J. R., Wong, C. H., *J. Org. Chem. 51* (1986) 2388.

[116] Hilhorst, R., Laane, C. Veeger, C., *FEBS Lett. 31* (1983) 159.

[117] Van Elsacker, P. C., Lemire, G. L., Lepoivre, J. A., Alderweireldt, F. C., *Bioorg. Chem. 17* (1989) 28.

[118] Simon, H., Günther, M., Bader, J., Tischer, W., *Angew. Chem., Int. Ed. Engl. 20* (1981) 861.

[119] Taschner, M. J., Black, D. J., *J. Am. Chem. Soc. 110* (1988) 6892.

[120] Skopan, H., Günther, H., Simon, H., *Angew. Chem. 99* (1987) 139.

[121] Vasic-Racki, D., Jonas, H., Wandrey, C., Hummel, W., Kula, M. R., *Appl. Microb. Biotechnol. 31* (1989) 215.

[122] Yamazaki, Y., Maeda, H., *Agric. Biol. Chem. 50* (1986) 2621.

[123] Nakamura, K., Yoneda, T., Miyai, T., Ushio, K., Oka, S., Ohno, A., *Tetrahedron Lett. 29* (1988) 2453.

[124] Oshima, T., Wandrey, C., Conrad, D., *Biotechnol. Bioeng. 34* (1989) 394.

[125] Vidal-Cros, A., Gaudry, M., Marquet, A., *J. Org. Chem. 54* (1989) 498.

[126] Furuyoshi, S., Kawataba, N., Nagata, S., Tanaka, H., Soda, K., *Agric. Biol. Chem. 53* (1989) 3075.

[127] Wichmann, R., Wandrey, C., *Biotechnol. Bioeng. 23* (1981) 2789.

[128] Wandrey, C., Wichmann, R., *Biotechnol. 1* (1987) 85.

[129] Kula, M. R., Wandrey, C., *Meth. Enzymol. 136* (1987) 9.

[130] Zamir, L. D., Sauriol, F., Nguyen, C. D., *Tetrahedron Lett. 28* (1987) 3057.

[131] Grünwald, J., Wirtz, B., Scollar, M., Klibanov, A. M., *J. Am. Chem. Soc. 108* (1986) 6734.

[132] Boland, W., Niedermayer, U., *Synthesis* (1987) 28.

[133] Yamazaki, Y., Hosono, K., *Tetrahedron Lett. 30* (1989) 5313.

[134] Wong, C. H., Matos, J. R., *J. Org. Chem. 50* (1985) 1992.

[135] Jones, J. B., Schwartz, H. M., *Can. J. Chem. 59* (1981) 1574.

[136] Belan, A., Bolte, J., Fauve, A., Gourcy, J. G., Veschambre, H., *J. Org. Chem. 52* (1987) 256.

[138] Irwin, A. J., Jones, J. B., *J. Am. Chem. Soc. 99* (1977) 556.

[139] Ng, G. S. Y., Yuan, L. C., Jakovac, I. J., Jones, J. B., *Tetrahedron 40* (1984) 1235.

[140] Jakovac, I. J., Ng, G., Lok, K. P., Jones, J. B., *Chem. Commun.* (1980) 515.

[141] Jakovac, I. J., Goodbrandt, H. B., Lok, K. P., Jones, J. B., *J. Am. Chem. Soc. 104* (1982) 4659.

[142] Goodbrandt, H. B., Jones, J. B., *Chem. Commun.* (1977) 469.

[143] Lok, L. K. P., Jakovac, I. J., Jones, J. B., *J. Am. Chem. Soc. 107* (1985) 2521.

[144] Jones, J. B., Francis, C. J., *Can. J. Chem. 62* (1984) 2578.

[145] Bridges, A. J., Roman, P. S., Ng, G. S., Jones, J. B., *J. Am. Chem. Soc. 106* (1984) 1461.

[146] Jones, J. B., Finch, M. A. W., *Can. J. Chem. 60* (1982) 2007.

[147] Ikeda, T., Hutchinson, C. R., *J. Org. Chem. 49* (1984) 2837.

[148] Riva, S., Bovara, R., Pasta, P., Carrea, G., *J. Org. Chem. 51* (1986) 2902.

[149] Carrea, G., Bovara, R., Cremonesi, P., Lodi, R., *Biotechnol. Bioeng. 26* (1984) 560.

[151] Helmchen-Zeier, R. E., Ph.D. Thesis, 4991, ETH, Zürich (1973) .

[152] Takemura, T., Jones, J. B., *J. Org. Chem. 48* (1983) 791.

[153] Haslegrawe, J. A., Jones, J. B., *J. Am. Chem. Soc. 104* (1982) 4666.

[154] Irwin, A. J., Jones, J. B., *J. Am. Chem. Soc. 98* (1976) 8476.

[155] Nakazaki, M., Chikamatsu, H., Fujii, T., Sasaki, Y., Ao, S., *J. Org. Chem. 48* (1983) 4337.

[156] Krawczyk, A. R., Jones, J. B., *J. Org. Chem. 54* (1989) 1795.

[157] Lam, L. K. P., Gair, I. A., Jones, J. B., *J. Org. Chem. 53* (1988) 1611.

[158] Graves, J. M. H., Clarck, A., Ringold, H. J., *Biochemistry 4* (1965) 2655.

[159] Baxter, A. D., Roberts, S. M., *Chem. Ind.* (1986) 510.

[160] Butt, S., Davies, H. G., Dawson, M. J., Lawrence, G. D., Leaver, J., Roberts, S. M., Turner, M. K., Wakefield, B. J., Wall, W. F., Winders, J. A., *Tetrahedron Lett. 26* (1985) 5077.

[161] Davies, H. G., Gartenmann, T. C. C., Leaver, J., Roberts, S. M., Turner, M. K., *Tetrahedron Lett. 27* (1986) 1093.

[162] Nakazaki, M., Chikamatsu, H., Naemura, K., Suzuki, T., Iwasaki, M., Sasaki, Y., Fujii, T., *J. Org. Chem. 46* (1981) 2726.

[163] Nakazaki, M., Chikamatsu, H., Sasaki, Y., *J. Org. Chem. 48* (1983) 2056.

[164] Dawies, J., Jones, J. B., *J. Am. Chem. Soc. 101* (1979) 5405.

[165] Prelog, V., *Pure Appl. Chem. 9* (1964) 119.

[166] Dodds, D. R., Jones, J. B., *Chem. Commun.* (1982) 1080.

[167] Jones, J. B., Dodds, D. R., *Can. J. Chem. 65* (1987) 2397.

[168] Nakazaki, M., Chikamatsu, H., Taniguchi, M., *Chem. Lett.* (1982) 1761.

[169] Nakajima, N., Esaki, N., Soda, K., *Chem. Commun.* (1990) 947.

[170] Pandey, G., Muralikrishna, C., Bhalerao, B. H., *Tetrahedron Lett. 21* (1990) 3771.

[171] Shen, G. J., Wang, Y. F., Bradshaw, C., Wong, C. H., *Chem. Commun.* (1990) 677.

[172] Nakamura, K., Shigara, T., Miyai, T., Ohno, A., *Bull. Chem. Soc., Japan. 63* (1990) 1735.

[173] Asano, Y., Yamada, A., Kato, Y., Yamaguchi, K., Hibino, Y., Hirai, K., Kondo, K., *J. Org. Chem. 55* (1990) 5567.

[174] Yamazaki, Y., Uebayashi, M., Someya, J., Hosono, K., *Agric. Biol. Chem. 54* (1990) 1781.

Chapter 5

[1] Harrison J. S., Rose, A. H. (Eds.), *The Yeasts*, 2nd Ed., New York: Academic Press, 1987.

[2] Pederson, C. S., *Microbiology of Food Fermentations*, Westport: AVI Publishing Co., 1971; Wood, B. J. B. (Ed.), *Microbiology of Fermented Foods*, London-New York: Elsevier, 1983.

[3] Gramatica, P., *Chimicaoggi* (1988) 17.

[4] Servi, S., *Synthesis* (1990) 1.

[5] Levene, P. A., Walti, A., *Org. Synth., Coll. Vol. 2* (1943) 545.

[6] Seebach, D., Sutter, M. A., Weber, R. M., Züger, M. F., *Org. Synth. 63* (1985) 1.

[7] Trivedi, N. B., Jacobson, G. K., *Prog. Ind. Microbiol. 23* (1986) 45.

[8] Simon, H., Bader, J., Günther, H., Neumann, S., Thanos, J., *Angew. Chem. Int. Ed. Engl. 24* (1985) 539.

[9] Deol, B. S., Ridley, D. D., Simpson, G. W., *Aust. J. Chem. 29* (1976) 2459.

[10] Kosmol, H., Kieslich, K., Vössing, R, Koch, H. J., Petzold, K., Gibian, H., *Liebigs Ann. Chem. 701* (1967) 199.

[11] Takaishi, Y., Yang, Y. L., Di Tullio, D., Sih, C. J., *Tetrahedron Lett. 23* (1982) 5489.

[12] Raddatz, P., Radunz, H. E., Schneider, G., Schwartz, H., *Angew. Chem., Int. Ed. Engl. 27* (1988) 426.

[13] Manzocchi, A., Casati, R., Fiecchi, A., Santaniello, E., *J. Chem. Soc., Perkin I* (1987) 2753.

[14] Bucciarelli, M., Forni, A., Moretti, I., Torre, G., *Synthesis* (1983) 897.

[15] Seebach, D., Roggo, S., Maetzke, T., Braunschweiger, H., Cercus, J., Kriger, M., *Helv. Chim. Acta 70* (1987) 1605.

[16] Fadnavis, N. W., Reddy, N. P., Bhalerao, U. T., *J. Org. Chem. 54* (1989) 3281.

[17] Hochköppler, A., Pfamatter, N., Luisi, P. L., *Chimia 43* (1989) 348.

[18] Haag, T., Arslan, T., Seebach, D., *Chimia 43* (1989) 351.

[19] Nakamura, K., Inoue, K., Ushio, K., Oka, S., Ohno, A., *J. Org. Chem. 53* (1988) 2589.

[20] Naoshima, Y., Nishiyama, T., Munakata, Y, *Chem. Lett.* (1989) 1517.

[21] Sih, C. J., Chen, C. S., *Angew. Chem. Int. Ed. Engl. 23* (1984) 570.

[22] Neuberg, C., *Adv. Carbohydr. Chem. 4* (1949) 75.

[23] Mc Leod, R., Prosser, H., Fiskentscher, L., Lányi, J., Mosher, H. S., *Biochemistry 3* (1964) 383.

[24] Shieh, W. R., Gopalan, A. S., Sih, C. J., *J. Am. Chem. Soc. 107* (1985) 2993.

[25] Sih, C. J., Zhou, B. N., Gopalan, A. S., Shieh, W. R., Van Middlesworth, F., "Selectivity - A Goal for Synthetic Efficiency" in : *Workshop Conferences Hoechst*, Vol. 14, Bartmann, W., Trost, B. (Eds.), Weinheim: VCH Verlagsgesellschaft, 1983.

[26] Chen, C. S., Zhou, B. M., Girdaukas, G., Shieh, W. R., Van Middlesworth, F., Gopalan, A. S., Sih, C. J., *Bioorg. Chem. 12* (1984) 98.

[27] Van Eys, J., Kaplan, N. O., *J. Am. Chem. Soc. 79* (1956) 2782.

[28] Wong, C. H., Drueckhammer, D. G., Sweers, H. M., *J. Am. Chem. Soc. 107* (1985) 4028.

[29] Furuichi, A., Akita, H., Matsukura, H., Oishi, T., Horikoshi, K., *Agric. Biol. Chem. 49* (1985) 2563.

[30] Nakamura, K., Miyai, T., Kawai, Y., Nakajima, N., Ohno, A., *Tetrahedron Lett. 31* (1990) 1159.

[31] Heidlas, J., Engel, K. H., Tressl, R., *Eur. J. Biochem. 172* (1988) 633.

[32] Lagunas, R., Dominguez, C., Busturia, A., Saez, M. J., *J. Bacteriol. 152* (1982) 19.

[33] Augermaier, L., Bader, J., Simon, H., *Hoppe Seyler's Z. Physiol. Chem. 362* (1981) 33; Bader, J., Kim, M. A., Simon, H., *ibid 362* (1981) 809.

[34] Kometani, T., Kitatsuji, E., Matsuno, R., *Chem. Lett.* (1989) 1465.

[35] Ehrler, J., Giovanni, F., Lamatsch, B., Seebach, D., *Chimia 40* (1986) 172.

[36] Sakai, T., Nakamura, T., Fukuda, K., Amano, E., Utaka, M., Takeda, A., *Bull. Chem. Soc., Japan 59* (1986) 3185.

[37] Fuganti, C., Grasselli, P., Spreafico, F., Zirotti, C., *J. Org. Chem. 49* (1984) 543.

[38] Wipf, B., Kupfer, E., Bertazzi, R., Leuenberger, H. G. W., *Helv. Chim. Acta 66* (1983) 485.

[39] Nakamura, K., Higaki, M., Ushio, K., Oka, S., Ohno, A., *Tetrahedron Lett. 26* (1985) 4213.

[40] Zhou, B. N., Gopalan, A. S., Van Middlesworth, F., Shieh, W. R., Sih, C. J., *J. Am. Chem. Soc. 105* (1983) 5925.

[41] Nakamura, K., Kawai, Y., Oka, S., Ohno, A., *Tetrahedron Lett. 30* (1989) 2245.

[42] Nakamura, K., Inoke, K., Ushio, K., Oka, S., Ohno, A., *Chem. Lett.* (1987) 679.

[43] Nakamura, K., Kawai, Y., Oka, S., Ohno, A., *Bull. Chem. Soc., Japan 62* (1989) 875.

[44] Nakamura, K., Kawai, Y., Ohno, A., *Tetrahedron Lett. 31* (1990) 267.

[45] Glänzer, B. I., Faber, K., Griengl, H., *Tetrahedron 43* (1987) 5791.

[46] Christen, M., Crout, D. H. G., *Chem. Commun.* (1988) 264.

[47] Manzocchi, A., Fiecchi, A., Santaniello, E., *J. Org. Chem. 53* (1988) 4405.

[48] Gillois, J., Buisson, D., Azerad, R., Jaouen, G., *Chem. Commun.* (1988) 1224.

[49] Aragozzini, F., Maconi, E., Pontenza, C., Scolastico, C, *Synthesis* (1989) 225.

[50] Fuganti, C., Grasselli, P., Casati, P., Carmeno, M., *Tetrahedron Lett. 26* (1985) 101.

[51] Hirama, M., Nakamine, T., Ito, S., *Chem. Lett.* (1986) 1381.

[52] Hirama, M., Shimizu, M., Iwashita, M., *Chem. Commun.* (1983) 599.

[53] Mori, K, Mori, H., Sugai, T., *Tetrahedron 41* (1985) 919.

[54] Nakamura, K., Ushio, K., Oka, S., Ohno, A., Yasui, S., *Tetrahedron Lett. 25* (1984) 3979.

[55] Gopalan, A. S., Jacobs, H. K., *Tetrahedron Lett. 30* (1989) 5705.

[56] Nakamura, K., Miyai, T., Nozaki, K., Ushio, K., Oka, S., Ohno, A., *Tetrahedron Lett. 27* (1986) 3155.

[57] Nakamura, K., Miyai, T., Ushio, K., Oka, S., Ohno, A., *Bull. Chem. Soc., Japan 61* (1988) 2089.

[58] Nakamura, K., Miyai, T., Nagar, A., Oka, S., Ohno, A., *Bull. Chem. Soc., Japan 62* (1989) 1179.

[59] Ohta, H., Kobayashi, N., Sugai, T., *Agric. Biol. Chem. 54* (1990) 489.

[60] Sato, T., Maeno, H., Noro, T., Fujisawa, T., *Chem. Lett.* (1988) 1739.

[61] Neuberg, C., Kerb-Etzdorf, E., *Biochem. Z. 92* (1918) 96.

[62] Neuberg, C., Vercellone, A., *Biochem. Z. 279* (1935) 140.

[63] Levene, P. A., Walti, A., Haller, H. L., *J. Biol. Chem. 71* (1927) 466.

[64] Poppe, L., unpublished results.

[65] Tsuboi, S., Kohara, N., Doi, K., Utaka, M., Takeda, A., *Bull. Chem. Soc., Japan 61* (1988) 3205.

[66] Kawahara, K., Matsumoto, M., Hashimoto, H., Miyano, S., *Chem. Lett.* (1988) 1163.

[67] Top, S., Jaouen, G., Gillois, J., Baldoli, C., Maiorana, S., *Chem. Commun.* (1988) 1284.

[68] Neuberg, C., Levite, A., *Biochem. Z. 91* (1918) 257.

[69] Färber, E., Nord, F. F., Neuberg, C., *Biochem. Z. 112* (1920) 313.

[70] Levene, P. A., Walti, A., *J. Biol. Chem. 98* (1932) 735.

[71] Zervinka, O., Hub, L., *Collect. Czech., Chem. Commun. 31* (1966) 2615.

[72] Le Drian, C., Greene, A. E., *J. Am. Chem. Soc. 104* (1982) 5473.

[73] Fuganti, C., Grasselli, P., Servi, S., Spreafico, F., Zirotti, C., Casati, P., *J. Chem. Res.(S)* (1984) 112.

[74] Nakamura, K., Inoue, Y., Shibahara, J., Oka, S., Ohno, A., *Tetrahedron Lett. 29* (1988) 4769.

[75] Hafner, T, Reissig, H. U., *Liebigs Ann. Chem.* (1989) 937.

[76] Fujisawa, T., Hayashi, H., Kishioka, Y., *Chem. Lett.* (1987) 129.

[77] Pondaven-Raphalen, A., Sturtz, G., *Bull. Soc. Chim. France* (1978) 215.

[78] Iriuchijima, S., Kojima, N., *Agric. Biol. Chem. 42* (1978) 451.

[79] Kozikowski, A. P., Mugrage, D. B., Li, C. S., Felder, L., *Tetrahedron Lett. 27* (1986) 4817.

[80] Manzocchi, A., Fiecchi, A., Santaniello, E., *Synthesis* (1987) 1007.

[81] Eichberger, G., Faber, K., Griengl, H., *Monatsch. Chem. 116* (1985) 1233.

[82] Slydatk, C., Stoffregen, A., Wattke, F., Tacke, R., *Biotechnol. Lett. 10* (1988) 731.

[83] Yamazaki, Y., Hosono, K., *Agric. Biol. Chem. 52* (1988) 3239.

[84] Ghiringhelli, D., *Tetrahedron Lett. 24* (1983) 287.

[85] Deschenaux, P. F., Kallimopoulos, T., Jacot-Guillarmod, A., *Helv. Chim. Acta 72* (1989) 1259.

[86] Guanti, G., Banfi, L., Narisano, E., *Tetrahedron Lett. 27* (1986) 3547.

[87] Takeshita, M., Terada, K., Akutsu, N., Yoshida, S., Sato, T., *Heterocycles 26* (1987) 3051.

[88] Rasor, P., Rüchardt, C., *Chem. Ber. 122* (1989) 1375.

[89] Bianchi, G., Comi, G., Venturini, I., *Gazz. Chim. Ital. 114* (1984) 285.

[90] Bucciarelli, M., Forni, A., Moretti, I., Torre, G., *Chem. Commun.* (1978) 456.

[91] Guanti, G., Banfi, L., Guaragna, A., Narisano, E., *Chem. Commun.* (1986) 138.

[92] Guette, J. P., Spassky, N., *Bull. Soc. Chim. France* (1972) 4217.

[93] Barry, J., Kagan, H. B., *Synthesis* (1981) 453.

[94] Ridley, D. D., Stralow, M., *Chem. Commun.* (1975) 400.

[95] Fujisawa, T., Itoh, T., Nakai, M., Sato, T., *Tetrahedron Lett. 26* (1985) 771.

[96] Sato, T., Okumura, Y., Itai, J., Fujisawa, T., *Chem. Lett.* (1988) 1537.

[97] Sato, T., Mizutani, T., Okumura, Y., Fujisawa, T., *Tetrahedron Lett. 30* (1989) 3701.

[98] Tanikaga, R., Hosoya, K., Kaji, A., *Synthesis* (1987) 389.

[99] Fujisawa, T., Kojima, E., Itoh, T., Sato, T., *Chem. Lett.* (1985) 1751.

[100] Kitazume, T., Ishikawa, N., *Chem. Lett.* (1983) 237.

[101] Belan, A., Bolte, J., Fauve, A., Gourcy, J. G., Veschambre, H., *J. Org. Chem. 52* (1987) 256.

[102] Nakamura, K., Kitayama, T., Inoue, Y., Ohno, A., *Bull. Chem. Soc., Japan 63* (1990) 91.

[103] Bernardi, R., Cardillo, R., Ghiringhelli, D., de Pava, O. V., *J. Chem. Soc., Perkin I* (1987) 1607.

[104] Bernardi, R., Cardillo, R., Ghiringehelli, D., *Chem. Commun.* (1984) 460.

[105] Chikashita, H., Kittaka, E., Kimura, Y., Itoh, K., *Bull. Chem. Soc., Japan 62* (1989) 833.

[106] Noda, Y., Kikuchi, M., *Chem. Lett.* (1989) 1755.

[107] Crumbie, R. L., Ridley, D. D., Simpson, G. W., *Chem. Commun.* (1977) 315.

[108] Fujisawa, T., Fujimura, A, Sato, T., *Bull. Chem. Soc., Japan 61* (1988) 1273.

[109] Hirai, K., Naito, A., *Tetrahedron Lett. 30* (1989) 1107.

[110] Kertesz, D. J., Kluge, A. F., *J. Org. Chem. 53* (1988) 4962.

[111] Xie, Z. F., Suemune, H., Nakamura, I., Sakai, K., *Chem. Pharm. Bull. 35* (1987) 4454.

[112] Ticozzi, C., Zanarotti, A., *Tetrahedron Lett. 29* (1988) 6167.

[113] Lowe, G., Swain, S., *J. Chem. Soc., Perkin I* (1985) 391.

[114] Dawson, M.J., Lawrence, G., Lilley, G., Todd, M., Noble, D., Green, S., Roberts, S., Wallace, T., Newton, R., Carter, M., Hallett, P., Paton, J., Reynolds, D., Young, S., *J. Chem. Soc., Perkin I* (1983) 2119.

[115] Newton, R. F., Paton, J., Reynolds, D. P., Young, S., Roberts, S. M., *Chem. Commun.* (1979) 908.

[116] Kurosawa, Y., Shimojima, H., Osawa, Y., *Steroids. Suppl. 1* (1965) 185.

[117] Eignerova, L., Prochazka, Z., *Coll. Czech. Chem. Comm. 39* (1974) 2828.

[118] von Falkenhausen, F., *Biochem. Z. 219* (1930) 241.

[119] Fuganti, C., Grasselli, P., Servi, S., Spreafico, F., Zirotti, C., Casati, P., *J. Org. Chem. 49* (1984) 4087.

[120] Ramaswamy, S., Oehlschlager, A. C., *J. Org. Chem. 54* (1989) 255.

[121] Terashima, S., Tamoto, K., *Tetrahedron Lett. 23* (1982) 3715.

[122] Fronza, G., Fuganti, C., Pedrocchi-Fantoni, G., Servi, S., *J. Org. Chem. 52* (1987) 1141.

[123] Fronza, G., Fuganti, C., Grasselli, P., Servi, S., *Tetrahedron Lett. 27* (1986) 4363.

[124] Fronza, G., Fuganti, C., Grasselli, P., Servi, S., *J. Org. Chem. 52* (1987) 2086.

[125] Fronza, G., Fuganti, C., Grasselli, P., Servi, S., *Tetrahedron Lett. 26* (1985) 4961.

[126] Fuganti, C., in : *Enzymes as Catalysts in Organic Synthesis*, Schneider, M. (Ed.), Dordrecht: D. Reidel Publishing, 1986; pp 3-17.

[127] Fuganti, C., Servi, S., in : *Bioflavour '87*, Schreier, P. (Ed.), Berlin: de Gruyter, 1988; p 555.

[128] Fuganti, C., Grasselli, P., *Chem. Commun.* (1978) 299.

[129] Fronza, G., Fuganti, C., Grasselli, P., *Chem. Commun.* (1980) 442.

[130] Fuganti, C., Grasselli, P., Servi, S., *Chem. Commun.* (1982) 1285.

[131] Pedrocchi-Fantoni, G., Servi, S., *J. Chem. Res., (S)* (1986) 199.

[132] Fuganti, C., Grasselli, P., Servi, S., Zirotti, C., *Tetrahedron Lett. 23* (1982) 4269.

[133] Servi, S., *Tetrahedron Lett. 24* (1983) 2023.

[134] Fronza, G., Fuganti, C., Högberg, H. E., Pedrocchi-Fantoni, G., Servi, S., *Chem. Lett.* (1988) 385.

[135] Fujise, S., *Biochem. Z. 236* (1931) 241.

[136] Neuberg, C., Nord, F. F., *Ber. 52* (1919) 2248.

[137] Neuberg, C., Lustig, H., Cagan, R. N., *Arch. Biochem. 1* (1943) 391.

[138] Deschamps, I., King, W. S., Novd, F. F., *J. Org. Chem. 14* (1949) 184.

[139] Takeshita, M., Sato, T., *Chem. Pharm. Bull. 37* (1989) 1085.

[140] Chénevert, R., Thiboutot, S., *Chem. Lett.* (1988) 1191.

[141] Bel-Rhlid, R., Fauve, A., Veschambre, H., *J. Org. Chem. 54* (1989) 3221.

[142] Fujisawa, T., Kojima, E., Sato, T., *Chem. Lett.* (1987) 2227.

[143] Fujisawa, T., Kojima, E., Itoh, T., Sato, T., *Tetrahedron Lett. 26* (1985) 6089.

[144] Ohta, H., Ozaki, K., Tsuchihashi, G., *Agric. Biol. Chem. 50* (1986) 2499.

[145] Bolte, J., Gourcy, J. G., Veschambre, H., *Tetrahedron Lett. 27* (1986) 565.

[146] Fauve, A., Veschambre, H., *J. Org. Chem. 53* (1988) 5215.

[147] Dauphin, G., Fauve, A., Veschambre, H., *J. Org. Chem. 54* (1989) 2239.

[148] Ohta, H., Ozaki, K., Tsuchihashi, G., *Chem. Lett.* (1987) 2225.

[149] Brooks, D. W., Grothaus, P. G., Irwin, W. L., *J. Org. Chem. 47* (1982) 2820.

[150] Brooks, D. W., Mazdiyashni, H., Grothaus, P. G., *J. Org. Chem. 52* (1987) 3223.

[151] Gulaya, V. E., Ananchenko, S. N., Torgov, I. V., Koshcheyenko, K. A., Bychkova, G. G., *Bioorg. Khim. 5* (1979) 768.

[152] Mori, K., Mori, H., *Tetrahedron 41* (1985) 5487.

[153] Mori, K., Fujiwhara, M., *Tetrahedron 44* (1988) 343.

[154] Brooks, D. W., Mazdiyashni, H., Chakrabarti, S., *Tetrahedron Lett. 25* (1984) 1241.

[155] Brooks, D. W., Mazdiyashni, H., Grothaus, P. G., *J. Org. Chem. 50* (1985) 3411.

[156] Rehm, H. J., Reed, G. (Eds.), *Biotechnology, Vol. 3: Biomass, Microorganisms for Special Applications, Microbial Products I, Energy from Renewable Resources*, Weinheim: VCH Verlagsgesellschaft, 1983.

[157] Lanzilotta, R. P., Bradley, D. G., Beard, C. C., *Appl. Microbiol. 29* (1975) 427.

[158] Brooks, D. W., Grothaus, P. G., Palmer, J. T., *Tetrahedron Lett. 23* (1982) 4187.

[159] Brooks, D. W., Grothaus, P. G., Masdyashni, H., *J. Am. Chem. Soc. 105* (1983) 4472.

[160] Brooks, D. W., Woods, K. W., *J. Org. Chem. 52* (1987) 2036.

[161] Sugai, T., Toyo, H., Mori, K., *Agric. Biol. Chem. 50* (1986) 3127.

[162] Mori, K., Komatsu, M., *Tetrahedron 43* (1987) 3409.

[163] Mori, K., Komatsu, M., *Liebigs Ann. Chem.* (1988) 107.

[164] Mori, K., Mori, H., Yanai, M., *Tetrahedron 42* (1986) 291.

[165] Mori, K., Nakazono, Y., *Tetrahedron 42* (1986) 283.

[166] Mori, K., Watanabe, H., *Tetrahedron 42* (1986) 273.

[167] Mori, K., Mori, H., *Tetrahedron 43* (1987) 4097.

[168] Yamamoto, H., Oritani, T., Yamashita, K., *Agric. Biol. Chem. 52* (1988) 2203.

[169] Mori, K., Takaishi, H., *Liebigs Ann. Chem.* (1989) 695.

[170] Mori, K., Takaishi, H., *Liebigs Ann. Chem.* (1989) 939.

[171] Mori, K., Suzuki, N., *Liebigs Ann. Chem.* (1990) 287.

[172] Mori, K., Tamura, H., *Liebigs Ann. Chem.* (1990) 361.

[173] Mori, K., Fujiwhara, M., *Liebigs Ann. Chem.* (1989) 41.

[174] Mori. K., Fujiwhara, M., *Liebigs Ann. Chem.* (1990) 369.

[175] Ito, M., Masahara, R., Tsukida, K., *Tetrahedron Lett.* (1977) 2767.

[176] Leiser, J. K., *Synth. Commun. 13* (1983) 765.

[177] Short, R. P., Kennedy, R. M., Masamune, S., *J. Org. Chem. 54* (1989) 1755.

[178] Hoffmann, G., Wiartalla, R., *Tetrahedron Lett. 23* (1982) 3887.

[179] Neubauer, O., Frommherz, K., *Hoppe-Seylers Z. Physiol. Chem. 70* (1911) 326.

[180] Fischer, F. G., Wiedmann, O., *Liebigs Ann. Chem. 513* (1934) 261.

[181] Suemune, H., Mizuhara, Y., Akita, H., Oishi, T., Sakai, K, *Chem. Pharm. Bull. 35* (1987) 3112.

[182] Tsuboi, S., Furutani, H., Takeda, A., *Bull. Chem. Soc., Japan 60* (1987) 833.

[183] Tsuboi, S., Nishiyama, E., Furutani,H., Utaka, M., Takeda, A., *J. Org. Chem. 52* (1987) 1359.

[184] Iriuchijima, S., Ogawa, M., *Synthesis* (1982) 41.

[185] Tsuboi, S., Nishiyama, E., Utaka, M., Takeda, A., *Tetrahedron Lett. 27* (1986) 1915.

[186] Fráter, Gy., *Helv. Chim. Acta 62* (1979) 2829.

[187] Crump, D. R., *Aust. J. Chem. 35* (1982) 1945.

[188] Lemieux, R. U., Giguere. J., *Can. J. Chem. 29* (1951) 678.

[189] Utaka, M., Higashi, H., Takeda, A., *Chem. Commun.* (1987) 1368.

[190] Hirama, M., Nakamine, T., Ito, S., *Tetrahedron Lett. 27* (1986) 5281.

[191] Bennett, F., Knight, D. W., *Tetrahedron Lett. 29* (1988) 4625.

[192] Seebach, D., Renaud, P., Schweizer, W. B., Züger, M. F., Btienne, M. J., *Helv. Chim. Acta 67* (1984) 1843.

[193] Seebach, D, Eberle, M., *Synthesis* (1986) 37.

[194] Brooks, D. W., Kellog, R. P., Cooper, C. S., *J. Org. Chem. 52* (1987) 192.

[195] Chikashita, H., Ohkawa, K., Itoh, K., *Bull. Chem. Soc., Japan 62* (1989) 3513.

[196] Friedmann, E., *Biochem. Z. 243* (1931) 125.

[197] Alfonso, C. M., Barros, M. T., Godinho, L., Maycook, C. D., *Tetrahedron Lett. 30* (1989) 2707.

[198] Sugai, T., Ohta, H., *Agric. Biol. Chem. 53* (1989) 2009.

[199] Mori, K., Watanabe, H., *Tetrahedron 41* (1985) 3423.

[200] Seebach, D., Züger, M., *Helv. Chim. Acta 65* (1982) 495.

[201] Seebach, D., Züger, M., *Tetrahedron Lett. 25* (1984) 2747.

[202] Ikunaka, M., Mori, K., *Agric. Biol. Chem. 51* (1987) 565.

[203] Ushio, K., Inouye, K., Nakamura, K., Oka, S., Ohno, A., *Tetrahedron Lett. 27* (1986) 2657.

[204] Hasegawa, J., Hamaguchi, S., Ogura, M., Watanabe, K., *J. Ferment. Technol. 59* (1981) 257.

[205] Mori, K., *Tetrahedron 45* (1989) 3233.

[206] Katsuki, T., Yamaguchi, M., *Tetrahedron Lett. 28* (1987) 651.

[207] Brandänge, S., Leijonmarck, H., Ölund, J., *Acta Chem. Scand. 43* (1989) 193.

[208] Brooks, D. W., Castro De Lee, N., Peevey, R., *Tetrahedron Lett. 25* (1984) 4623.

[209] Oguni, N., Ohkawa, Y., *Chem. Commun.* (1988) 1376.

[210] Kitahara, T., Koseki, K., Mori, K., *Agric. Biol. Chem. 47* (1983) 389.

[211] Meyers, A. I., Amos, R. A., *J. Am. Chem. Soc. 102* (1980) 870.

[212] Mori, K., Sakai, T., *Liebigs Ann. Chem.* (1988) 13.

[213] Mori, K., Maemoto, S., *Liebigs Ann. Chem.* (1987) 863.

[214] Masoni, C., Deschenaux, P. F., Kallimopoulos, T., Jacot-Guillarmod, A., *Helv. Chim. Acta 72* (1989) 1284.

[215] Deschenaux, P. F., Kallimopulos, T., Stoeckli-Evans, H., Jacot-Guillarmod, J., *Helv. Chim. Acta 72* (1989) 731.

[216] Ernst, B., Wagner, B., *Helv. Chim. Acta 72* (1989) 165.

[217] Bennett, F., Knight, D. W., *Tetrahedron Lett. 29* (1988) 4865.

[218] Hirama, M., Noda, T., Ito, S., *J. Org. Chem. 50* (1985) 127.

[219] Mori, K., Sugai, T., *Synthesis* (1982) 752.

[220] Rösslein, L., Tamm, C., *Helv. Chim. Acta 71* (1988) 47.

[221] Fráter, Gy., *Helv. Chim. Acta 62* (1979) 2825.

[222] Fráter, Gy., *Helv. Chim. Acta 63* (1980) 1383.

[223] Kitazume, T., Kobayashi, T., *Synthesis* (1987) 187.

[224] Züger, M. F., Giovanni, F., Seebach, D., *Angew. Chem. Int. Ed. Engl. 22* (1983) 1012.

[225] Akita, H., Furuichi, A., Koshiji, H., Horikoshi, H., Oishi, T., *Chem. Pharm. Bull. 31* (1983) 4376.

[226] Fráter, Gy., Müller, U., Günther, W., *Tetrahedron 40* (1984) 1269.

[227] Sato, T., Tsurumaki, M., Fujisawa, T., *Chem. Lett.* (1986) 1367.

[228] Fujisawa, T., Itoh, T., Sato, T., *Tetrahedron Lett. 25* (1984) 5083.

[229] Chikashita, H., Motozawa, T., Itoh, K., *Synth. Commun. 19* (1989) 1119.

[230] Akita, H., Furuichi, A., Koshiji, H., Horikoshi, K., Oishi, T., *Chem. Pharm. Bull. 31* (1983) 4384.

[231] Buisson, D., Henrot, S., Larchevoque, M., Azerad, R., *Tetrahedron Lett. 28* (1987) 5033.

[232] Itoh, T., Fukuda, T., Fujisawa, T., *Bull. Chem. Soc., Japan 62* (1989) 3851.

[233] Itoh, T., Yonekawa, Y., Sato, T., Fujisawa, T., *Tetrahedron Lett. 27* (1986) 5405.

[234] Watabu, H., Ohkubko, M., Matsubara, H., Sakai, T., Tsuboi, S., Utaka, M., *Chem. Lett.* (1989) 2183.

[235] Horikoshi, K., Furuichi, A., Koshiji, H., Akita, H., Oishi, T., *Agric. Biol. Chem. 47* (1983) 453.

[236] Akita, H., Furuichi, A., Koshiji, H., Horikoshi, K., Oishi, T., *Tetrahedron Lett. 23* (1982) 4051.

[237] Akita, H., Koshiji, H., Furuichi, A., Horikoshi, K., Oishi, T., *Tetrahedron Lett. 24* (1983) 2009.

[238] Hoffmann, R. W., Ladner, W., Stenbach, K., Massa, W., Schmidt, R., Snatzke, G., *Chem. Ber. 114* (1981) 2786.

[239] Hoffman, R. W., Helbig, W., Ladner, W., *Tetrahedron Lett. 23* (1982) 3479.

[240] Hoffmann, R. W., Ladner, W., *Chem. Ber. 116* (1983) 1631.

[241] Hoffmann, R. W., Ladner, W., Helbig, W., *Liebigs Ann. Chem.* (1984) 1170.

[242] Kitahara, T., Mori, K., *Tetrahedron Lett. 26* (1985) 451.

[243] Kitahara, T., Kurata, H., Mori, K., *Tetrahedron 44* (1988) 4339.

[244] Buisson, D., Azerad, R., *Tetrahedron Lett. 27* (1986) 2631.

[245] Herradon, B., Seebach, D., *Helv. Chim. Acta 72* (1989) 690.

[246] Kitahara, T., Toruhara, K., Watanabe, H., Mori, K., *Tetrahedron 45* (1989) 6387.

[247] Mori, K., Ikunaka, M., *Tetrahedron 43* (1987) 45.

[248] Mori, K., Tsuji, M., *Tetrahedron 42* (1986) 435.

[249] Brooks, D. W., Wilson, M., Webb, M., *J. Org. Chem. 52* (1987) 2244.

[250] Mori, K., Tsuji, M., *Tetrahedron 44* (1988) 2835.

[251] Xie, Z. F., Funakoshi, K., Suemune, H., Oishi, T., Akita, H., Sakai, A., *Chem. Pharm. Bull 34* (1986) 3058.

[252] Okano, K., Mizuhara, Y., Suemune, H., Akita, H., Sakai, K., *Chem. Pharm. Bull. 36* (1988) 1358.

[253] Muys, G. T., Van der Ven, B., De Jonge, A. P., *Appl. Microbiol. 11* (1963) 389; Muys, G. T., Van der Ven, B., de Jonge, A. P., U.S. Pat. 3'076'750 (1959); *Chem. Abstr. 58* (1963) 11'928d.

[254] Francke, A., *Nature (London) 197* (1963) 384; Francke, A., *Biochem. J. 95* (1965) 633.

[255] Keppler, J. G., *J. Am. Oil. Chem. Soc. 54* (1987) 474.

[256] Utaka, M., Watabu, H., Takeda, A., *J. Org. Chem. 52* (1987) 4363.

[257] Moriuchi, F., Muroi, H., Aibe, H., *Chem. Lett.* (1987) 1141.

[258] Naoshima, Y., Ozawa, H., Kondo, H., Hayashi, S., *Agric. Biol. Chem. 47* (1983) 1431.

[259] Naoshima, Y., Hasegawa, H., Saeki, T., *Agric. Biol. Chem. 51* (1987) 3417.

[260] Utaka, M., Watabu, H., Takeda, A., *Chem. Lett.* (1985) 1475.

[261] Sehgal, S. N., Venzina, C., *Bacteriol. Proc. A15* (1969) 3.

[262] Dasaradhi, L., Fadnavis, N. W., Bhalerao, U. T., *IUPAC Int. Symp. Chem. Nat. Prod.* (17th), symposium paper, New Delhi, India, 1990; p 355.

[263] Naoshima, Y., Hasegawa, H., *Chem. Lett.* (1987) 2379.

[264] Naoshima, Y., Hasegawa, H., Nishiyama, T., Nakamura, A., *Bull. Chem. Soc., Japan 62* (1989) 608.

[265] Naoshima, Y., Nakamura, A., Nishiyama, T., Haramaki, T., Mende, M., Munaka, Y., *Chem. Lett.* (1989) 1023.

[266] Fischer, F. G., Wiedmann, O., *Liebigs Ann. Chem. 520* (1935) 53.

[267] Fuganti, C., Ghiringhelli, D., Graselli, P., *Chem. Commun.* (1975) 846.

[268] Fuganti, C., Grasselli, P., Servi, S., Högberg, H. E., *J. Chem. Soc., Perkin I* (1988) 3061.

[269] Ferraboschi, P., Grisenti, P., Casati, R., Fiecchi, A., Santaniello, E., *J. Chem. Soc., Perkin I* (1987) 1743.

[270] Gramatica, P., Manitto, P., Monti, D., Speranza, G., *Tetrahedron 44* (1988) 1299.

[271] Fuganti, C., Graselli, P., *Chem. Commun.* (1979) 995.

[272] Gramatica, P., Manitto, P., Poli, L., *J. Org. Chem. 50* (1985) 4625.

[273] Gramatica, P., Manitto, P., Monti, D., Speranza, G., *Tetrahedron 42* (1986) 6687.

[274] Gramatica, P., Giardina, G., Spranza, G., Manitto, P., *Chem. Lett.* (1985) 1395.

[275] Gramatica, P., Manitto, P., Monti, D., Speranza, G., *Tetrahedron 43* (1987) 4481.

[276] Leuenberger, H. G. W., Boguth, W., Barner, R., Schmid, M., Zell, R., *Helv. Chim. Acta 62* (1979) 455.

[277] Sato, T., Hanayama, K., Fujisawa, T., *Tetrahedron Lett. 29* (1988) 2197.

[278] Ferraboschi, P., Fecchi, A., Grisenti, P., Santaniello, E., *J. Chem. Soc., Perkin I* (1987) 1749.

[279] Gramatica, P., Manitto, P., Ranzi, B. M., Delbianco, A., Francavilla, M., *Experientia 38* (1982) 775.

[280] Ohta, H., Kobayashi, N., Ozaki, K., *J. Org. Chem. 54* (1989) 1809.

[281] Eiter, K., Austrian Pat. 181'926 (1952) *Chem. Abstr. 50* (1956) 2672q; Eiter, K., Letnansky, K., *Monatsch. Chem. 85* (1954) 822.

[282] Utaka, M., Koinishi, S., Mizuoka, A., Ohkubko, T., Sakai, T., Tsuboi, S., Takeda, A., *J. Org. Chem. 54* (1989) 4989.

[283] Utaka, M., Onoue, S., Takeda, A., *Chem. Lett.* (1987) 917.

[284] Utaka, M., Koinishi, S., Takeda, A., *Tetrahedron Lett. 27* (1986) 4737.

[285] Sebek, O. K., Lincoln, F. H., Schneider, W. P., *5th Int. Ferment. Congr.*, West Berlin, (Abstr.), 17.05, 1976, .

[286] Leuenberger, H. G. W., Boguth, W., Widmer, E., Zell, R., *Helv. Chim. Acta 59* (1976) 1832.

[287] Zell, R., Widmer, E., Lukac, T., Leuenberger, H. G. W., Schönholzer, P., Broger, E. A., *Helv. Chim. Acta 64* (1981) 2447.

[288] Protiva, M., Capek, A., Jilek, J. O., Kakac, B., Tadra, B., *Collect. Czech. Chem. Commun. 30* (1961) 2236.

[289] Gil, G., Ferre, E., Barre, M., Le Petit, J., *Tetrahedron Lett. 29* (1988) 3797.

[290] Lüers, H., Mengele, J., *Biochem. Z. 179* (1926) 238; Neuberg, C., Simon, E., *Biochem. Z. 171* (1926) 256.

[291] Neuberg, C., Welde, E., *Biochem. Z. 60* (1914) 427.

[292] Neuberg, C., Reinfurth, E., *Biochem. Z. 138* (1923) 561.

[293] Neuberg, C., Welde, E., *Biochem. Z. 67* (1914) 18.

[294] Takeshita, M., Yoshida, S., Kiya, R., Higuchi, N., Kobayashi, Y., *Chem. Pharm. Bull. 37* (1989) 615.

[295] Buist, P. H., Dallmann, H. G., *Tetrahedron Lett. 29* (1988) 258.

[296] Buist, P. H., Dallmann, H. G., Seigel, P. M., Rymerson, R. T., *Tetrahedron Lett. 28* (1987) 857.

[297] Glänzer, B. I., Faber, K., Griengl, H., Roehr, M., Wöhrer, W., *Enzyme Microb. Technol. 10* (1988) 745.

[298] Glänzer, B. I., Faber, K., Griengl, H., *Tetrahedron 43* (1987) 771.

[299] Glänzer, B. I., Faber, K., Griengl, H., *Tetrahedron Lett. 27* (1986) 4293.

[300] Mori, K., Akao, H., *Tetrahedron 36* (1980) 91; Sugai, T., Kuwahara, S., Hoshino, C., Matsuno, N., Mori, K., *Agric. Biol. Chem. 46* (1982) 3579.

[301] Glänzer, B. I., Faber, K., Griengl, H., *Enzyme Microb. Technol. 10* (1988) 689.

[302] Chen, S. L., Peppler, H. J., *J. Biol. Chem. 221* (1956) 101.

[303] Kitazume, T., Ishkawa, N., *Chem. Lett.* (1984) 1815.

[304] Fronza, G., Fuganti, C., Grasselli, P., Poli, G., Servi, S., *J. Org. Chem. 53* (1988) 6153.

[305] Fronza, G., Fuganti, C., Pedrocchi-Fantoni, G., Servi, S., *Chem. Lett.* (1989) 2141.

[306] Fuganti, C., Graselli, P., *Chem. Ind.* (1977) 983.

[307] Ohta, H., Ozaki, K., Konishi, J., Tsuchihashi, G., *Agric. Biol. Chem. 50* (1986) 1261.

[308] Becvarova, H., Hanc, O., Mauk, K., *Folia Microbiol. 8* (1963) 165.

[309] Smith, P. F., Heudlin, D., *J. Bacteriol. 65* (1953) 440.

[310] Gröger, D., Schmandler, H. P., Mothes, K., *Z. Allg. Microbiol. 6* (1966) 275.

[311] Vojtisek, V., Netrval, J., *Folia Microbiol. 27* (1982) 173.

[312] Fuganti, C., Grasselli, P., Poli, G., Servi, S., Zorzella, A., *Chem. Commun.* (1988) 1619.

[313] Neuberg, C., Hirsch, J., *Z. Allg. Microbiol. 113* (1921) 282.

[314] Hanc, O., Kakac, B., Zvacek, J., Tuma, J., Czech. Pat. 231'401 (1986) (Appl. 1960); *Chem. Abstr. 106* (1987) 3871.

[315] Gröger, D., Schmander, H. P., Frömmel, H., Ger. (East) Pat. 62'544 (1967); *Chem. Abstr. 70* (1969) 86'297g.

[316] Long, A., Ward, O. P., *Biotechnol. Bioeng. 34* (1989) 933.

[317] Bertolli, G., Fronza, G., Fuganti, C., Grasselli, P., Majori, L., Spreafico, F., *Tetrahedron Lett. 22* (1981) 965.

[318] Behrens, M., Iwanoff, N. N., *Biochem. Z. 169* (1926) 478.

[319] Neuberg, C., Liebermann, L., *Biochem. Z. 121* (1921) 311.

[320] Merck and Co., US Pat. 3'338'796 (1964) *Chem. Abstr. 67* (1967) 89'797.

[321] Long, A., James, P., Ward, O. P., *Biotechnol. Bioeng. 33* (1989) 657.

[322] Fuganti, C., Grasselli, P., *Chem. Commun.* (1982) 205.

[323] Bujons, J., Guajardo, R., Kyler, K. S., *J. Am. Chem. Soc. 110* (1988) 604.

[324] Medina, J. C., Guajardo, R., Kyler, K. S., *J. Am. Chem. Soc. 111* (1989) 2310.

[325] Medina, J. C., Kyler, K. S., *J. Am. Chem. Soc. 110* (1988) 4818.

[326] Kamal, A., Sattur, P. B., *Chem. Commun.* (1989) 835.

[327] Rama Rao, K., Bhanumathi, N., Srinivasan, T. N., Sattur, P. B., *Tetrahedron Lett. 31* (1990) 899.

[328] Bernardi, R., Fuganti, C., Grasselli, P., Marinoni, G., *Synthesis* (1980) 50.

[329] Csuk, R., Glänzer, B. I., *Chem. Rev. 91* (1991) 49.

[330] Spiliotis, V., Papahatjis, D., Ragoussis, N., *Tetrahedron Lett. 31* (1990) 1615.

[331] Naoshima, Y., Maede, J., Munakata, Y., Nishiyama, T., Kamezawa, M., Tachibana, H., *Chem. Commun.* (1990) 964.

[332] Nakamura, K., Kitayama, T., Inoue, Y., Ohno, A., *Tetrahedron 46* (1990) 7471.

[333] Gopalan, A. S., Jacobs, H. K., *Tetrahedron Lett. 31* (1990) 5575.

[334] Ramaswamy, S., Oehlschlager, A. C., *Tetrahedron 47* (1991) 1145.

[335] Kodama, M., Minami, H., Mima, Y., Fukuyama, Y., *Tetrahedron Lett. 31* (1990) 4025.

[336] Sato, M., Sakai, J., Sugita, Y., Nakano, T., Kaneko, C., *Tetrahedron Lett. 31* (1990) 7463.

[337] Kawamura, K., Ohta, T., Otani, G., *Chem. Pharm. Bull. 38* (1990) 2092.

[338] Takemura, T., Hosoya, Y., Mori, N., *Can. J. Chem. 68* (1990) 523.

[339] Yamamoto, H., Oritani, T., Koga, H., Horiuchi, T., Yamashita, K., *Agric. Biol. Chem. 54* (1990) 1915.

[340] Burnier, G., Vogel, P., *Helv. Chim. Acta 73* (1990) 985.

[341] Sakaki, J., Suzuki, M., Kobayashi, S., Sato, M., Kanako, C., *Chem. Lett.* (1990) 901; Sakai, J., Kobayashi, S., Sato, S., Kaneko, C., *Chem. Pharm. Bull. 37* (1989) 2952.

[342] Fujisawa, T., Yamanaka, K., Mobele, B. J., Shimizu, M., *Tetrahedron Lett. 32* (1991) 399.
[343] Brenelli, E. C. S., Moran, P. J. S., Rodrigues, J. A. R., *Synth. Commun. 20* (1990) 261.
[344] Kitahara, T., Miyake, M., Kido, M., Mori, K., *Tetrahedron: Asymmetry 1* (1990) 775.
[345] Mori, K., Nagano, E., *Biocatalysis 3* (1990) 25.
[346] Utaka, M., Watabu, H., Higashi, H., Sakai, T., Tsuboi, S., Torii, S., *J. Org. Chem. 55* (1990) 3917.
[347] Kahn, M., Fujita, K., *Tetrahedron 47* (1991) 1137.
[348] Wahhab, A., Tavares., D. F., Rauk., A., *Can. J. Chem. 68* (1990) 1558.
[349] Jacobs, H., Berryman, K., Jones, J., Gopalan, A., *Synth. Commun. 20* (1990) 999.
[350] Dasaradhi, L., Fadnavis, N. W., Bhalareo, U. T., *Chem. Commun.* (1990) 729.
[351] Naoshima, Y., Nakamura, A., Munakata, Y., Kamezawa, M., Tachibana, H., *Bull. Chem. Soc., Japan. 63* (1990) 1263.
[352] Gibbs, D. E., Barnes, D., *Tetrahedron Lett. 31* (1990) 5555.
[353] Buist, P. H., Marecak, D. M., Partington, E. T., Skala, P., *J. Org. Chem. 55* (1990) 5667.
[354] Kamal, A., Rao, M. V., Rao, A. B., *Chem. Lett.* (1990) 655.
[355] Rama Rao, K., Bhanumathi, N., Sattur, P. B., *Tetrahedron Lett. 21* (1990) 3201.
[356] Fuganti, C., Pedrocchi-Fantoni, G., Servi, S., *Tetrahedron Lett. 31* (1990) 4195.
[357] Itho, T., Takagi, Y., Fujisawa, T., *Tetrahedron Lett. 53* (1989) 6153.

Chapter 6

[1] Toone, E. J., Simon, E. S., Bednarski, M. D., Whitesides, G. M., *Tetrahedron 45* (1989) 5365.
[2] Billhardt, U. M., Stein, P., Whitesides, G. M., *Bioorg. Chem. 17* (1989) 1.
[3] International Union of Biochemistry, *Enzyme Nomenclature 1984*, New York: Academic Press, 1984.
[4] Szejtli, J., *Cyclodextrins and Their Inclusion Complexes*, Budapest: Akadémiai Kiadó, 1982.
[5] Ooi, Y., Hashimoto, T., Mitsuno, N., Satoh, T., *Tetrahedron Lett. 25* (1984) 2241.
[6] Mitsuo, N., Takeichi, H., Satoh, T., *Chem. Pharm. Bull. 32* (1984) 1183.
[7] Ooi, Y., Mitsuno, N., Satoh, T., *Chem. Pharm. Bull. 33* (1985) 5547.
[8] Boos, W., *Meth. Enzymol. 89* (1982) 59.
[9] Itano, K., Yamasaki, K., Kirahara, C., Tanaka, O., *Carbohydr. Res. 87* (1980) 27.
[10] Björkling, F., Godtfredsen, S. E., *Tetrahedron 44* (1988) 2957.
[11] Esaki, N., Shimoi, H., Tanaka, H., Soda, K., *Biotechnol. Bioeng. 34* (1989) 1231.
[12] Passerat, N., Bolte, J., *Tetrahedron Lett. 28* (1987) 1277.
[13] Calton, G. J., Wood, L. L., Updike, M. H., Lantz, L., Hamman, J. P., *Bio/Technology 4* (1986) 317.
[14] Crans, D. C., Whitesides, G. M., *J. Am. Chem. Soc. 107* (1985) 7019.
[15] Crans, D. C., Whitesides, G. M., *J. Am. Chem. Soc. 107* (1985) 7008.
[16] Yamada, H., Shimizu, S., *Angew. Chem. Int. Ed. Engl. 27* (1988) 622.
[17] Whitesides, G. M., Wong, C. H., *Angew. Chem. Int. Ed. Engl. 24* (1985) 617.
[18] Simon, E. S., Grabowski, S., Whitesides, G. M., *J. Am. Chem. Soc. 111* (1989) 8920.
[19] Walt, D. R., Rios-Mercadillo, V. M., Augé, J., Whitesides, G. M., *J. Am. Chem. Soc. 102* (1980) 7805; Walt, D. R., Findeis, M. A., Rios-Mercadillo, V. M., Augé, J., Whitesides, G. M., *J. Am. Chem. Soc. 106* (1984) 234.
[20] Utagawa, T., Morisawa, H., Nakamatsu, T., Yamazaki, A., Yamanaka, S., *FEBS Lett. 119* (1980) 101.
[21] Utagawa, T., Morisawa, H., Yamanaka, S., Yamazaki, A., Yoshinaga, F., Hirose, Y., *Agric. Biol. Chem. 49* (1985) 3239.
[22] Krenitsky, T. A., Koszalka, G. W., Tuttle, J. V., Rideout, J. L., Elion. G. B., *Carbohydr. Res. 97* (1981) 139.

[23] Krenitsky, T. A., Rideout, J. L., Chao, E. Y., Koszalka, G. W., Gurney, F., Crouch, E. C., Cohn, N. K., Wolberg, G., Vinegar, R., *J. Med. Chem. 29* (1986) 138.

[24] Hennen, W. J., Wong, C. H., *J. Org. Chem. 54* (1989) 4692.

[25] Rehm, H. J., Reed, G. (Eds.), *Biotechnology, Vol. 7B: Gene Technology*, Weinheim: VCH Verlagsgesellschaft, 1989.

[26] Walker, J. M. (Ed.), *Nucleic Acids: Methods in Molecular Biology*, Vol. 2, Clifton, NJ: Humana Press, 1984; Walker, J. M. (Ed.), *New Nucleic Acid Techniques: Methods in Molecular Biology*, Vol. 4, Clifton, NJ: Humana Press, 1988.

[27] Hindley, J., *DNA Sequencing*, Amsterdam: Elsevier, 1983.

[28] Gait, M. J. (Ed.), *Oligonucleotide Synthesis: A Practical Approach*, Oxford: IRL Press, 1984; Gassen, H. G., Lang, A. (Eds.), *Chemical and Enzymatic Synthesis of Gene Fragments*, Weinheim: VCH Verlagsgesellschaft, 1982.

[29] Stevenson, D. E., Akhtar, M., Gani, D., *Tetrahedron Lett. 27* (1986) 5661.

[30] Crout, D. H. G., Rathbone, D. L., *Chem. Commun.* (1988) 98.

[31] Yamamoto, K., Tosa, T., Yamashita, K., Chibata, I., *Eur. J. Appl. Microbiol. 3* (1976) 169.

[32] Chibata, I., Tosa, T., Sato, T., *Meth. Enzymol. 44* (1976) 739.

[33] Findeis, M. A., Whitesides, G. M., *J. Org. Chem. 52* (1987) 2838.

[34] Akhtar, M., Botting, M. P., Cohen, M. A., Gani, T., *Tetrahedron 43* (1987) 5899.

[35] Koyama, T., Ogura, K., Baker, F. C., Jamienson, G. C., Scooley, D. A., *J. Am. Chem. Soc. 109* (1987) 2853.

[36] Ohnuma, S., Koyama, T., Ogura, K., *Bull. Chem. Soc., Japan 62* (1989) 2742.

[37] Nógrádi, M., *Stereoselective Synthesis*, Weinheim: VCH Verlagsgesellschaft, 1986.

[38] Von der Osten, C. H., Sinskey, A. J., Barbas, C. J., Pederson, R. L., Wang, Y. F., Wong, C. H., *J. Am. Chem. Soc. 111* (1989) 3924.

[39] Barbas, C. F., Wang, Y. F., Wong, C. H., *J. Am. Chem. Soc. 112* (1990) 2013.

[40] Bednarski, M. D., Waldmann, H. J., Whitesides, G. M., *Tetrahedron Lett. 27* (1986) 5807.

[41] Bednarski, M. D., Simon, E. S., Bischoffberger, N., Fessner, W. D., Kim, M. J., Lees, W., Saito, T., Waldmann, H., Whitesides, G. M., *J. Am. Chem. Soc. 111* (1989) 627.

[42] Effenberger, F., Straub, A., *Tetrahedron Lett. 28* (1987) 1641.

[43] Wong, C. H., Mazenod, F. P., Whitesides, G. M., *J. Org. Chem. 48* (1983) 3493.

[44] Durrwachter, J. R., Drueckhammer, D. G., Nozaki, K., Sweers, H. M., Wong, C. H., *J. Am. Chem. Soc. 108* (1986) 7812.

[45] Shultz, M., Waldmann, H., Vogt, W., Kunz, H., *Tetrahedron Lett. 31* (1990) 867.

[46] Wong, C. H., Whitesides, G. M., *J. Org. Chem. 48* (1983) 3199.

[47] Borsyenko, C. W., Spaltenstein, A., Straub, J. A., Whitesides, G. M., *J. Am. Chem. Soc. 111* (1989) 9275.

[48] Durrwachter, J. R., Wong, C. H., *J. Org. Chem. 53* (1988) 4175.

[49] Ziegler, T., Straub, A., Effenberger, F., *Angew. Chem. Int. Ed. Engl. 27* (1988) 716.

[50] Pederson, R. L., Kim, M. J., Wong, C. H., *Tetrahedron Lett. 29* (1988) 4645.

[51] Turner, N. J., Whitesides, G. M., *J. Am. Chem. Soc. 111* (1989) 624.

[52] Drueckhammer, D. G., Durrwachter, J. R., Pederson, R. L., Crans, D. C., Daniels, L., Wong, C. H., *J. Org. Chem. 54* (1989) 70.

[53] Ohta, H., Miyamae, Y., Tsuchihashi, G. I., *Agric. Biol. Chem. 50* (1986) 3181.

[54] Ohta, H., Miyamae, Y., Tsuchihashi, G. I., *Agric. Biol. Chem. 53* (1989) 215.

[55] Ohta, H., Miyamae, Y., Tsuchihashi, G. I., *Agric. Biol. Chem. 53* (1989) 281.

[56] Ohta, H., Miyamae, Y., Kimura, Y., *Chem. Lett.* (1989) 379.

[57] Ohta, H., Kimura, Y., Sugano, Y., *Tetrahedron Lett. 29* (1988) 6957.

[58] Ohta, H., Hiraga, S., Miyamoto, K., Tsuchihashi, G. I., *Agric. Biol. Chem. 52* (1988) 3023.

[59] Becker, W., Pfeil, E., *J. Am. Chem. Soc. 88* (1966) 4299.

[60] Becker, W., Pfeil, E., *Biochem. Z. 346* (1966) 301; Becker, W., Freund, H., Pfeil, E., *Angew. Chem. Int. Ed. Engl. 4* (1965) 1079; Becker, W., Pfeil, E., *Naturwissenschaften 51* (1964) 193.

[61] Hochuli, E., *Helv. Chim. Acta 66* (1983) 489.

[62] Brousse, J., Roos, E. C., Van der Gren, A., *Tetrahedron Lett. 29* (1988) 4485.

[63] Moran, J. R., Whitesides, G. M., *J. Org. Chem. 49* (1984) 704.

[63] Effenberger, F., Hörsch, B., Förster, S., Ziegler, T., *Tetrahedron Lett. 31* (1990) 1249.

[64] Effenberger, F., Ziegler, T., Förster, S., *Angew. Chem. 99* (1987) 491.

[65] Brusse, J., Loos, W. T., Kruse, C. G., Van der Gen, A., *Tetrahedron 46* (1990) 979.

[66] Kieslich, J. K., *Microbial Transformations of Non-Steroid Cyclic Compounds*, Stuttgart: G. Thieme Verlag, 1976.

[67] Oritani, T., Yamashita, K., *Agric. Biol. Chem. 38* (1974) 1965.

[68] Oritani, T., Kundo, S., Yamashita, K., *Agric. Biol. Chem. 46* (1982) 757.

[69] Ziffer, H., Kawai, M., Imuta, M., Froussios, C., *J. Org. Chem. 48* (1983) 3017.

[70] Oritani, T., Ichimura, M., Hanyu, Y., Yamashita, K., *Agric. Biol. Chem. 47* (1983) 2613.

[71] Kasai, M., Ziffer, H., Silverton, J. V., *Can. J. Chem. 63* (1985) 1287.

[72] Ohta, H., Ikemoto, M., Ii, H., Okamoto, Y., Tsuchihashi, G., *Chem. Lett.* (1986) 1169.

[73] Hirai, K., Miyakoshi, S., Naito, A., *Tetrahedron Lett. 30* (1989) 2555.

[74] Kato, Y., Ohta, H., Tsuchihashi, G., *Tetrahedron Lett. 28* (1987) 1303.

[75] Wu, S. H., Zhang, L. Q., Chen, C. S., Girdaukas, G., Sih, C. J., *Tetrahedron Lett. 26* (1985) 4323.

[76] Kotani, H., Kuze, Y., Uchida, S., Minabe, T., Iimori, T., Okano, K., Kobayashi, S., Ohno, M., *Agric. Biol. Chem. 47* (1983) 1363.

[77] Gopalan, A. S., Sih, C. J., *Tetrahedron Lett. 25* (1984) 5235.

[78] Patel, D. V., Van Middlesworth, F., Donaubauer, J., Gaunnett, P., Sih, C. J., *J. Am. Chem. Soc. 108* (1986) 4603.

[79] Xie, Z. F., Suemune, H., Sakai, K., *Chem. Commun.* (1987) 838.

[80] Claridge, C. A., Schmitz, H., *Appl. Environ. Microbiol. 36* (1978) 63; Claridge, C. A., Schmitz, H., *Appl. Environ. Microbiol. 37* (1979) 693.

[81] Asano, Y., Yasuda, T., Tani, Y., Yamada, H., *Agric. Biol. Chem. 46* (1982) 1183.

[82] Nagasawa, T., Nauba, H., Ruyno, K., Takeuchi, K., Yamada, H., *Eur. J. Biochem. 162* (1987) 691.

[83] Nagasawa, T., Yamada, H., *IUPAC Int. Symp. Chem. Nat. Prod.* (17th), symposium paper, New Delhi, India, 1990; p 54.

[84] Mauger, J., Nagasawa, T., Yamada, H., *Tetrahedron 45* (1989) 1347.

[85] Bengis-Gaber, G., Gutman, A. L., *Tetrahedron Lett. 29* (1988) 2589.

[86] Bengis-Gaber, G., Gutman, A. L., *Appl. Microbiol. Biotechnol. 32* (1989) 11.

[87] Kabuto, K., Imuta, M., Kempner, E. S., *J. Org. Chem. 43* (1978) 2357.

[88] Trincone, A., Lama, L., Lanzotti, V., Nicolaus, B., Rosa, M. De, Rossi, M., Gambacorta, A., *Biotechnol. Bioeng. 35* (1990) 559.

[89] Kataoka, M., Shimizu, S., Yamada, H., *Agric. Biol. Chem. 54* (1990) 177.

[90] Matzinger, P. K., Wirtz, B., Leuenberger, H. G. W., *Appl. Microbiol. Biotechnol. 32* (1990) 533.

[91] Nakazaki, M., Chikamatsu, H., Naemura, K., Nishino, M., Murakami, H., Asao, M., *Chem. Commun.* (1978) 667.

[92] Yamazaki, Y., Hosono, K., *Biotechnol. Lett. 11* (1989) 679.

[93] Adlercreutz, P., *Appl. Microbiol. Biotechnol. 30* (1989) 257.

[94] Ohta, H., Yamada, H., Tsuchihashi, G. I., *Chem. Lett.* (1987) 2325.

[95] Uskokovic, M. R., Lewis, L. R., Patridge, J. J., Despreaux, C. W., Pruess, D. L., *J. Am. Chem. Soc. 101* (1979) 6742.

[96] Buisson, D., Azerad, R., Revial, D., d'Angelo, J., *Tetrahedron Lett. 25* (1984) 6005.

[97] D'Angelo, J., Revial, G., Azerad, R., Buisson, D., *J. Org. Chem. 51* (1986) 40.

[98] Wang, N. Y., Hsu, C. T., Sih, C. J., *J. Am. Chem. Soc. 103* (1981) 6538.

[99] Hsu, C. T., Wang, N. Y., Latimer, L. H., Sih, C. J., *J. Am. Chem. Soc. 105* (1983) 593.

[100] Sih, C. J., Heather, J. B., Peruzzotti, G. P., Price, P., Sood, R., Hsu-Lee, L. F., *J. Am. Chem. Soc. 97* (1975) 865.

[101] Sih, C. J., Heather, J. B., Peruzzotti, G. P., Price, P., Sood, R., Hsu-Lee, L. F., *J. Am. Chem. Soc. 95* (1973) 1676.

[102] Okano, K., Suemune, H., Sakai, K., *Chem. Pharm. Bull. 37* (1989) 1995.

[103] Davies, H. G., Roberts, S. M., Wakefield, B. J., Winders, J. A., *Chem. Commun.* (1985) 1166.

[104] Fourneron, J. D., Archelas, A., Furstoss, R., *J. Org. Chem. 54* (1989) 4686.

[105] Ohta, H., Matsumoto, S., Okamoto, Y., Sugai, T., *Chem. Lett.* (1989) 625.

[106] Weijers, C. A. G. M., Van Gingel, C. G., de Bont, J. A. M., *Enzyme Microb. Technol. 10* (1988) 214.

[107] May, S. W., Schwartz, R. D., *J. Am. Chem. Soc. 96* (1974) 4031.

[108] Takahashi, O., Umezawa, J., Furuchashi, K., Takagi, M., *Tetrahedron Lett. 30* (1989) 1583.

[109] Smet, M. J., Witholt, B., Wynberg, H., *J. Org. Chem. 46* (1981) 3128.

[110] Ohta, H., Tetsukawa, H., *Chem. Commun.* (1978) 849.

[111] May, S. W., Steltenkamp, H. S., Schwartz, R. D., McCoy, C. J., *J. Am. Chem. Soc. 98* (1976) 7856.

[112] Ballard, D. H. G., Courtis, A., Shirlei, I. M., Taylor, R. C., *Chem. Commun.* (1983) 954.

[113] Van den Tweel, W. J. J., de Bont, J. A. M., Vorage, M. J. A. W., Marsman, E. H., Tramper, J., Koppejan, J., *Enzyme Microb. Technol. 10* (1988) 134.

[114] Gibson, D. T., Hensley, M., Yoshioka, H., Mabry, T. J., *Biochemistry 9* (1970) 849.

[115] Kobal, V. M., Gibson, D. T., Davis, R. E., Garza, A., *J. Am. Chem. Soc. 95* (1973) 4420.

[116] Hudlicky, T., Seoane, G., Pettus, T., *J. Org. Chem. 54* (1989) 4239.

[117] Taylor, S. J. C., Ribbons, D. W., Slawin, A. M. Z., Widdowson, D. A., Williams, D. J., *Tetrahedron Lett. 28* (1987) 6391.

[118] Boyd, D. R., Mc Mordie, R. A., Sharma, N. D., Dalton, H., Williams, P., Jenkins, R. O., *Chem. Commun.* (1989) 339.

[119] Ley, S. V., Strenfeld, F., Taylor, S., *Tetrahedron Lett. 28* (1987) 225.

[120] Ley, S. V., Sternfield, F., *Tetrahedron 45* (1989) 3463.

[121] Ley, S. V., Sternfield, F., *Tetrahedron Lett. 29* (1988) 5303.

[122] Ley, S. V., Parra, M., Redgrave, A. J., Sternfeld, F., Vidal, A., *Tetrahedron Lett. 30* (1989) 3557.

[123] Hudlicky, T., Luna, H., Barbieri, G., Kwart, L. D., *J. Am. Chem. Soc. 110* (1988) 4735.

[124] Kobayashi, M., Koyama, T., Ogura, K., Seto, S., Ritter, F. J., Brüggemann-Rotgans, I. E. M., *J. Am. Chem. Soc. 106* (1984) 6602.

[125] Yamada, H., Ruyno, K., Nagasawa, T., Enomoto, K., Watanabe, I., *Agric. Biol. Chem. 50* (1986) 2859; Ruyno, K., Nagasawa, T., Yamada, H., *Agric. Biol. Chem. 52* (1988) 1813.

[126] Thiem, J., Stangier, P., *Liebigs Ann. Chem.* (1990) 1101.

[127] Ozaki, A., Toone, E. J., Osten, C. H., Sinskey, A. J., Whitesides, G. M., *J. Am. Chem. Soc. 112* (1990) 4970.

[128] Schultz, M., Waldmann, H., Kunz, H., Vogt, W., *Liebigs Ann. Chem.* (1990) 1019.

[129] Straub, A., Effenberger, F., Fischer, P., *J. Org. Chem. 55* (1990) 3926.

[130] Ziegler, T., Hörsch, B., Effenberger, F., *Synthesis* (1990) 575.

[131] Niedermeyer, U., Kula, M. R., *Angew. Chem. Int. Ed. Engl. 29* (1990) 386.

[132] Klempier, N., Raadt, A., Faber, K., Griengl, H., *Tetrahedron Lett. 32* (1991) 341.

[133] Hönike-Schmidt, P., Schneider, M. P., *Chem. Commun.* (1990) 648.

[134] Cohen, M. A., Sawden, J., Turner, N. J., *Tetrahedron Lett. 31* (1990) 7223.

[135] Kakeya, H., Sakai, N., Sugai, T., Ohta, H., *Tetrahedron Lett. 32* (1991) 1343.

[136] Geary, P. J., Pryce, R. J., Roberts, S. M., Ryback, G., Winders, J. A., *Chem. Commun.* (1990) 204.

[137] Howard, P. W., Stephenson, G. R., Taylor, S. C., *Chem. Commun.* (1990) 1182.

[138] Hudlicky, T., Luna, H., Price., J. D., *J. Org. Chem. 55* (1990) 4683.

[139] Carless, H. A. J., Oak, O. Z., *Chem. Commun.* (1991) 61.

Index